Handbook of Optimization

Handbook of Optimization

Edited by **Calanthia Wright**

NY RESEARCH PRESS
New York

Published by NY Research Press,
23 West, 55th Street, Suite 816,
New York, NY 10019, USA
www.nyresearchpress.com

Handbook of Optimization
Edited by Calanthia Wright

International Standard Book Number: 978-1-63238-259-7 (Hardback)

Printed in China.

Contents

Preface

Optimization is a mathematical procedure that is also termed as the act of rendering something optimal or the synchronized augmentation of profitability and growth. It is helpful in the determination of optimal allocation of resources that are scarce. The goal of optimization is to find an alternative with the highest possible performance or most cost effective performance under specific given guidelines, by the process of maximising the wanted factors and reducing the undesired ones. This is of course a highly specialized field that requires a sharp mind and study. One should remember that the term 'programming' means something different in mathematics than it does in the usual usage of the term in computing. In the case of mathematical optimization, it refers to writing instructions for performing calculations. There are factors which affect the practice of optimization and it can be hampered by the lack of complete information as well as lack of time to evaluate the information and data already available. The logical and scientific foundation of this field ensures that there is a constant demand for those who are proficient in optimization. The need for graduates in this field is always present and accounted for as well as new research methodologies.

This book is an attempt to compile and collate the available data on the various methods of optimization. I am thankful to those who put their hard work and effort into the data in this book. I am also thankful to the support given by my family and friends during the compilation of this book.

Editor

Optimization of Integer Order Integrators for Deriving Improved Models of Their Fractional Counterparts

Maneesha Gupta and Richa Yadav

Advanced Electronics Lab, Department of ECE, Netaji Subhas Institute of Technology, Dwarka, New Delhi 110078, India

Correspondence should be addressed to Richa Yadav; yadav.richa1@gmail.com

Academic Editor: Manuel Lozano

Second and third order digital integrators (DIs) have been optimized first using Particle Swarm Optimization (PSO) with minimized error fitness function obtained by registering mean, median, and standard deviation values in different random iterations. Later indirect discretization using Continued Fraction Expansion (CFE) has been used to ascertain a better fitting of proposed integer order optimized DIs into their corresponding fractional counterparts by utilizing their refined properties, now restored in them due to PSO algorithm. Simulation results for the comparisons of the frequency responses of proposed 2nd and 3rd order optimized DIs and proposed discretized mathematical models of half integrators based on them, with their respective existing operators, have been presented. Proposed integer order PSO optimized integrators as well as fractional order integrators (FOIs) have been observed to outperform the existing recently published operators in their respective domains reasonably well in complete range of Nyquist frequency.

1. Introduction

In fractional calculus (FC) initially the development of mathematical models of fractional order integrators (FOIs) and fractional order differentiators (FODs) [1, 2] started presumably for searching different generalizing approaches for switching from integer order to the fractional order domain. The frequency response of ideal fractional order differ-integrator ($s^{\pm\alpha}$) is

$$H_{\text{fractional}}(j\omega) = (j\omega)^{\pm\alpha}, \quad (1)$$

where $j = \sqrt{-1}$ and ω gives angular frequency in radians. Variable "α" defines the order of fractional order operators and its value lies in between 0 and 1.

Due to its exotic nature FC has strengthened its grasp over different research areas such as statistical modeling, mechanical system analysis [3], control [4], automated control [5], instrumentation [6], signal processing [7], and radio engineering and image processing [8, 9]. At present time, FC is also extensively spreading its niche in different design methods for improving fractional order controllers by fractional integral and derivative functions [10, 11]. An important design method which has undoubtedly helped in sharpening up the focus of an efficient formulation of new blocks of fractional operators is *"linear interpolation"* method [12–16]. This most vastly followed tri-step method involves the following steps. (1) First step deals with linear interpolation of two existing digital integrators. (2) In second step, a new fractional order integrator is formulated by discretization of integer order integrator developed in first step by anyone of the two existing discretization (direct and indirect) schemes. (3) In the third step, this expanded series is truncated up to some arbitrary order (N), which defines the order of the new fractional operator ($s^{\pm\alpha}$). Direct discretization [17] works with the help of different expansion techniques such as Taylor Series Expansion (TSE) [18], Power Series Expansion (PSE) [19, 20], and Continued Fraction Expansion (CFE) [21, 22], whereas for indirect discretization [23–26] different rational approximations have been developed by many mathematicians, namely, Carlson and Halijak [27], Steiglitz [28], Khovanskii [29], Roy [30], Oustaloup et al. [31], and Maione [32], which proved like boon for the development of fractional integrators/differentiators.

In spite of the efficient and systematic nature of different previously mentioned approximations, still there existed a dissent about performance of these approximations, when these were later adapted by well-known discretization schemes for obtaining FOIs. Moreover development of new approximations required many exhaustive mathematical applications to make them perform in finite dimensional form in proper range of frequencies during discretization. Instead of deriving new methods of rational approximations, authors have suggested here an alternate solution in which the integrator operator (used as *s-z* transform) itself should be pushed near boundaries of an optimal solution (near ideal response) before converting it to fractional system. Reason behind this is that every minute dispensation in properties of this *integer* order operator will directly propel into its *noninteger* order counterpart. The main novelty proposed by this paper is to put together unique combination of optimized integer order operators and an accurate approximation technique for configuring their efficient fractional order (fo) operators. The current scenario of design methods for deriving optimal operators [33–41], is cumulatively proceeding towards "not so distant" phase of perfect optimization with negligible errors. So, another motivating point behind the approach proposed in this paper is to trespass all disadvantages of existing conventional design methods of FOIs by using refined capabilities of an efficient optimization scheme.

The frequency response of ideal integer order digital integrators (DIs) is

$$H_{\text{int}}(\omega) = \left(\frac{1}{j\omega}\right), \quad \text{where } |\omega| \le \pi, \tag{2}$$

and the frequency response of ideal integer order digital differentiators (DDs) is

$$H_{\text{diff}}(\omega) = j\omega, \quad \text{where } |\omega| \le \pi, \tag{3}$$

where $j = \sqrt{-1}$, ω is the angular frequency in radians, and "T" is the sampling period.

In recent novel advancements, different optimization algorithms, namely, Linear Programming (LP) [33, 34], Genetic Algorithm (GA) [35–37], Simulated Annealing (SA) [38, 39], and pole-zero (PZ) optimization [40, 41], have been used for obtaining more refined integer order differ-integrators. Jain et al. [35] have recently derived optimal models of recursive DIs and DDs by GA algorithm while Upadhyay [41] has developed these operators by analyzing poles and zeros of existing recursive DDs by PZ optimization technique for obtaining better results for frequency responses over wideband of complete frequency spectrum. Al-Alaoui [39] has done a credible work in improving design process of DIs and DDs based on Newton-Cotes integration rules by using linear interpolation and SA.

This paper attempts to find the optimal solutions of DIs by using an efficient modified application of Particle Swarm Optimization (PSO) technique [42–44], before discretizing them to their fractional mathematical models. Values of mean, median, and standard deviation in different random iterations have been continuously registered using PSO for the optimization of 2nd and 3rd order DIs with minimized error fitness function. Later these optimized DIs are discretized by CFE of indirect discretization scheme [23], for finding half integrator models of different orders. The simulation results of magnitude responses, phase responses, and relative magnitude errors for the proposed optimized DIs have been compared with all the recently published integrators optimized by different optimization algorithms, namely, LP [34], GA [35], SA [39] and PZ [41]. Frequency responses of the proposed FOIs based on 2nd and 3rd order optimized DIs have been also compared with their existing fractional models [2, 17] in complete frequency range.

The paper is organized as follows. Section 2 deals with the brief description of original and modified PSO algorithm for optimizing DIs along with its trade-off conditions, and the resultant 2nd and 3rd order optimized integrators thus derived have been presented in this section. Comparisons of proposed and existing DIs are also given in the same section. These optimized operators have been discretized by indirect discretization using CFE technique of indirect discretization for deriving models of FOIs in Section 3. Section 4 presents the simulation results of comparisons of proposed half integrators with responses of ideal and the existing models. Section 5 concludes the paper.

2. PSO Algorithm for Optimizing 2nd and 3rd Order Integer Order Integrators

2.1. Outline of Basic Functionality of Original PSO Algorithm. In conventional PSO algorithm [42–44], first a population of random solutions has been initialized and different generations are updated for searching optimal values. In this algorithm, the potential solutions, called particles, fly through the problem space by following the current optimum particles. The position of each particle encodes a possible solution to the given problem. The velocity of the particle is the parameter to be added to the current position to find the new position of the next generation. The velocity and position are updated for all particles in every generation unless an optimal solution is found. The particles keep track of its coordinates in the problem space which are associated with the best solution (fitness) it has achieved so far. These values are called pbest and the best element among all the pbest values of all the particles is called gbest. The velocities are initialized to zero and after finding the two best values, the particle updates its velocity and positions in accordance with the following update equations:

$$\begin{gathered} v(i) = v(i-1) + c_1 r_1 \left(p_{\text{pbest}} - p(i)\right) + c_2 r_2 \left(p_{\text{gbest}} - p(i)\right), \\ p(i+1) = p(i) + v(i), \end{gathered} \tag{4}$$

where $v(i)$ is the current particle velocity, $v(i-1)$ is previous particle velocity, $p(i)$ is current particle position, $p(i+1)$ is the position of particle in next generation, p_{pbest} is the best

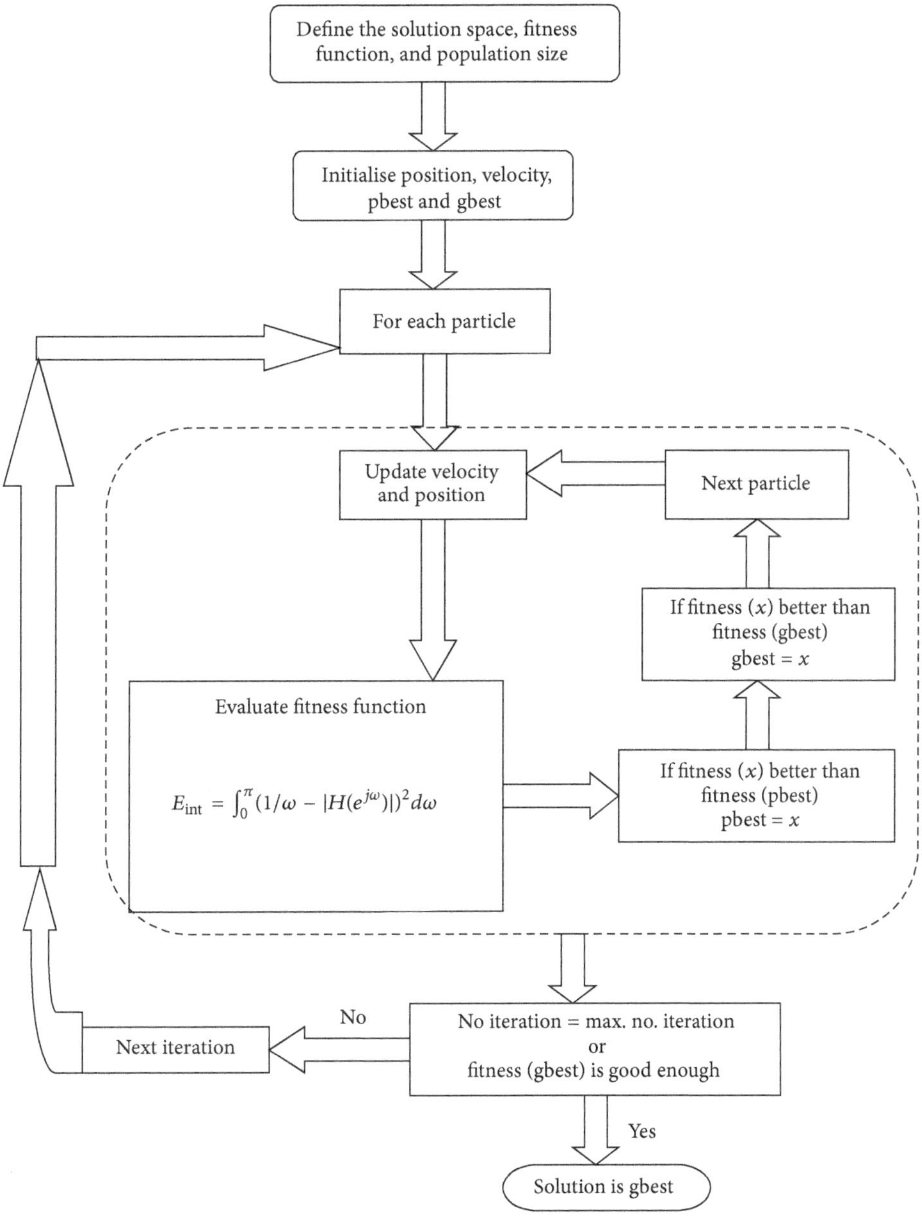

FIGURE 1: Flowchart of modified PSO algorithm.

position the particle has seen, p_{gbests} is the best position the swarm has seen, c_1, c_2 are constants set to 2, and r_1, r_2 are the random numbers between 0 and 1.

2.2. Parameters Used for Optimization and Trade-Offs between Multiple Parameters during Convergence of Optimized Digital Integrators. A mathematical model of transfer function (TF) of optimized digital integrator (DI) can be described in generalized form as

$$H(z) = \left(\frac{B_M (z)^{b_M} + \cdots + B_1 (z)^{b_1} + B_0 (z)^{b_0}}{A_N (z)^{a_N} + \cdots + A_1 (z)^{a_1} + A_0 (z)^{a_0}} \right), \quad (5)$$

where A_i and B_j, are arbitrary constants whereas a_i and b_j are real numbers with ($i = 0, 1, \ldots, N$) and ($j = 0, 1, \ldots, M$).

Transfer functions (TFs) of 2nd and 3rd order DIs which have been optimized using PSO in this paper are given in (6) and (7), respectively:

$$H_{\text{PSO\_2nd\_opt}}(z) = \left(\frac{B_2 z^2 + B_1 z + B_0}{A_2 z^2 + A_1 z + A_0} \right), \quad (6)$$

$$H_{\text{PSO\_3rd\_opt}}(z) = \left(\frac{B_3 (z)^3 + B_2 (z)^2 + B_1 z + B_0}{A_3 (z)^3 + A_2 (z)^2 + A_1 z + A_0} \right). \quad (7)$$

The coefficients of the previously mentioned TFs are generated by random iterations of PSO algorithm as it takes real numbers as particles.

TABLE 1: Transfer functions and fitness values of worst and average cases of 2nd order PSO optimized DIs.

TFs of worst cases of 2nd order DIs along with their fitness function values (F.V.)		TFs of average cases of 2nd order DIs along with their fitness function values (F.V.)	
Transfer function	F.V.	Transfer function	F.V.
$H_1(z)=\left(\frac{0.373324z^2+0.427523z+0.397575}{z^2-0.042561z-0.999916}\right)$	0.643052	$H_1(z)=\left(\frac{0.22896z^2-0.825829z-0.502494}{z^2-0.776284z-0.181729}\right)$	0.382107
$H_2(z)=\left(\frac{-0.929251z^2+0.381076z-0.75834}{z^2-0.725805z-0.306558}\right)$	3.90016	$H_2(z)=\left(\frac{-0.336478z^2-0.82224z+0.101197}{z^2-0.818946z-0.147828}\right)$	0.407093
$H_3(z)=\left(\frac{0.541699z^2+0.241831z+0.506633}{z^2-0.0552516z-1}\right)$	4.73234	$H_3(z)=\left(\frac{0.0849835z^2+0.752164z+0.756218}{z^2-0.167219z-0.882869}\right)$	0.50102
$H_4(z)=\left(\frac{-0.354583z^2-0.833923z-0.813869}{z^2-0.0637125z-1}\right)$	5.89578	$H_4(z)=\left(\frac{-0.39729z^2-0.795585z-0.130415}{z^2-0.722962z-0.318656}\right)$	0.618717
$H_5(z)=\left(\frac{0.353852z^2+0.203374z+0.466221}{z^2-0.0406162z-1}\right)$	18.799	$H_5(z)=\left(\frac{0.284382z^2+0.949776z+0.363676}{z^2-0.401952z-0.648204}\right)$	0.737534

Magnitude response of the optimized DI is

$$H_{\text{mag}} = |H(w)| = \frac{\text{abs (numerator)}}{\text{abs (denominator)}}. \quad (8)$$

Phase response of the optimized DI is

$$H_{\text{phase}} = \arg |H(w)|. \quad (9)$$

Relative magnitude error (dB) or RME value is given by

$$\begin{aligned} \text{RME} &= 20\log_{10}\left|\frac{H_{\text{ideal}}(\omega)-H_{\text{approx}}(\omega)}{H_{\text{ideal}}(\omega)}\right| \\ &= 20\log_{10}\left|\frac{(1/j\omega)-H_{\text{approx}}(\omega)}{(1/j\omega)}\right|, \end{aligned} \quad (10)$$

where $H_{\text{ideal}}(\omega)$ is the magnitude of ideal value of the operator and $H_{\text{approx}}(\omega)$ is the value of the digital integrator that is approximated.

In the recently published literature on digital and fractional operators most of the researchers have focused on formulating the fitness function for optimization technique based on only single parameter [38–41] that is magnitude error because practically if phase and magnitude are optimized for a multiparameter problem the solution is always stuck in a trade-off due to their contradictory nature. The reason behind this is that when the optimal solution starts converging towards the best results of one parameter (magnitude response) it simultaneously degrades optimality of second parameter (phase responses) after certain number of iterations. Solution for such trade-offs is to choose dominating parameter in problem under consideration. In this paper the design strategy to find optimized DIs is kept more concerned for minimum magnitude response error, which is comparatively more significant characteristic. The fitness function used to approximate a digital integrator is taken as the mean square error of the transfer function and it is given as

$$E_{\text{int}} = \int_0^{\pi}\left(\frac{1}{\omega}-\left|H\left(e^{j\omega}\right)\right|\right)^2 d\omega. \quad (11)$$

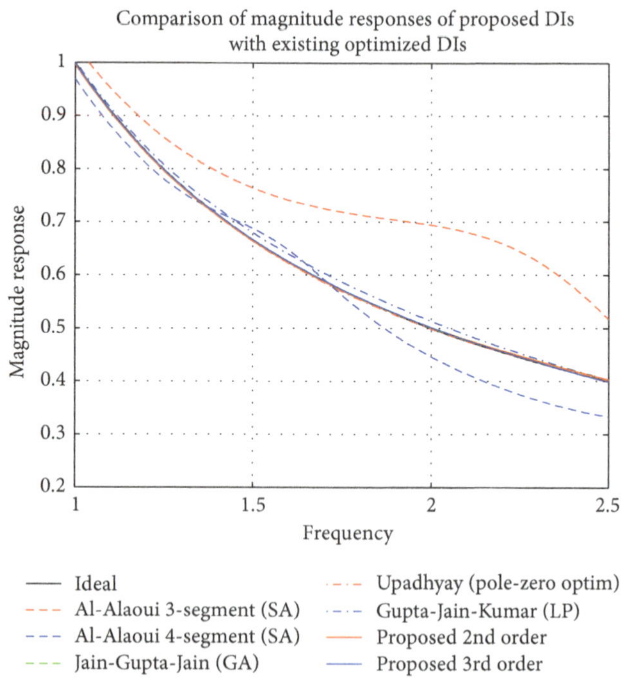

FIGURE 2: Comparison of magnitude responses of proposed 2nd and 3rd order optimized DIs with existing [26, 34, 39, 41] optimized models.

Aim of PSO optimization is to search the best optimal solution (TF of DI) with minimized mean square error.

2.3. Modified PSO Algorithm for Derivation of Optimized DIs. In this paper, a slight modification in the original PSO algorithm has been introduced in velocity, taking into account the given problem of deriving the best feasible optimized DIs with very small magnitude response errors *without* worsening their corresponding phase responses. In modified PSO used in this paper, the maximum velocity which constraints the movement of swarms to the range is

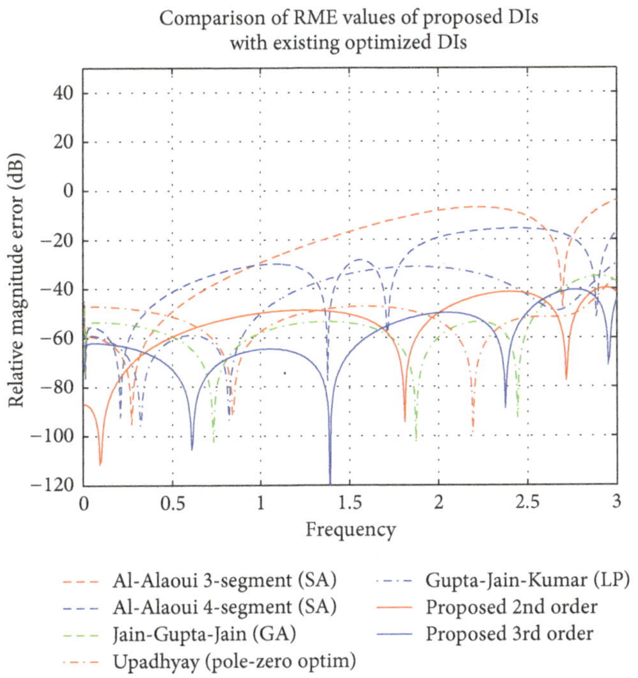

Figure 3: Comparison of RME values of proposed 2nd and 3rd order optimized DIs with existing [26, 34, 39, 41] optimized models.

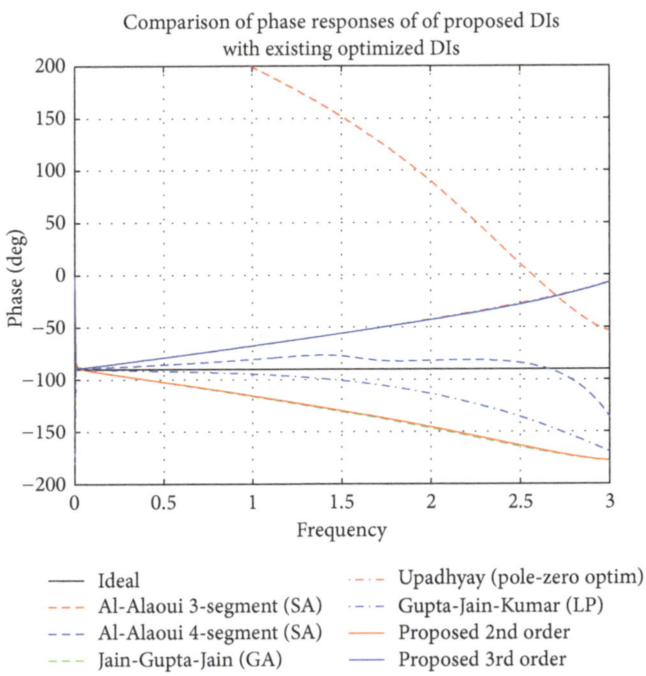

Figure 4: Comparison of phase responses of proposed 2nd and 3rd order optimized DIs with existing [26, 34, 39, 41] optimized models.

divided by a factor of ten after a certain number of iterations. Here we have used dynamic velocity control variant of PSO in which velocity is scaled down after every specific number of generations. This modification results in improved performance as compared to the original algorithm. In this paper the parameters used for PSO algorithm for finding

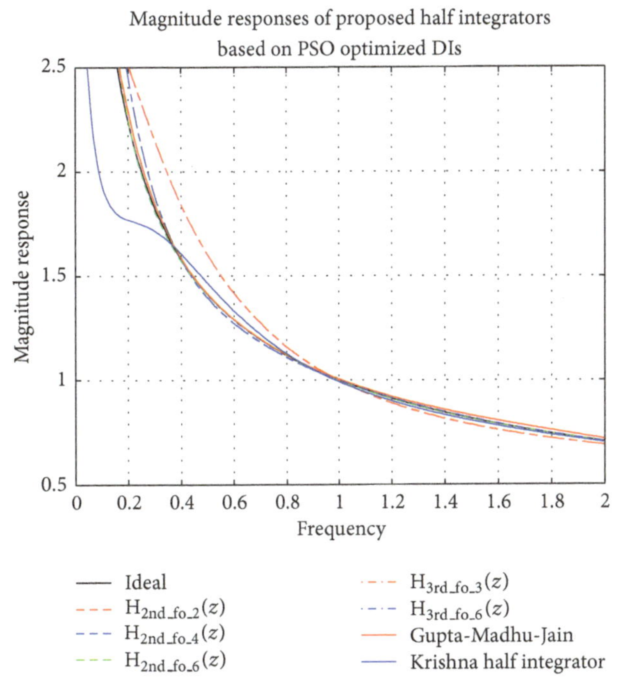

Figure 5: Comparison of magnitude responses of proposed half integrators based on 2nd and 3rd order optimized DIs with existing [2, 17] models.

optimized solutions of 2nd and 3rd order DIs are as follows: particle range = (−1, 1), learning factors (c_1 and c_2) = 2, diversity factor = 1, maximum velocity factor = 0.5, swarm size = 200, and number of generations = 200. The flowchart of modified PSO has been given in Figure 1.

Intel Core 2 Duo CPU T6600 @2.20 GHz (Installed Memory (RAM) of 4 GB) has been used for simulations. Transfer functions of worst and average cases of 2nd and 3rd order DIs have been given in Tables 1 and 2, respectively, along with their corresponding fitness function values (F.V.).

In this paper when the code was run in C++ for 100 iterations, the best optimized (with least mean square error) results for 2nd and 3rd order proposed DIs have been observed for 200 generations using 200 particles. Transfer functions (TFs) of proposed optimized integrators have been given next in pole zero form in (12) and (13), respectively.

TF of proposed 2nd order optimized integrator is

$$H_{\mathrm{PSO\_2nd\_opt}}(z) = \left(\frac{(0.8651\,(z+0.1042)\,(z+0.6276))}{((z-1)\,(z+0.5547))}\right). \tag{12}$$

TF of proposed 3rd order optimized integrator is

$$H_{\mathrm{PSO\_3rd\_opt}}(z) = \left(\frac{(0.8656\,(z+0.1057)\,(z+0.2021)\,(z+0.503))}{((z+0.38)\,(z+0.2522)\,(z-0.9997))}\right). \tag{13}$$

Proposed 2nd and 3rd order DIs optimized here by PSO have been compared with the recently published integrators optimized by different optimization algorithms, namely, LP

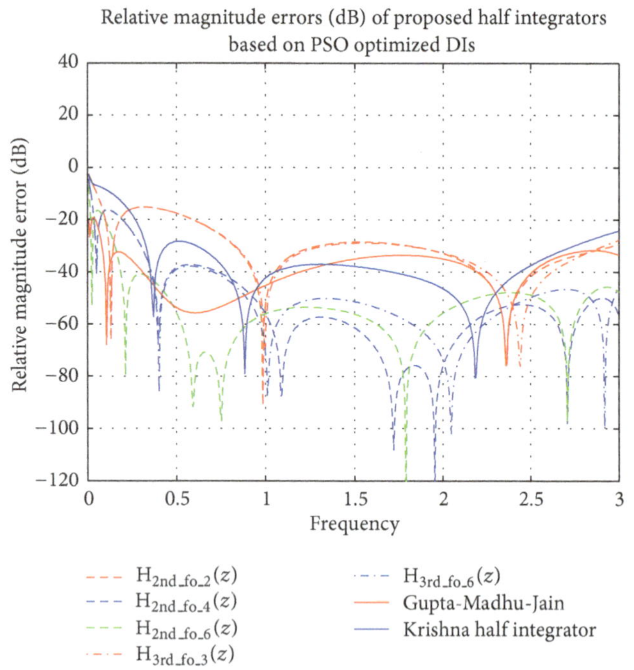

Figure 6: Comparison of RME values of proposed half integrators based on 2nd and 3rd order optimized DIs with existing [2, 17] models.

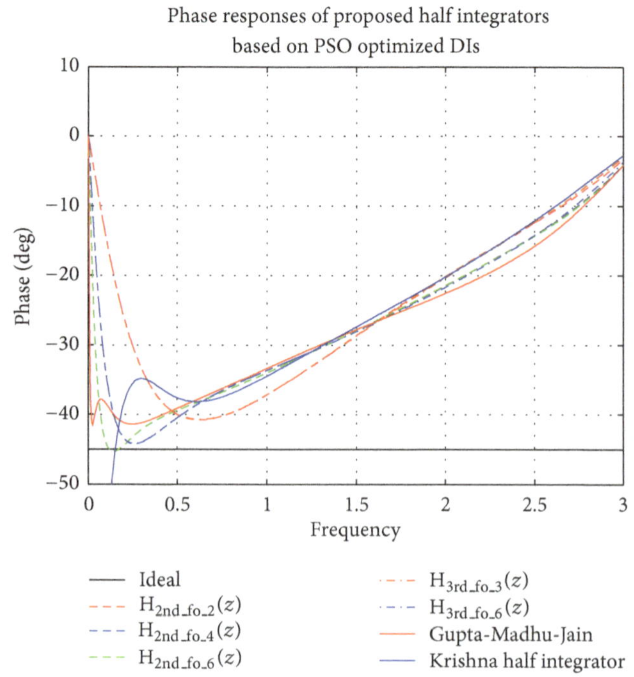

Figure 7: Comparison of phase responses of proposed half integrators based on 2nd and 3rd order optimized DIs with existing [2, 17] models.

[34], GA [35], SA [39], and PZ [41], for validating efficient performance of proposed PSO. Comparisons of bode diagrams (i.e., magnitude versus the frequency (w) and phase (degrees) versus frequency (w)) and RME responses (in dB) have been presented in Figures 2, 3 and 4 respectively.

Results for performance values (RME) for each optimization method have been compared in Table 3. It can be observed from Figure 3 that the proposed PSO method clearly outperforms existing SA [39] and LP [34] techniques and gives comparable results for PZ [41] and GA [35] methods. The 2nd order DI, namely, $H_{\text{PSO\_2nd\_opt}}(z)$ given in (12), outperforms Al-Alaoui 3-segment and 4-segment optimized DIs in complete spectrum. $H_{\text{PSO\_2nd\_opt}}(z)$ excels Jain-Gupta-Jain DI [35] in frequency range over $0 \leq \omega \leq 0.5\pi$ radians and $2.58 \leq \omega \leq 3\pi$ radians. PSO optimized proposed 3rd order DI, namely, $H_{\text{PSO\_3rd\_opt}}(z)$, clearly outperforms the existing Al-Alaoui optimized 3-segment, 4-segment [39], and Upadhyay [41] DIs in almost complete range with RME of ≤ -41 dB. $H_{\text{PSO\_3rd\_opt}}(z)$ given in (13) gives comparable results with GA optimized [35] operator. Proposed optimized DIs are observed to satisfy stability criterion and show linearly decreasing phase responses (see Figure 4).

Proposed DIs optimized by PSO clearly excel operators optimized by SA [39], LP [34], and PZ [41] techniques, but lie in close vicinity of the GA [35] responses. So a statistical test has been done for comparing performances of PSO and GA for validating superiority of the proposed modified algorithm. The products of number of particles and number of generations ($N_p * N_g$) have been varied from ($50 * 50$) to ($100 * 100$) and then to ($200 * 200$) for different iterations and values of mean, median, and standard deviation have been registered for all the worst, average, and best results of 2nd and 3rd order optimized DIs (see Table 4). From this table it is clear that PSO technique performs better than GA method and also it was noticed that with the increase in values of N_p, N_g, and number of iterations, the response of resultant DIs comes closer to the ideal response as mean, median, and standard deviation values go on decreasing with efficient minimization of error function. In this paper we have considered values of root mean square error.

3. Discretization of Proposed 2nd and 3rd Order Optimized DIs for Half Integrator Models

The optimized integer order DIs derived in Section 2 are used as s-to-z transformations for CFE based indirect discretization [2, 23, 24] scheme. Formula given by Khovanskii in [29] has been used for different rational approximations of FOIs. That is

$$(s)^{\pm\alpha} = \frac{1}{1-}\,\frac{\alpha s}{1+(1+\alpha)\,s-}\,\frac{(1+\alpha)\,s\,(1+s)}{2+(3+\alpha)\,s-}\,\frac{2\,(2+\alpha)\,s\,(1+s)}{3+(5+\alpha)\,s-}$$
$$\times\frac{3\,(3+\alpha)\,s\,(1+s)}{4+(9+\alpha)\,s-}\,\frac{4\,(4+\alpha)\,s\,(1+s)}{5+(9+\alpha)\,s-}$$
$$\times\frac{5\,(5+\alpha)\,s\,(1+s)}{6+(11+\alpha)\,s-}\,\frac{6\,(6+\alpha)\,s\,(1+s)}{7+(13+\alpha)\,s-}. \quad (14)$$

Infinite series of (14) can be terminated up to 2, 4, and 6 terms, and CFE based indirect discretization used in

Table 2: Transfer functions and fitness values of worst and average cases of 3rd order PSO optimized DIs.

TFs of worst cases of 3rd order DIs along with their fitness function values (F.V.)		TFs of average cases of 3rd order DIs along with their fitness function values (F.V.)	
Transfer function	F.V.	Transfer function	F.V.
$H_1(z)=\left(\frac{0.142394z^3+0.428511z^2-0.306611z+0.217758}{z^3-0.903135z^2-0.674295z+0.568774}\right)$	0.666364	$H_1(z)=\left(\frac{-0.65685z^3-0.806245z^2-0.820988z-0.19093}{z^3+0.100534z^2-0.473273z-0.550301}\right)$	0.548217
$H_2(z)=\left(\frac{-0.738721z^3-0.425561z^2-0.787252z+0.588187}{z^3-0.0845531z^2-0.714103z-0.24424}\right)$	2.38383	$H_2(z)=\left(\frac{-0.171166z^3+0.795027z^2+0.339133z+0.481229}{z^3-0.740329z^2-0.0821587z-0.130453}\right)$	0.651617
$H_3(z)=\left(\frac{0.524852z^3+0.958813z^2+0.585323z+0.779817}{z^3-0.462779z^2-0.380859z-0.247247}\right)$	3.91385	$H_3(z)=\left(\frac{0.522989z^3+0.163603z^2-0.0262705z+0.483852}{z^3-0.979861z^2-0.140418z+0.0831846}\right)$	0.673745
$H_4(z)=\left(\frac{0.979099z^3-0.852921z^2+0.960434z+0.875076}{z^3-0.827092z^2-0.246012z+0.135716}\right)$	4.06169	$H_4(z)=\left(\frac{0.225804z^3+0.780081z^2+0.758662z-0.0994224}{z^3-0.237348z^2-0.575985z-0.13299}\right)$	0.713815
$H_5(z)=\left(\frac{0.0594656z^3+0.700018z^2+0.346537z+0.314347}{z^3-0.791867z^2-0.573836bz+0.410591}\right)$	4.68934	$H_5(z)=\left(\frac{0.123855z^3+0.491766z^2+0.727444z+0.568523}{z^3-0.319642z^2-0.287981z-0.331056}\right)$	0.976067

Table 3: Comparisons of RME values of the proposed PSO optimized DIs with other integer order existing integrators optimized by different optimization techniques.

ω (π rad)	Al-Alaoui 3-segment (SA)	Al-Alaoui 4-segment (SA)	Jain-Gupta-Jain (GA)	Upadhyay (P-Z)	Gupta-Jain-Kumar (LP)	Proposed 2nd order DI (PSO)	Proposed 3rd order DI (PSO)
0.25	−72.03	−60.18	−54.46	−48.31	−71.06	−72.03	−63.82
0.5	−49.15	−41.21	−59.47	−51.65	−59.89	−59.68	−73.04
0.75	−37.25	−34.02	−80.04	−61.71	−63.91	−53.27	−71.45
1.0	−30.29	−30.34	−57.28	−57.28	−49.94	−49.88	−64.63
1.25	−22.72	−32.96	−53.84	−48.71	−39.33	−48.71	−68.2
1.75	−11.64	−35.05	−53.85	−47.91	−31.46	−62.62	−54.2
2.0	−8.24	−19.11	−60.53	−52.89	−31.43	−49.98	−49.94
2.25	−6.92	−15.99	−53.84	−61.5	−34.5	−42.77	−53.66
2.75	−18.19	−19.73	−37.54	−37.76	−45.88	−55.55	−40.61
3.0	−4.17	−18.11	−35.52	−36.44	−30.14	−41.34	−45.01

Table 4: Statistical test for comparison of proposed PSO and GA techniques for approximations of 2nd and 3rd order digital integrators.

	Genetic algorithm			Modified PSO algorithm		
	Approximating 2nd order digital integrator for comparing PSO and GA					
$N_p * N_g$	50 * 50	100 * 100	200 * 200	50 * 50	100 * 100	200 * 200
Mean	0.915405	0.230096	0.0874599	0.614370	0.201256	0.0936231
Median	0.482815	0.194402	0.075695	0.491575	0.114377	0.0488172
Standard deviation	0.737526	0.131617	0.0400587	0.43579	0.137471	0.042671
	Approximating 3rd order digital integrator for comparing PSO and GA					
$N_p * N_g$	50 * 50	100 * 100	200 * 200	50 * 50	100 * 100	200 * 200
Mean	0.915406	0.230096	0.0874599	0.633554	0.221398	0.0969881
Median	0.1478565	0.069402	0.02937085	0.1801123	0.1207528	0.0861992
Standard deviation	0.737526	0.131617	0.0400587	0.491451	0.152073	0.0530145

[24] has been adopted for deriving half integrators fitted in continuous-time domain (see Table 5). Also 2nd order optimized DI. resulted in 2nd, 4th, and 6th order FOIs whereas 3rd order optimized DI when substituted for different terms resulted in 3rd and 6th order half integrators. In the TFs of derived proposed half integrators given in Table 5, order of optimized DIs and order of FOIs are used as subscripts of symbol "H".

4. Comparisons of Proposed FOIs Based on 2nd and 3rd Order Optimized DIs with the Existing Models

Simulation results of comparison of magnitude responses, RME and phase responses of half integrators (see Table 5) based on different 2nd and 3rd order optimized DIs (given in (12) and (13)) with the existing half integrators given in [2, 17] and ideal half integrator have been presented in Figures 5–7.

Comparison of RME values of proposed and existing FOIs has been presented in Table 6. It is observed from Figure 6 that the proposed 2nd and 3rd order FOIs, namely, $H_{2nd\_fo\_2}(z)$ and $H_{3rd\_fo\_3}(z)$, only lag in performance but still these outperform the existing [2, 17] models in higher frequency spectrum. The 6th order operator $H_{2nd\_fo\_6}(z)$ is the best among the proposed FOIs and it clearly outperforms all the existing models [2, 17] in complete frequency spectrum with RME of ≤−40 dB. Proposed $H_{2nd\_fo\_2}(z)$ and $H_{3rd\_fo\_3}(z)$ outperform Gupta-Madhu-Jain half integrator [17] in frequency range over $0.88 \le \omega \le 3\pi$ radians. $H_{2nd\_fo\_4}(z)$, $H_{2nd\_fo\_6}(z)$, and $H_{3rd\_fo\_6}(z)$ half integrators excel Krishna half integrator [2] in almost complete range. All the proposed FOIs are observed to be stable and follow linear phase curve in almost complete frequency spectrum (see Figure 7).

5. Conclusions

This paper attempts to focus on a simple yet unaddressed fact which till date has remained in the backdrop as the optimization of an integer order operator before converting it into its fractional the order (fo) counterparts. Proposed FOIs validate that properties of an efficiently optimized DI pass directly into its fractional domain when it is used as s-to-z transformation during indirect discretization. Here, 2nd and 3rd order integer order integrators have been explored by PSO algorithm for finding their optimal solutions for a minimized error fitness function. The simulation results of integer order operators have revealed the effectiveness of the proposed modified PSO algorithm, because of their low

TABLE 5: Mathematical models of proposed half integrators based on 2nd and 3rd order PSO optimized DIs.

Mathematical models of half integrators based on 2nd and 3rd order PSO optimized DIs	
Name of PSO optimized DIs	Transfer functions of FOIs derived by indirect discretization of CFE
$H_{\text{PSO\_2nd\_opt}}(z)$	$H_{\text{2nd-fo-2}}(z) = \left(\frac{0.9302\,(z-0.1825)\,(z+0.5869)}{(z-0.7422)\,(z+0.5604)}\right)$ $H_{\text{2nd-fo-4}}(z) = \left(\frac{0.93011\,(z-0.003372)\,(z+0.5633)\,(z+0.6071)\,(z-0.6396)}{(z-0.9034)\,(z-0.293)\,(z+0.5788)\,(z+0.5566)}\right)$ $H_{\text{2nd-fo-6}}(z) = \left(\frac{\left(\begin{array}{c}0.93011\,(z-0.8073)\,(z-0.3427)\,(z+0.616)\\(z+0.5758)\,(z+0.5587)\,(z+0.04899)\end{array}\right)}{\left(\begin{array}{c}(z+0.5936)\,(z+0.5649)\,(z+0.5556)\\(z-0.5924)\,(z-0.9502)\,(z-0.1107)\end{array}\right)}\right)$
$H_{\text{PSO\_3rd\_opt}}(z)$	$H_{\text{3rd-fo-3}}(z) = \left(\frac{0.93046\,(z-0.1852)\,(z+0.2354)\,(z+0.433)}{(z-0.742)\,(z+0.3892)\,(z+0.2492)}\right)$ $H_{\text{3rd-fo-6}}(z) = \left(\frac{\left(\begin{array}{c}0.93038\,(z-0.6395)\,(z+0.4678)\,(z+0.3939)\\(z+0.2476)\,(z+0.223)\,(z-0.008033)\end{array}\right)}{\left(\begin{array}{c}(z-0.9031)\,(z-0.2946)\,(z+0.2512)\\(z+0.2397)\,(z+0.383)\,(z+0.4194)\end{array}\right)}\right)$

TABLE 6: Comparisons of RME values of the proposed half integrators based on optimized DIs with the existing models.

ω (π rad)	Proposed half integrators based on 2nd order PSO optimized DI			Proposed half integrators based on 3rd order PSO optimized DI		Gupta-Madhu-Jain half integrator	Krishna half integrator
	$H_{\text{2nd-fo-2}}(z)$	$H_{\text{2nd-fo-4}}(z)$	$H_{\text{2nd-fo-6}}(z)$	$H_{\text{3rd-fo-3}}(z)$	$H_{\text{3rd-fo-6}}(z)$		
0.25	−15.79	−24.55	−43.25	−15.79	−29.26	−35.53	−17.42
0.5	−17.52	−38.74	−55.85	−17.52	−38.97	−53.62	−28.31
0.75	−25.82	−41.56	−85.14	−25.81	−42.61	−53.65	−35.56
1.0	−58.3	−58.3	−55.44	−64.01	−74.75	−45.26	−42.05
1.25	−31.25	−57.34	−53.48	−31.13	−50.67	−38.93	−37.13
1.75	−28.94	−81.15	−75.32	−29.82	−58.25	−33.59	−40.57
2.0	−32.62	−75.83	−55.54	−32.62	−73.25	−34.88	−46.46
2.25	−41.43	−57.13	−48.99	−41.15	−58.93	−43.69	−55.28
2.75	−32.99	−59.6	−59.6	−36.33	−46.94	−32.6	−29.8
3.0	−29.56	−56.03	−47.75	−28.27	−46.74	−33.74	−24.3

orders and very less magnitude errors as compared to existing ones. Proposed FOIs based on the optimized DIs present very less RME values of the order of −40 dB and show linear phase curves in almost full band of Nyquist frequency.

References

[1] L. Debnath, "Recent applications of fractional calculus to science and engineering," *International Journal of Mathematics and Mathematical Sciences*, vol. 2003, no. 54, pp. 3413–3442, 2003.

[2] B. T. Krishna, "Studies on fractional order differentiators and integrators: a survey," *Signal Processing*, vol. 91, no. 3, pp. 386–426, 2011.

[3] N. M. F. Ferreira, F. B. Duarte, M. F. M. Lima, M. G. Marcos, and J. A. T. Machado, "Application of fractional calculus in the dynamical analysis and control of mechanical manipulators," *Fractional Calculus and Applied Analysis*, vol. 11, pp. 91–113, 2008.

[4] S. Manabe, "The noninteger integral and its application to control systems," *English Translation Journal Japan*, vol. 6, pp. 83–87, 1961.

[5] J. A. T. Machado and A. M. S. Galhano, "A new method for approximating fractional derivatives: application in nonlinear control," in *Proceedings of the 6th EUROMECH Nonlinear Dynamics Conference (ENOC '08)*, pp. 4–8, Saint Petersburg, Russia, July 2008.

[6] Y. Ferdi, J. P. Herbeuval, A. Charef, and B. Boucheham, "R wave detection using fractional digital differentiation," *ITBM-RBM*, vol. 24, no. 5-6, pp. 273–280, 2003.

[7] J. Mocak, I. Janiga, M. Rievaj, and D. Bustin, "The use of fractional differentiation or integration for signal improvement," *Measurement Science Review*, vol. 7, pp. 39–42, 2007.

[8] B. Mathieu, P. Melchior, A. Oustaloup, and C. Ceyral, "Fractional differentiation for edge detection," *Signal Processing*, vol. 83, no. 11, pp. 2421–2432, 2003.

[9] Y. Pu, W. Wang, J. Zhou, Y. Wang, and H. Jia, "Fractional differential approach to detecting textural features of digital

image and its fractional differential filter implementation," *Science in China F*, vol. 51, no. 9, pp. 1319–1339, 2008.

[10] S. Das, I. Pan, S. Das, and A. Gupta, "Improved model reduction and tuning of fractional-order $PI^{\lambda}D^{\mu}$ controllers for analytical rule extraction with genetic programming," *ISA Transactions*, vol. 51, no. 2, pp. 237–261, 2012.

[11] S. Das, I. Pan, K. Halder, S. Das, and A. Gupta, "Impact of fractional order integral performance indices in LQR based PID controller design via optimum selection of weighting matrices," in *Proceedings of the International Conference on Computer Communication and Informatics (ICCCI '12)*, pp. 1–6, Coimbatore, India, January 2012.

[12] Y. Chen and B. M. Vinagre, "A new IIR-type digital fractional order differentiator," *Signal Processing*, vol. 83, no. 11, pp. 2359–2365, 2003.

[13] Y. Chen, B. M. Vinagre, and I. Podlubny, "Continued fraction expansion approaches to discretizing fractional order derivatives-an expository review," *Nonlinear Dynamics*, vol. 38, no. 1–4, pp. 155–170, 2004.

[14] Y. Q. Chen and K. L. Moore, "Discretization schemes for fractional-order differentiators and integrators," *IEEE Transactions on Circuits and Systems I*, vol. 49, no. 3, pp. 363–367, 2002.

[15] M. A. Al-Alaoui, "Novel class of digital integrators and differentiators," *IEEE Transactions on Signal Processing*, p. 1, 2009.

[16] M. A. Al-Alaoui, "Al-Alaoui operator and the new transformation polynomials for discretization of analogue systems," *Electrical Engineering*, vol. 90, no. 6, pp. 455–467, 2008.

[17] M. Gupta, M. Jain, and N. Jain, "A new fractional order recursive digital integrator using continued fraction expansion," in *Proceedings of the India International Conference on Power electronics (IICPE)*, pp. 1–3, New Delhi, India, 2010.

[18] M. Gupta, P. Varshney, and G. S. Visweswaran, "Digital fractional-order differentiator and integrator models based on first-order and higher order operators," *International Journal of Circuit Theory and Applications*, vol. 39, no. 5, pp. 461–474, 2011.

[19] F. Leulmi and Y. Ferdi, "An improvement of the rational approximation of the fractional operator s^{α}," in *Proceedings of the Saudi International Electronics, Communications and Photonics Conference (SIECP '11)*, pp. 1–6, Riyadh, Saudi Arabia, 2011.

[20] Y. Ferdi, "Computation of fractional order derivative and integral via power series expansion and signal modelling," *Nonlinear Dynamics*, vol. 46, no. 1-2, pp. 1–15, 2006.

[21] G. S. Visweswaran, P. Varshney, and M. Gupta, "New approach to realize fractional power in z-domain at low frequency," *IEEE Transactions on Circuits and Systems II*, vol. 58, no. 3, pp. 179–183, 2011.

[22] P. Varshney, M. Gupta, and G. S. Visweswaran, "Switched capacitor realizations of fractional-order differentiators and integrators based on an operator with improved performance," *Radioengineering*, vol. 20, no. 1, pp. 340–348, 2011.

[23] B. T. Krishna and K. V. V. Reddy, "Design of fractional order digital differentiators and integrators using indirect discretization," *An International Journal for Theory and Applications*, vol. 11, pp. 143–151, 2008.

[24] B. T. Krishna and K. V. V. Reddy, "Design of digital differentiators and integrators of order $1/2$," *World Journal of Modelling and Simulation*, vol. 4, pp. 182–187, 2008.

[25] R. Yadav and M. Gupta, "Design of fractional order differentiators and integrators using indirect discretization approach," in *Proceedings of the 2nd International Conference on Advances in Recent Technologies in Communication and Computing (ARTCom '10)*, pp. 126–130, Kottayam, India, October 2010.

[26] R. Yadav and M. Gupta, "Design of fractional order differentiators and integrators using indirect discretization scheme," in *Proceedings of the India International Conference on Power Electronics (IICPE '10)*, New Delhi, India, January 2011.

[27] G. E. Carlson and C. A. Halijak, "Simulation of the fractional derivative operator $(1/s)^{1/2}$ and the fractional integral operator $(1/s)^{1/2}$," *Kansas State University Bulletin*, vol. 45, pp. 1–22, 1961.

[28] S. K. Steiglitz, "Computer-aided design of recursive digital filters," *IEEE Transactions Audio and Electroacoustics*, vol. AU-18, no. 2, pp. 123–129, 1970.

[29] A. N. Khovanskii, *The Application of Continued Fractions and their Generalizations to Problems in Approximation Theory*, P. Noordhoff Ltd., 1963, Translate by Peter Wynn.

[30] S. C. Roy, "Some exact steady-state sinewave solutions of nonuniform RC lines," *British Journal of Applied Physics*, vol. 14, pp. 378–380, 1963.

[31] A. Oustaloup, F. Levron, B. Mathieu, and F. M. Nanot, "Frequency-band complex noninteger differentiator: characterization and synthesis," *IEEE Transactions on Circuits and Systems I*, vol. 47, no. 1, pp. 25–39, 2000.

[32] G. Maione, "Laguerre approximation of fractional systems," *Electronics Letters*, vol. 38, no. 20, pp. 1234–1236, 2002.

[33] N. Papamarkos and C. Chamzas, "A new approach for the design of digital integrators," *IEEE Transactions on Circuits and Systems I*, vol. 43, no. 9, pp. 785–791, 1996.

[34] M. Gupta, M. Jain, and B. Kumar, "Recursive wideband digital integrator and differentiator," *International Journal of Circuit Theory and Applications*, vol. 39, no. 7, pp. 775–782, 2011.

[35] N. Jain, M. Gupta, and M. Jain, "Linear phase second order recursive digital integrators and differentiators," *Radioengineering Journal*, vol. 21, pp. 712–717, 2012.

[36] C.-C. Hse, W.-Y. Wang, and C.-Y. Yu, "Genetic algorithm-derived digital integrators and their applications in discretization of continuous systems," in *Proceedings of the CEC Congress on Evolutionary Computation*, pp. 443–448, Honolulu, Hawaii, USA, 2002.

[37] S. Das, B. Majumder, A. Pakhira, I. Pan, S. Das, and A. Gupta, "Optimizing continued fraction expansion based iir realization of fractional order differ-integrators with genetic algorithm," in *Proceedings of the International Conference on Process Automation, Control and Computing (PACC '11)*, pp. 1–6, Coimbatore, India, July 2011.

[38] G. K. Thakur, A. Gupta, and D. Upadhyay, "Optimization algorithm approach for error minimization of digital integrators and differentiators," in *Proceedings of the International Conference on Advances in Recent Technologies in Communication and Computing (ARTCom '09)*, pp. 118–122, Kottayam, India, October 2009.

[39] M. A. Al-Alaoui, "Class of digital integrators and differentiators," *IET Signal Processing*, vol. 5, no. 2, pp. 251–260, 2011.

[40] D. K. Upadhyay and R. K. Singh, "Recursive wideband digital differentiator and integrator," *Electronics Letters*, vol. 47, no. 11, pp. 647–648, 2011.

[41] D. K. Upadhyay, "Class of recursive wide band digital differentiators and integrators," *Radioengineering Journal*, vol. 21, no. 3, pp. 904–910, 2012.

[42] J. Kennedy and R. Eberhart, "Particle swarm optimization," in *Proceedings of the IEEE International Conference on Neural Networks*, pp. 1942–1948, December 1995.

[43] Y.-T. Kao and E. Zahara, "A hybrid genetic algorithm and particle swarm optimization for multimodal functions," *Applied Soft Computing Journal*, vol. 8, no. 2, pp. 849–857, 2008.

[44] Y. Shi and R. C. Eberhart, "Empirical study of particle swarm optimization," in *Proceedings of the IEEE international Conference Evolution Computer*, pp. 1945–1950, Washington, DC, USA, 1999.

2

Physics-Inspired Optimization Algorithms: A Survey

Anupam Biswas, K. K. Mishra, Shailesh Tiwari, and A. K. Misra

Department of Computer Science & Engineering, Motilal Nehru National Institute of Technology Allahabad, Allahabad 211004, India

Correspondence should be addressed to K. K. Mishra; mishrakrishn@gmail.com

Academic Editor: Qingsong Xu

Natural phenomenon can be used to solve complex optimization problems with its excellent facts, functions, and phenomenon. In this paper, a survey on physics-based algorithm is done to show how these inspirations led to the solution of well-known optimization problem. The survey is focused on inspirations that are originated from physics, their formulation into solutions, and their evolution with time. Comparative studies of these noble algorithms along with their variety of applications have been done throughout this paper.

1. Introduction

Leonid Kantorovich introduced linear programming for optimizing production in plywood industry in 1939 and probably it was the first time the term optimization of a process was used, though Fermat and Lagrange used calculus for finding optima and Newton and Gauss proposed methods for moving towards an optimum. Every technological process has to achieve optimality in terms of time and complexity and this led the researchers to design and obtain best possible or better solutions. In previous studies, several mathematical solutions were provided by various researchers such as LP [1], NLP [2] to solve optimization problems. The complexity of the proposed mathematical solutions is very high which requires enormous amount of computational work. Therefore, alternative solutions with lower complexity are appreciated. With this quest, nature-inspired solutions are developed such as GA [3], PSO [4], SA [5], and HS [6]. These nature-inspired metaheuristic solutions became very popular as the algorithms provided are much better in terms of efficiency and complexity than mathematical solutions. Generally, these solutions are based on biological, physical, and chemical phenomenon of nature.

In this paper, the algorithms inspired by the phenomenon of physics are reviewed, surveyed, and documented. This paper mainly focuses on the following issues:

(i) most inspirational facts and phenomena,

(ii) their formulation into a solution,

(iii) parameters considered for this formulation,

(iv) effectiveness of these parameters,

(v) variation with time in inspiration,

(vi) other biological influences,

(vii) convergence, exploration, and exploitation,

(viii) Various applications.

The rest of the paper is organized as follows. Section 2 overviews the history of physics-inspired algorithms and also the description of few major algorithms. In Section 3 a correlative study of these major algorithms is done on the basis of their inspirational theory and formulation method. Various parameters used in these algorithms along with their variants and respective applications are also discussed in this section. In Section 4, finally the overall study is concluded.

2. Historical Study

Both simplicity and efficiency attract researchers towards natural phenomenon, resulting in some popular algorithms such as GA [3] based on Darwin's principle of survival of the fittest, SA [5] in 1983 based on the annealing process of metal, PSO [4] in 1995 based on the behavior of fishes and birds swarms, and HS [6] in 2001 based on the way a musician adjusts instruments to obtain good harmony. Richard Feynman's proposal of quantum computing system [7, 8], inspired

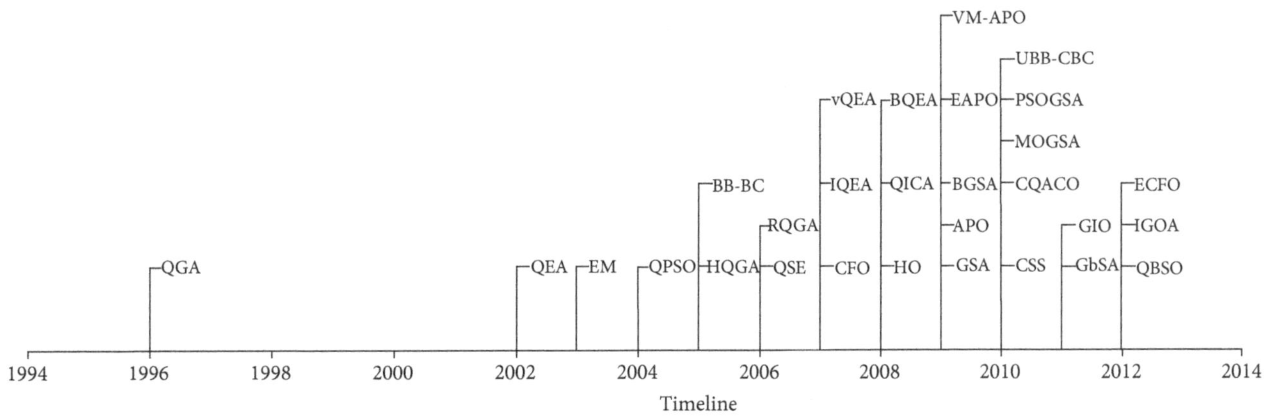

FIGURE 1: Evolution of physics-inspired optimization algorithms.

by quantum mechanics in 1982, paved way for physics-inspired optimization algorithms. With this, the concept of quantum computing was developed and in 1995 Narayanan and Moore [9] proposed Quantum-Inspired Genetic Algorithm (QGA). This is the beginning of physics-inspired optimization algorithms. After half a decade later, in 2002, Han and Kim proposed Quantum-Inspired Evolutionary Algorithm (QEA). In 2004 Quantum-Inspired Particle Swarm Optimization (QPSO) was proposed by Sun et al. [10] and in 2007 another swarm-based Quantum Swarm Evolutionary Algorithm (QSE) was proposed by Wang et al. [11]. Apart from the quantum mechanics, other principles and theorems of physics also begun to draw the attention of researchers. In 2003, Birbil and Fang [12] proposed Electromagnetism-like (EM) mechanism based on the superposition principle of electromagnetism. Big Bang-Big Crunch (BB-BC) [13] based on hypothetical theorem of creation and destruction of the universe was proposed in 2005. Based on Newton's gravitational law and laws of motion algorithms emerged such as CFO [14] by Formato in 2007, GSA by Rashedi et al. [15], APO by Xie et al. [16] in 2009, and GIO by Flores et al. [17] in 2011. Hysteretic Optimization (HO) [18] based on demagnetization process was proposed in 2008. In 2010, Kaveh and Talatahari proposed CSS [19] based on electrostatic theorems such as Coulomb's law, Gauss's law, and superposition principle from electrostatics and Newton's laws of motion. In 2011, Shah-Hosseini proposed Spiral Galaxy-Based Search Algorithm (GbSA) [20]. Jiao et al. [21] proposed QICA in 2008 based on quantum theory and immune system. Li et al. [22] proposed CQACO based on quantum theory and ant colony in 2010. Most recently in 2012, Zhang et al. [23] proposed IGOA based on gravitational law and immune system, and Jinlong and Gao [24] proposed QBSO based on quantum theory and bacterial forging. These major algorithms along with their modified, improved, and hybrid versions along with the year of proposal are shown in Figure 1. We have categorized these algorithms with their variants and their notion of inspiration as follows:

(A) Newton's gravitational law
(i) Pure physics
(1) CFO
(a) Variant (pure physics)
(1) ECFO
(2) APO
(a) Variant (pure physics)
(1) EAPO
(2) VM-APO
(3) GSA
(a) Variant (pure physics)
(1) BGSA
(2) MOGSA
(b) Variant (Semiphysics)
(1) PSOGSA
(4) GIO
(ii) Semiphysics
(1) IGOA

(B) Quantum mechanics
(i) Pure physics
(1) QGA
(a) Variant (pure physics)
(1) RQGA
(2) QGO
(b) Variant (semiphysics)
(1) HQGA
(2) QEA
(a) Variant (pure physics)
(1) BQEA
(2) vQEA
(3) IQEA
(ii) Semiphysics
(1) QPSO
(2) QSE
(3) QICA
(4) CQACO
(5) QBSO

(C) Universe theory
(i) Pure physics
(1) BB-BC
(a) Variant (pure physics)
(1) UBB-CBC
(2) GbSA

(D) Electromagnetism
  (i) Pure physics
    (1) EM
(E) Glass demagnetization
  (i) Pure physics
    (1) HO
(F) Electrostatics
  (i) Pure physics
    (1) CSS.

3. Algorithms

3.1. Newton's-Gravitation-Law-Based Algorithms

3.1.1. CFO. CFO [14] is inspired by the theory of particle kinematics in gravitational field. Newton's universal law of gravitation implies that larger particles will have more attraction power as compared to smaller particles. Hence, smaller ones will be attracted towards the larger ones. As a result, all smaller particles will be attracted towards the largest particle. This largest particle can be resembled as global optimum solution in case of optimization. To mimic this concept in CFO, a set of solutions is considered as probes on the solution space. Each probe will experience gravitational attraction due to the other. Vector acceleration experienced by probe p with respect to other probes at iteration t is given by the equation below:

$$a_t^p = G\sum_{\substack{k=1 \\ k\neq p}}^{N} U\left(F_t^k - F_t^p\right)\cdot\left(F_t^k - F_t^p\right)^{\alpha} \times \frac{\left(R_t^k - R_t^p\right)}{\left\|R_t^k - R_t^p\right\|^{\beta}}. \quad (1)$$

Here, G is CFO's gravitational constant, R_t^p and F_t^p are the position of a probe p and objective function value at that position, respectively, at iteration t, R_t^k and F_t^k are the position of all other probe, and objective function value at that position, respectively, at iteration t, $U(\cdot)$ is the unit step function. The CFO exponents α and β, by contrast, have no analogues in nature but these exponents provide flexibility to the algorithm. These parameters have drastic effect on overall exploration and convergence of the algorithm. The algorithm does not have any apparent mechanism for exploitation.

In this equation, $U(F_t^k - F_t^p)\cdot(F_t^k - F_t^p)^{\alpha}$ defines CFO's mass which is analogous to real objects mass in space.

The a_t^p causes the probe p to move from position R_t^p to R_{t+1}^p and the new location is obtained by the following equation:

$$R_{t+1}^p = R_t^p + \frac{1}{2}a_t^p\Delta t^2. \quad (2)$$

Here, Δt is the time interval between iterations. Recently, Ding et al. proposes an extended version of CFO, namely, ECFO [25]. Applications of this algorithm are neural network [26] and antenna applications [27, 28].

3.1.2. APO. APO [16] is based on the concept of artificial physics or physicomimetics [29], which was applied to robots. Analogous to Newton's gravitation law a new kind of force law is defined as follows:

$$F = G\frac{m_1 m_2}{r^p}, \quad (3)$$

where F is the force exerted between particles m_1 and m_2 in a hypothetical universe, G is gravitational constant, and r is the distance between particles m_1 and m_2. Unlike real universe, the value of p is not always equal to 2; instead it varies from −5 to +5.

Mass is defined in APO as follows:

$$m_i = e^{(f(x_{\text{best}})-f(x_i))/(f(x_{\text{worst}})-f(x_{\text{best}}))}. \quad (4)$$

Considering value of $p = -1$ in (3), the force in APO is defined as follows:

$$F_{ij}^k = \begin{cases} Gm_i m_j\left(x_j^k - x_i^k\right) & \text{if } f\left(X_j\right) < f\left(X_i\right) \\ -Gm_i m_j\left(x_j^k - x_i^k\right) & \text{if } f\left(X_j\right) \geq f\left(X_i\right), \end{cases} \quad (5)$$

where F_{ij}^k is the kth component of force exerted on particle i by particle j, x_i^k and x_j^k are the kth dimension of particles i and j, respectively. The kth component of the *total* force F_i^k exerted on particle i by all other particles is given by the following:

$$F_i^k = \sum_{j=1}^{N} F_{ij}^k \quad \forall i \neq \text{best}. \quad (6)$$

Velocity and positions of particles are updated with following equation:

$$v_i^k(t+1) = wv_i^k(t) + \frac{\lambda F_i^k}{m_i},$$
$$x_i^k(t+1) = x_i^k(t) + v_i^k(t+1), \quad (7)$$

where λ is uniformly distributed random variable in $[0, 1]$, w is user-defined weight $0 < w < 1$.

Main exploitation and convergence component of APO algorithm is the computation of force exerted on each particle by others. Overall exploration of algorithm is controlled by the weight parameter w. The parameter λ is actually for putting limitation to convergence. But, due to randomness, it also serves for exploration. To overcome the lack of convergence component, an extended version of APO is proposed in [30], where individual particle's best position is tracked in iteration and utilized in velocity updating. A vector model of APO is defined in [31].

3.1.3. GSA. GSA [15] is inspired by Newton's law of universal gravitation and law of motion. In addition to this, another fact of physics is also considered, according to which the actual value of gravitational constant G depends on the actual age of the universe. So G at time t can be expressed as follows:

$$G(t) = G(t_0) \times \left(\frac{t_0}{t}\right)^{\beta}, \quad \beta < 1, \quad (8)$$

where $G(t_0)$ is the value of the gravitational constant at the first cosmic quantum-interval of time t_0, β is a time-dependent exponent.

In GSA, the solution space is considered as an imaginary universe. Every point in solution space is considered as an agent having mass. To compute mass of any agent i, a parameter $m_i(t)$ is computed. The parameter $m_i(t)$ and mass $M_i(t)$ of agent i are computed as follows:

$$m_i(t) = \frac{\text{fit}_i(t) - \text{worst}(t)}{\text{best}(t) - \text{worst}(t)}, \quad M_i(t) = \frac{m_i(t)}{\sum_{j=1}^{N} m_j(t)}, \tag{9}$$

where $\text{fit}_i(t)$ is the fitness value of the agent i at time t.

Force exerted on each considered agent is computed as follows:

$$F_{ij}^d = G(t)\frac{M_{pi}(t) \times M_{aj}(t)}{R_{ij}(t) + \varepsilon}\left(x_j^d(t) - x_i^d(t)\right), \quad R_{ij} = \left\|X_i(t)\,X_j(t)\right\|_2, \quad F_i^d(t) = \sum_{j=1, j\neq i}^{N} \text{rand}_j F_{ij}^d(t), \tag{10}$$

where F_{ij}^d is the force acting on mass M_i and mass M_j at time t, M_{aj} is the active gravitational mass related to agent j, M_{pi} is the passive gravitational mass related to agent i, $G(t)$ is gravitational constant at time t, ε is a small constant, $R_{ij}(t)$ is the Euclidian distance between two agents i and j, $F_i^d(t)$ is the total force that acts on agent i in a dimension d at time t, and rand_j is a random number in the interval $[0, 1]$.

Acceleration of any agent i at time t in direction d is computed with equation given below:

$$a_i^d(t) = \frac{F_i^d}{M_i(t)}. \tag{11}$$

The next position of each agent and at which velocity they will move is calculated as follows:

$$v_i^d(t+1) = \text{rand}_i \times v_i^d(t) + a_i^d(t), \quad x_i^d(t+1) = x_i^d(t) + v_i^d(t+1). \tag{12}$$

The concept of variable gravitational constant G provides a good mechanism for convergence to the algorithm. As in subsequent iterations the value of G gradually increases, attraction force experienced by each agent also increases. Thus, agents converge towards the better agents with incremental attraction. However, the effect of attraction force is controlled by a random parameter. This random control of force ensures exploitation as well as exploration. Another random parameter used in velocity updating also implies exploration of search space.

In [32] binary version of GSA is proposed, multiobjective GSA [33] is proposed by Mirjalili and Hashim, and a hybrid of PSO and GSA is proposed in [34].

Applications of GSA algorithm are in power system [35–42], economic dispatch problem [43, 44], Wessinger's equation [45], fuzzy system [46, 47], forecasting of future oil demand [48], slope stability analysis [49], clustering [50–52], prototype classification [53], feature selection [54], web services [55], PID controller [56], antenna application [47], and so forth.

3.1.4. IGOA. IGOA [23] algorithm is an improved version of GSA [15]. The gravitation-law-based algorithm GSA can easily fall into local optimum solution and convergence rate is also comparatively slow [57]. To overcome these problems, IGOA introduces new operators which are inspired from biological immune system (BIS) [58]. In BIS mainly two kinds of activities take place, activities of antigens and activities of antibody. An antigen can only be negotiated with corresponding right antibody which comes from mother during birth. But for an unknown antigen, BIS also can act accordingly by learning. That means BIS has immune memory and antibody diversity. IGOA mimics this mechanism to avoid falling into local optimum. In this case, local optimum is similar to the unknown antigen in BIS. In IGOA vaccination and memory antibody replacement is used to improve the convergence speed and antibody diversity mechanism to catch the diversity of solution space along with GSA. IGOA is a newly proposed algorithm and not yet applied in any real-life application.

3.1.5. GIO. GIO [17] algorithm is similar to GSA [15] and CSS [19] where each point in the search space is assigned mass and charges, respectively. Although perspective of assigning masses or charges to each point is similar, the way of assignment and notion is different. CSS is inspired from electrostatic dynamics law, whereas GSA is inspired from Newton's gravitational laws and laws of motion. GIO is also inspired from Newton's law but unlike GSA this algorithm keeps hypothetical gravitational constant G as constant. Force exerted between two bodies is computed as follows:

$$F_{ij} = \frac{M(f(B_i)) \cdot M(f(B_j))}{\left|B_i - B_j\right|^2}\widehat{B}_{ij}, \tag{13}$$

where B_i is the position of the ith body and B_j is that of jth body, F_{ij} is exerting force on the mass B_i; $|B_i - B_j|$ is the Euclidean distance and $\widehat{B}_{ij}$ is the unit vector between bodies B_i and B_j; $f(B_i)$ is the fitness of body B_i, $M(f(B_i))$ is corresponding mass of body B_i and is computed as follows:

$$M(f(B_i)) = \left(\frac{f(B_i) - \min f(B)}{\max f(B) - \min f(B)} \times (1 - \text{mapMin}) + \text{mapMin}\right)^2, \tag{14}$$

where $\min f(B)$ is the minimum fitness value of the positions of the bodies so far, $\max f(B)$ is the maximum fitness value of the positions of the bodies so far. mapMin is a constant used to limit the fitness value $f(B_i)$ to a mass in the interval [mapMin, 1). As each body are interacts with other bodies so resultant force acting on body B_i is computed as follows:

$$F_i = \sum_{j=1}^{n} \frac{M(f(B_i)) \cdot M\left(f\left(B_j^{\text{best}}\right)\right)}{\left|B_i - B_j^{\text{best}}\right|^2} B_i \widehat{B}_j^{\text{best}}. \quad (15)$$

Velocity with which a body will move to its new position is computed as follows:

$$V_{t+1} = \chi\left(V_t + R \cdot C \cdot B_k\right), \quad (16)$$

where V_t is the current velocity of B_i, R is a random real number generated in the range of $[0, 1)$, C is the gravitational interaction coefficient, B_k is the displacement of body B_i and is computed with (17) and χ is the inertia constraint, and is computed with (18):

$$B_k = \sqrt{\frac{M(f(B_i))}{|F_i|}} \widehat{F}_i, \quad (17)$$

$$\chi = \frac{2k}{\left|2 - \phi - \sqrt{\phi^2 - 4\phi}\right|}. \quad (18)$$

In (18), k is an arbitrary value in the range $(0, 1]$, $\phi = C_1 + C_2 > 4$, where C_1 and C_2 are the cognitive and the gravitational interaction constants, respectively.

New position of a body is obtained by adding the computed velocity corresponding to it. Formula given above for computing velocity is for unimodal optimization problems, which is further modified for multimodal optimization problems as follows:

$$V_{t+1} = \chi\left(V_t + C_1 \cdot R_1 \cdot \left(B^{\text{best}} - B\right) + C_2 \cdot R_2 \cdot B_k\right). \quad (19)$$

Here, R_1 and R_2 are real random numbers in the range $[0, 1)$.

New position for next iteration is obtained by adding the updated velocity with current body as follows:

$$B_{t+1} = B_t + V_{t+1}. \quad (20)$$

Certain precaution has been taken during resultant force computation in order to avoid numerical errors. The force between masses $M(B_i)$ and $M(B_j^{\text{best}})$ is computed only if $|B_i - B_j| \geq 1 \times 10^{-5}$. In order to avoid division by 0, B_k is computed only if for a body B_i resultant force $F_i > 0$.

The concept of inertia constant χ is similar to the concept of constriction parameter in constricted PSO [59–61]. Exploration of a body in GIO is controlled by this parameter. Exploration, exploitation, and convergence are ensured by computation of mass and resultant force. The inertia constant χ also helps in convergence. Though Flores et al. [17] shows GIO's superiority over PSO in multimodal problems but it has not been yet applied in any real-life application.

3.2. Quantum-Mechanics-Based Algorithms

3.2.1. QGA. According to quantum mechanics, electrons are moving around the nucleus in an arc path, known as orbits. Depending on the angular momentum and energy level, electrons are located in different orbits. An electron in lower level orbit can jump to higher level orbit by absorbing certain amount of energy; similarly higher level electron can jump to lower energy level by releasing certain amount of energy. This kind of jumping is considered as discrete. There is no intermediate state in between two energy levels. The position where an electron lies on the orbit is unpredictable; it may lie at any position in orbit at a particular time. Unpredictability of electron's position is also referred as superposition of electron.

In classical computing, a bit is represented either by 0 or 1, but in quantum computing this is termed as qubit. State of a qubit can be 0 or 1 or both at the same time in superposition state. This superposition of qubit mimics the superposition of electrons or particles. State of qubit at any particular time is defined in terms of probabilistic amplitudes. The position of an electron is described in terms of qubits by a vector called quantum state vector. A quantum state vector can be described with the equation given below:

$$|\Psi\rangle = \alpha|0\rangle + \beta|1\rangle, \quad (21)$$

where α and β are complex numbers that specify the probability amplitudes of obtaining the qubit in "0" state and in "1" state, respectively. In this case, the value of α and β always satisfies the equation $|\alpha|^2 + |\beta|^2 = 1$. For n positions of electrons, states can be described by n state vectors. These n positions of an electron can be known simultaneously.

QGA [9] utilized the concept of parallel universe in GA [3] to mimic quantum computing. According to this parallel universe interpretation, each universe contains its own version of population. All populations follow the same rules, but one universe can interfere in population of other universe. This interference occurs as in the form of a different kind of crossover called interference crossover, which provides good exploration capability to the algorithm. In QGA, all the solutions are encoded using superposition and all of these solutions may not be valid, which creates problems during implementation of crossover. Udrescu et al. propose RQGA [62], which provides a mechanism to overcome this problem. Hybrid versions [63] merge QGA with permutation-based GA and [64] merge QGA with real-valued GA. Malossini and Calarco propose QGOA [65] very similar to QGA with special quantum-based selection and fitness evaluation methods.

Many applications have been developed in recent years on the basis of this algorithm such as structural aligning [66], clustering [67, 68], TSP [69], combinatorial optimization problem [70], web information retrieval [71], computational grid [72], software testing [73], dynamic economic dispatch [74], area optimization [75], operation prediction [76], computer networking [77, 78], PID controller [79], multivariate problem [80], course timetabling [81], minimal redact [82], image applications [83–86], smart antenna [87], hardware [88], fuzzy system [89, 90], neural network [91], and robot application [92].

3.2.2. QEA. Quantum bit and superposition of states are the main basis of this algorithm. QEA [93] is originally inspired by quantum computing, which itself is inspired by the quantum mechanics. In QEA, the state of a qubit or Q-bit is represented as pair of numbers (α, β) in a column matrix $\left[\begin{smallmatrix}\alpha\\ \beta\end{smallmatrix}\right]$, where $|\alpha|^2 + |\beta|^2 = 1$ and $|\alpha|^2$ gives the probability that the Q-bit will be found in the "0" state and $|\beta|^2$ gives the probability that the Q-bit will be found in the "1" state.

A Q-bit individual which is a string of Q-bits is defined as follows:

$$\left[\begin{array}{c|c|c|c} \alpha_1 & \alpha_2 & \cdots & \alpha_m \\ \beta_1 & \beta_2 & \cdots & \beta_m \end{array}\right], \quad (22)$$

where $|\alpha_i|^2 + |\beta_i|^2 = 1$, $i = 1, 2, \ldots, m$. With this Q-bit representation, a population set is formulated and operations are performed on that population. Zhang and Gao further improved this algorithm as IQEA [94], by introducing probability amplitude ratio $\gamma_\alpha = |\beta|/|\alpha|$ if $\alpha \neq 0$ and $\gamma_\beta = |\alpha|/|\beta|$ if $\beta \neq 0$ to define relative relationship between α and β. As quantum rotation gate is unable to cover the entire search space since it outputs discrete values, a mechanism for calculating rotation angle of quantum rotation gate is defined. Platel et al. propose versatile QEA [95], with introducing new concept of hitchhiking phenomenon into QEA with little bit elitism in updating parameters and P. Li and S. Li propose Bloch QEA [96] based on Bloch coordinates depicted by qubits. Here α and β are defined as $\cos(\theta/2)$ and $e^{i\varphi}\sin(\theta/2)$, respectively. This θ and φ define bloch points.

Applications of QEA-related algorithms are combinatorial optimization [97, 98], image segmentation [99], Knapsack Problems [100–102], resource optimization [103, 104], numerical optimization [105, 106], extrusion [107], unit commitment problem [108, 109], power system [110, 111], signaling [112], face identification [113, 114], financial data analysis [115], Option pricing model calibration [116, 117], stock market prediction [118], and so forth.

3.2.3. QSE. QSE [11] takes the concepts from both QEA [93] and PSO [4]. Similar to PSO's swarm intelligent concept, quantum swarms are represented using Q-bits. Unlike QEA, representation of Q-bit in QSE changes probabilistic parameters. α and β are replaced with angular parameters $\sin\theta$ and $\cos\theta$, here θ is quantum angle. Q-bit $[\theta]$ is represented as $\left[\begin{smallmatrix}\sin\theta\\ \cos\theta\end{smallmatrix}\right]$, where $|\sin\theta|^2 + |\cos\theta|^2 = 1$. For m Q-bits, this can be represented as $\left[\begin{smallmatrix}\sin\theta_1 & \sin\theta_1 & \cdots & \sin\theta_1\\ \cos\theta_1 & \cos\theta_1 & \cdots & \cos\theta_1\end{smallmatrix}\right]$. Each bit position x_{ij}^t of each individual, at time t, is determined with the following

$$x_{ij}^t = \begin{cases} 1 & \text{if random } [0,1] > \left|\cos\theta_{ij}\right|^2 \\ 0 & \text{otherwise.} \end{cases} \quad (23)$$

Velocity is updated as in PSO. Another quantum-swarm-based PSO called QPSO was proposed by Sun et al. [10]. Unlike QSE state of particle is not determined by the probabilistic angular parameters. Here, state of particle is determined by a wave function $\Psi(x)$ as follows:

$$\Psi(x) = \frac{1}{\sqrt{L}} e^{-(\|c-x\|/L)}. \quad (24)$$

Here, c and x are the center or current best and current location vector, L is called creativity or imagination parameter of particle. Location vector is defined as:

$$x(t) = c \pm \frac{L}{2} \ln\left(\frac{1}{R}\right) \quad (25)$$

Here, R is a random number in range $[0, 1]$. The creativity parameter L is updated as follows:

$$L(t+1) = 2 \times \alpha \times |c - x(t)|. \quad (26)$$

Here, α is the creative coefficient and acts as main ingredient for convergence towards the optima. Huang et al. [119] have improved this later on by considering global best instead of current best.

Applications of these algorithms are flow shop scheduling [120], unit commitment problem [121, 122], neural network [123], power system [124–126], vehicle routing problem [127–129], engineering design [130, 131], mining association rules [132], and so forth.

3.2.4. QICA. Basic concept of QICA [21] is Artificial Immune System's clonal selection, which is hybridized with the framework of quantum computing. Basic quantum representational aspect is similar to QEA [93]. QICA introduces some new operators to deal with premature convergence and diverse exploration. The clonal operator Θ is defined as follows:

$$\Theta(Q(t)) = \left[\Theta(q_1)\,\Theta(q_2)\cdots\Theta(q_n)\right]^T, \quad (27)$$

where $Q(t)$ is quantum population and $\Theta(q_i) = I_i q_i$, I_i is the identity matrix of dimensionality D_i, which is given by the following:

$$D_i = \left\lceil N_c \times \frac{A(q_i)}{\sum_{i=1}^{n} A(q_i)} \right\rceil. \quad (28)$$

Here, $A(\cdot)$ is function for adaptive self-adjustment and N_c is a given value relating to the clone scale. After cloning these are added to population.

The immune genetic operator consists of two main parts, that is, quantum mutation and recombination. Before performing quantum mutation, population is guided towards the best one by using following equation:

$$\begin{bmatrix} \alpha_i^u \\ \beta_i^u \end{bmatrix} = U(\theta_i) \times \begin{bmatrix} \alpha_i^p \\ \beta_i^p \end{bmatrix}, \quad (29)$$

where α_i^u and β_i^u are updated values, α_i^p and β_i^p are previous values of probabilistic coefficients, $U(\theta_i) = \left[\begin{smallmatrix} rl\sin\theta_i & -\sin\theta_i \\ \cos\theta_i & \cos\theta_i \end{smallmatrix}\right]$ is quantum rotation gate and θ_i is defined as follows:

$$\theta_i = k \times f(\alpha_i, \beta_i), \quad (30)$$

where k is a coefficient which determines the speed of convergence and the function $f(\alpha_i, \beta_i)$ determines the search direction. This updated population is mutated using quantum NOT gate as:

$$Q^u(t) = \sqrt{1 - |Q^u(t)|^2}. \quad (31)$$

Quantum recombination is similar to interference crossover in QGA [9]. Finally, the clonal selection operator selects the best one from the population observing the mutated one and original population. Clonal operator of QICA increases explorative power drastically in contrast to QEA.

3.2.5. CQACO. CQACO [22] merges quantum computing and ACO [133]. Ant's positions are represented by quantum bits. This algorithm also represents qubits similar to QEA [93] and QICA [21] and uses the concept of quantum rotation gate as in QICA. Similar to CQACO, Wang et al. [134] proposed quantum ant colony optimization. Another variant is proposed by You et al. [135] in 2010. Quantum concept with ACO provides good exploitation and exploration capability to these algorithms. Applications of these algorithms are fault diagnosis [136], robot application [137], and so forth.

3.2.6. QBSO. QBSO [24] is the newest among all the quantum-based algorithms. This algorithm is semiphysics-inspired, as it incorporates concepts of both bacterial forging and quantum theory. In other words, QBSO is an improved version of BFO [138]. As BFO is unable to solve discrete problems, QBSO deals with this problem by using quantum theory to adapt the process of the BFO to accelerate the convergence rate. BFO consists of chemotaxis, swarming, reproduction, elimination, and dispersal processes whereas QBSO consists mainly of three of them, chemotaxis, reproduction, and elimination dispersal. In QBSO also the qubit is defined as in (21). Quantum bacterium of S bacteria is represented in terms of the three processes, that is, bit position, chemotactic step, and reproduction loop.

The ith quantum bacterium's quantum jth bit at the mth chemotactic step of the nth reproduction loop in the pth elimination dispersal event is updated as follows:

$$\Psi_{ij}(m+1,n,p) = \begin{cases} \sqrt{1-\left(\Psi_{ij}(m,n,p)\right)^2}, & \text{if } \left(\theta_{ij}^{i+1} = 0 \text{ and } \rho(m+1,n,p) < c_1\right) \\ \text{abs}\left(\begin{array}{c} \Psi_{ij}(m,n,p) \times \cos\theta_{ij}^{i+1} \\ -\sqrt{1-\left(\Psi_{ij}(m,n,p)\right)^2} \times \sin\theta_{ij}^{i+1} \end{array}\right), & \text{otherwise}, \end{cases} \quad (32)$$

where θ_{ij}^{t+1} is the quantum rotation angle, which is calculated through (33), t is the iteration number of the algorithm, ρ_{ij} is uniform random number in range $[0, 1]$, and c_1 is mutation probability which is a constant in the range $[0, 1]$.

After updating quantum bacterium, the corresponding bit position in the population x_{ij} is updated with (34), where γ_{ij} is uniform random number between 0 and 1:

$$\theta_{ij}^{i+1} = e_1\left(b_j(m,n,p) - x_{ij}(m,n,p)\right). \quad (33)$$

Here, e_1 is attracting effect factor and b_j is the jth bit position of global optimal bit:

$$x_{ij}(m+1,n,p) = \begin{cases} 1, & \text{if } \gamma_{ij}(m+1,n,p) > \left(\Psi_{ij}(m+1,n,p)\right)^2 \\ 0, & \text{if } \gamma_{ij}(m+1,n,p) \le \left(\Psi_{ij}(m+1,n,p)\right)^2. \end{cases} \quad (34)$$

Fitness value of each point solution in population is represented as the health of that particular bacterium.

3.3. Universe-Theory-Based Algorithms

3.3.1. BB-BC. BB-BC [13] algorithm is inspired mainly from the expansion phenomenon of Big Bang and shrinking phenomenon of Big Crunch. The Big Bang is usually considered to be a theory of the birth of the universe. According to this theory all space, time, matter, and energy in the universe were once squeezed into an infinitesimally small volume and a huge explosion was carried out resulting in the creation of our universe. From then onwards, the universe is expanding. It is believed that this expansion of the universe is due to Big Bang. However, many scientists believe that this expansion will not continue forever and all matters would collapse into the biggest black hole pulling everything within it, which is referred as Big Crunch.

BB-BC algorithm has two phases, namely, Big Bang phase and Big Crunch phase. During Big Bang phase, new population is generated with respect to center of mass. During Big Crunch phase, the center of mass is computed which resembles black hole (gravitational attraction). Big Bang phase ensures exploration of solution space. Big Crunch phase fullfills necessary exploitation as well as convergence.

BB-BC algorithm suffers botching all candidates into a local optimum. If a candidate with best fitness value converges to an optima at the very beginning of the algorithm, then all remaining candidates follow that best solution and trapped into local optima. This happens because the initial population is not uniformly distributed in the solution space. So, this algorithm provides a methodology to obtain uniform initial population in BB-BC. Initially, that is, at level 1, two candidates C_1 and C_2 are considered; at level 2, C_1 and C_2 are subdivided into C_3 and C_4; at level 3, C_1 and C_2 are again divided into C_4, C_5, C_6, C_7, C_8, C_9, and C_{10} and so on. This kind of division continues until we get the required numbers of candidates for initial population. In this way at nth level, C_1 and C_2 are subdivided into $2^n - 2$ candidates and include in population. In addition to this, in Big Crunch phase chaotic map is introduced, which improves convergence speed of algorithm. In this Chaotic Big Crunch phase, next position of each candidate is updated as follows:

$$x_i^{\text{new}} = x_c \pm \frac{\alpha(t)\left(x_{\max} - x_{\min}\right)}{t}, \quad (35)$$

where $\alpha_{t+1} = cf(\alpha(t))$, $0 < \alpha(t) < 1$, here $cf(x)$ is a chaotic map or function. BB-BC with uniform population is called UBB-BC and with chaotic map is called BB-CBC. If both are used, then it is called UBB-CBC.

Applications of this algorithm are fuzzy system [139–141], target tracking [142, 143], smart home [144], course timetabling [145], and so forth.

3.3.2. GbSA. GbSA [20] is inspired by spiral arm of spiral galaxies to search its surrounding. This spiral movement recovers from botching into local optima. Solutions are adjusted with this spiral movement during local search as well. This algorithm has two components:

(1) SpiralChaoticMove,

(2) LocalSearch.

SpiralChaoticMove actually mimics the spiral arm nature of galaxies. It searches around the current solution by spiral movement. This kind of movement uses some chaotic variables around the current best solution. Chaotic variables are generated with formula $x_{n+1} = \lambda x_n(1 - x_n)$. Here, $\lambda = 4$ and $x_0 = 0.19$. In this way, if it obtains a better solution than the current solution, it immediately updates and goes for *LocalSearch* to obtain more suitable solution around the newly obtained solution. GbSA is applied to Principle Component Analysis (PCA). *LocalSearch* ensures exploitation of search space and *SpiralChaoticMove* provides good exploration mechanism of search space ensuring reachability of algorithm towards the global optimum solution.

3.4. Electromagnetism-Based Algorithms

3.4.1. Electromagnetism-Like: EM. EM [12] algorithm is based on the superposition principle of electromagnetism, which states that the force exerted on a point via other points is inversely proportional to the distance between the points and directly proportional to the product of their charges. Points in solution space are considered as particles. The charge of each point is computed in accordance with their objective function value. In classical physics, charge of a particle generally remains constant, but in this heuristic the charge of each point is not constant and changes from iteration to iteration. The charge q^i of each point i determines its power of attraction or repulsion. This charge of a particle is evaluated as follows:

$$q^i = e^{(-n((f(x^i)-f(x^{\text{best}}))/\sum_{k=1}^{m}(f(x^k)-f(x^{\text{best}}))))}, \quad \forall i, \tag{36}$$

where m is the total number of points and n is the number of dimensions. This formula shows that points having better objective values will possess higher charges. This heuristic does not use signs to indicate positive or negative charge as in case of electric charge. So, direction of force (whether attractive or repulsive force) is determined by the objective function values (fitness) of two particular points. If point x^j has better value than x^i, then corresponding force is considered as attractive otherwise repulsive. That means that x^{best} attracts all other points towards it. The total force F^i (attractive or repulsive) exerted on point i is computed by the following equation:

$$F^i = \sum_{j\neq i}^{m} \begin{cases} \left(x^j - x^i\right) \dfrac{q^i q^j}{\left\|x^j - x^i\right\|^2} \\ \quad \text{if } f\left(x^j\right) < f\left(x^i\right) \\ \left(x^i - x^j\right) \dfrac{q^i q^j}{\left\|x^j - x^i\right\|^2} \\ \quad \text{if } f\left(x^j\right) \geq f\left(x^i\right). \end{cases}, \quad \forall i. \tag{37}$$

After evaluating the total force vector F^i, the point i is moved in the direction of the force with a random step length as given in (38). Here, RNG is a vector whose components denote the allowed feasible movement towards the upper bound or the lower bound

$$x^i = x^i + \lambda \frac{F^i}{\left\|F^i\right\|}(\text{RNG}) \qquad i = 1, 2, \ldots, m. \tag{38}$$

EM algorithm provides good exploration and exploitation mechanism with computation of charge and force. Exploration and convergence of EM are controlled by the random parameter λ. Exploration is also controlled with RNG, by limiting movements of particles.

Debels et al. [146] propose a hybrid version of EM combining the concept of GA with EM.

Numerous applications are developed on the basis of this algorithm such as scheduling problems [147–150], course timetabling [151], PID controller [152], fuzzy system [153–155], vehicle routing problem [156], networking [157], inventory control [158], neural network [159, 160], TSP [161, 162], feature selection [163], antenna application [164], robotics application [165], flow path designing [166], and vehicle routing [167].

3.5. Glass-Demagnetization-Based Algorithms

3.5.1. HO. HO [18] is inspired by the demagnetization process of a magnetic sample. A magnetic sample comes to a very stable low-energy state called ground state, when it is demagnetized by an oscillating magnetic field of slowly decreasing amplitude. After demagnetization, the system are shakeup repeatedly to obtain improved result. HO simulates these two processes of magnetic sample to get low-energy state by repeating demagnetization followed by a number of shakeups.

The process of demagnetization is mainly for exploration and convergence. After exploring, better solutions are searched by performing a number of shake-up operations. The algorithm possesses two kinds of stopping conditions, firstly fixed number of shakeups n_s for each instance of a given size N and secondly required number of shakeups M_{req} to obtain the current low-energy state or global optimum. Besides this M_{req} repetition of shakeup in current low-energy state, minimum number of shakeups $n_{s\,\min}$ is set to ensure that algorithm does not accept suboptimum too early. Similarly, the maximum number of shakeups $n_{s\max}$ is also set to avoid wasting time on hard situation. HO algorithm is applied to TSP [168], spin glasses [169], vehicle routing problem [170], protein folding [171], and so forth.

3.6. Electrostatics-Based Algorithms

3.6.1. CSS. This algorithm inherits Coulomb's law, Gauss's law and superposition principle from electrostatics, and the Newtonian laws of mechanics. CSS [19] deploys each solution as a Charged Particle (CP). If two charged particles having charges q_i and q_j reside at distance r_{ij}, then according to Coulomb's law electric force F_{ij} exerted between them is as follows:

$$F_{ij} = k_e \frac{q_i q_j}{r_{ij}^2}. \tag{39}$$

Here, k_e is a constant called the Coulomb constant. Now if q_i amount of charges is uniformly distributed within a sphere of radius "a" then electric field E_{ij} at a point outside the sphere is as follows:

$$E_{ij} = k_e \frac{q_i}{r_{ij}^2}. \tag{40}$$

The electric field at a point inside the sphere can be obtained using Gauss's law as follows:

$$E_{ij} = k_e \frac{q_i}{a^3} r_{ij}. \tag{41}$$

The resultant force on a charge q_j at position r_j due to the electric field of a charge q_i at position r_i can be expressed in vector form as:

$$F_{ij} = E_{ij} q_j \frac{r_i - r_j}{\left\| r_i - r_j \right\|}. \tag{42}$$

For multiple charged particles, this equation can be expressed as follows:

$$F_j = k_e q_j \sum_{i,i \neq j} \left(\frac{q_i}{a^3} r_{ij} \cdot i_1 + \frac{q_i}{r_{ij}^2} \cdot i_2 \right) \times \frac{r_i - r_j}{\left\| r_i - r_j \right\|} \begin{cases} i_1 = 1,\ i_2 = 0 \Longleftrightarrow r_{ij} < a \\ i_1 = 0,\ i_2 = 1 \Longleftrightarrow r_{ij} \geq a. \end{cases} \tag{43}$$

To formulate the concept of charges into CSS algorithm, a set of charged particles are considered. Each point in the solution space is considered as possible positions of any charged particle. Charges in each charged particle is computed as follows:

$$q_i = \frac{\text{fit}(i) - \text{fitworst}}{\text{firbest} - \text{fitworst}}. \tag{44}$$

Distance among particles is computed with the following:

$$r_{ij} = \frac{\left\| X_i - X_j \right\|}{\left\| \left(X_i + X_j \right)/2 - X_{\text{best}} \right\| + \varepsilon}. \tag{45}$$

Radius of particle is computed with

$$a = 0.10 \times \max \left(\left\{ x_{i,\max} - x_{i,\min} \mid i = 1, 2, \ldots, n \right\} \right). \tag{46}$$

The value of the resultant electrical force acting on a charged particle is determined as follows:

$$F_j = q_j \sum_{i,i \neq j} \left(\frac{q_i}{a^3} r_{ij} \cdot i_1 + \frac{q_i}{r_{ij}^2} \cdot i_2 \right) P_{ij} \left(X_i - X_j \right). \tag{47}$$

Here, P_{ij} defines the attractiveness or repulsiveness of the force exerted. A good particle may attract a bad one and similarly bad one can also attract good one. So, if bad one attracts good one, then it is not suitable for an optimization problem. The parameter P_{ij} limits these kinds of attractions as follows:

$$P_{ij} = \begin{cases} 1, & \text{if } \left(\dfrac{\text{fit}(i) - \text{fitbest}}{\text{fit}(j) - \text{fit}(i)} > \text{rand} \right) \\ & \text{or } \left(\text{fit}(j) > \text{fit}(i) \right) \\ 0, & \text{otherwise.} \end{cases} \tag{48}$$

Again in Newtonian mechanics or classical mechanics the velocity v of a particle is defined as follows:

$$v = \frac{r_{\text{new}} - r_{\text{old}}}{t_{\text{new}} - t_{\text{old}}} = \frac{r_{\text{new}} - r_{\text{old}}}{\Delta t}. \tag{49}$$

Displacement from r_{old} to r_{new} position along with acceleration a can be expressed as follows:

$$r_{\text{new}} = \frac{1}{2} a \cdot \Delta t^2 + v_{\text{old}} \cdot \Delta t + r_{\text{old}}. \tag{50}$$

Newton's second law states that "the acceleration of an object is directly proportional to the net force acting on it and inversely proportional to its mass;" that is, $F = ma$, so r_{new} can be expressed as follows:

$$r_{\text{new}} = \frac{1}{2} \frac{F}{m} \cdot \Delta t^2 + v_{\text{old}} \cdot \Delta t + r_{\text{old}}. \tag{51}$$

In CSS movements due to the electric force exerted among those particles are measured and accordingly new positions of particles are updated. New position ($X_{j,\text{new}}$) of CP and with which velocity ($V_{j,\text{new}}$) will reach the position ($X_{j,\text{new}}$) is computed as follows:

$$X_{j,\text{new}} = \text{rand}_{j1} \cdot k_a \cdot \frac{F_j}{m_j} \cdot \Delta t^2 + \text{rand}_{j2} \cdot k_v \cdot \Delta t + X_{j,\text{old}},$$
$$V_{j,\text{new}} = \frac{X_{j,\text{new}} - X_{j,\text{old}}}{\Delta t}. \tag{52}$$

Here, k_a is the parameter related to the attracting forces and k_v is velocity coefficient. The effect of the pervious velocity and the resultant force acting on a charged particle can be decreased with parameter k_v or increased with parameter k_a. These parameters can be computed as follows:

$$k_v = 0.5\left(1 - \frac{\text{iter}}{\text{iter}_{\max}}\right),$$
$$k_a = 0.5\left(1 + \frac{\text{iter}}{\text{iter}_{\max}}\right), \quad (53)$$

where iter is the current iteration number and $\text{iter}_{\max}$ is the maximum number of iterations.

The CSS algorithm possesses good exploring as well as exploiting capability of solution domain. Exploitation of CP is mainly ensured by the resulting electric force F_j of any particle j. Handling of attractiveness and repulsiveness of resulting force of any CP with the noble concept of parameter P_{ij} is very effective for exploitation. However, whether CP is going to explore or exploit the search space depends on the parameters k_a and k_v. Higher value of k_a implies higher impact on resulting electric force, which results exploitation of search space. Whereas higher value of k_v implies high exploration. Initially, values of k_a and k_v are almost same, but gradually k_a increases and k_v decreases. Hence, at the beginning, the algorithm explores the search space. As in successive iterations k_a increases, gradually the effect of attraction of good solutions also increases. Thus, the algorithm ensures convergence towards better solutions. The algorithm does not suffer from premature convergence due high exploration at the beginning of the algorithm. However, since good solution attracts others, if initial set of CPs not uniformly distributed over solution space, then the algorithm may be trapped into any local optima,.

Applications of this algorithm are mainly related to structural engineering designs [172–175] and geometry optimization [176].

4. Conclusion

In this paper, we have categorically discussed various optimization algorithms that are mainly inspired by physics. Major areas covered by these algorithms are quantum theory, electrostatics, electromagnetism, Newton's gravitational law, and laws of motion. This study shows that most of these algorithms are inspired by quantum computing and significant numbers of applications are developed on the basis of them. Parallel nature of quantum computing perhaps attracts researchers towards quantum-based algorithms. Another most attractive area of physics for inspiration is Newton's gravitational laws and laws of motion. We have realized that hybridization of quantum computing and biological phenomenon draws most attention these days. As biological phenomenon suggests best strategies and quantum computing provide simultaneity to those strategies; so merging of both into one implies better result. In this paper, we have studied formational aspects of all the major algorithms inspired by physics. We hope, this study will definitely be beneficial for new researchers and motivate them to formulate great solutions from those inspirational theorems of physics to optimization problems.

Abbreviations

ACO:	Ant colony optimization
APO:	Artificial physics optimization
BB-BC:	Big bang-big crunch
BFO:	Bacterial forging optimization
BGSA:	Binary gravitational search algorithm
BIS:	Biological immune system
BQEA:	Binary Quantum-inspired evolutionary algorithm
CFO:	Central force optimization
CQACO:	Continuous quantum ant colony optimization
CSS:	Charged system search
EAPO:	Extended artificial physics optimization
ECFO:	Extended central force optimization
EM:	Electromagnetism-like heuristic
GA:	Genetic Algorithm
GbSA:	Galaxy-based search algorithm
GIO:	Gravitational interaction optimization
GSA:	Gravitational search algorithm
HO:	Hysteretic optimization
HQGA:	Hybrid quantum-inspired genetic algorithm
HS:	Harmony search
IGOA:	Immune gravitation inspired optimization algorithm
IQEA:	Improved quantum evolutionary algorithm
LP:	Linear programming
MOGSA:	Multiobjective gravitational search algorithm
NLP:	Nonlinear programming
PSO:	Particle swarm optimization
PSOGSA:	PSO gravitational search algorithm
QBSO:	Quantum-inspired bacterial swarming optimization
QEA:	Quantum-inspired evolutionary algorithm
QGA:	Quantum-inspired genetic algorithm
QGO:	Quantum genetic optimization
QICA:	Quantum-inspired immune clonal algorithm
QPSO:	Quantum-behaved particle swarm optimization
QSE:	Quantum swarm evolutionary algorithm
RQGA:	Reduced quantum genetic algorithm
SA:	Simulated annealing
TSP:	Travelling salesman problem
UBB-CBC:	Unified big bang-chaotic big crunch
VM-APO:	Vector model of artificial physics optimization
vQEA:	Versatile quantum-inspired evolutionary algorithm.

References

[1] http://en.wikipedia.org/wiki/Linear_programming#CITEREF-Vazirani2001.

[2] D. P. Bertsekas, *Nonlinear Programmingby*, Athena Scientific, Belmont, Mass, USA, 2nd edition, 1999.

[3] J. H. Holland, "Genetic algorithms and the optimal allocation of trials," *SIAM Journal on Computing*, vol. 2, no. 2, pp. 88–105, 1973.

[4] J. Kennedy and R. Eberhart, "Particle swarm optimization," in *Proceedings of the IEEE International Conference on Neural Networks. IV*, pp. 1942–1948, December 1995.

[5] S. Kirkpatrick and M. P. Vecchi, "Optimization by simulated annealing," *Science*, vol. 220, no. 4598, pp. 671–680, 1983.

[6] Z. W. Geem, J. H. Kim, and G. V. Loganathan, "A new heuristic optimization algorithm: harmony search," *Simulation*, vol. 76, no. 2, pp. 60–68, 2001.

[7] R. P. Feynman, "Simulating physics with computers," *International Journal of Theoretical Physics*, vol. 21, no. 6-7, pp. 467–488, 1982.

[8] R. P. Feynman, "Quantum mechanical computers," *Foundations of Physics*, vol. 16, no. 6, pp. 507–531, 1986.

[9] A. Narayanan and M. Moore, "Quantum-inspired genetic algorithms," in *Proceedings of the IEEE International Conference on Evolutionary Computation (ICEC '96)*, pp. 61–66, May 1996.

[10] J. Sun, W. Xu, and B. Feng, "A global search strategy of quantum-behaved particle swarm optimization," in *Proceedings of the 2004 IEEE Conference on Cybernetics and Intelligent Systems*, vol. 1, pp. 111–116, December 2004.

[11] Y. Wang, X. Feng, Y. Huang et al., "A novel quantum swarm evolutionary algorithm and its applications," *Neurocomputing*, vol. 70, no. 4–6, pp. 633–640, 2007.

[12] S. I. Birbil and S. Fang, "An electromagnetism-like mechanism for global optimization," *Journal of Global Optimization*, vol. 25, no. 3, pp. 263–282, 2003.

[13] O. K. Erol and I. Eksin, "A new optimization method: Big Bang-Big Crunch," *Advances in Engineering Software*, vol. 37, no. 2, pp. 106–111, 2006.

[14] R. A. Formato, "Central force optimization: a new metaheuristic with applications in applied electromagnetics," *Progress in Electromagnetics Research*, vol. 77, pp. 425–491, 2007.

[15] E. Rashedi, H. Nezamabadi-Pour, and S. Saryazdi, "GSA: a gravitational search algorithm," *Information Sciences*, vol. 179, no. 13, pp. 2232–2248, 2009.

[16] L. Xie, J. Zeng, and Z. Cui, "General framework of artificial physics optimization algorithm," in *Proceedings of the World Congress on Nature and Biologically Inspired Computing (NaBIC '09)*, pp. 1321–1326, IEEE, December 2009.

[17] J. Flores, R. López, and J. Barrera, "Gravitational interactions optimization," in *Learning and Intelligent Optimization*, pp. 226–237, Springer, Berlin, Germany, 2011.

[18] K. F. Pál, "Hysteretic optimization for the Sherrington-Kirkpatrick spin glass," *Physica A*, vol. 367, pp. 261–268, 2006.

[19] A. Kaveh and S. Talatahari, "A novel heuristic optimization method: charged system search," *Acta Mechanica*, vol. 213, no. 3, pp. 267–289, 2010.

[20] H. Shah-Hosseini, "Principal components analysis by the galaxy-based search algorithm: a novel metaheuristic for continuous optimisation," *International Journal of Computational Science and Engineering*, vol. 6, no. 1-2, pp. 132–140, 2011.

[21] L. Jiao, Y. Li, M. Gong, and X. Zhang, "Quantum-inspired immune clonal algorithm for global optimization," *IEEE Transactions on Systems, Man, and Cybernetics B*, vol. 38, no. 5, pp. 1234–1253, 2008.

[22] W. Li, Q. Yin, and X. Zhang, "Continuous quantum ant colony optimization and its application to optimization and analysis of induction motor structure," in *Proceedings of the IEEE 5th International Conference on Bio-Inspired Computing: Theories and Applications (BIC-TA '10)*, pp. 313–317, September 2010.

[23] Y. Zhang, L. Wu, Y. Zhang, and J. Wang, "Immune gravitation inspired optimization algorithm," in *Advanced Intelligent Computing*, pp. 178–185, Springer, Berlin, Germany, 2012.

[24] C. Jinlong and H. Gao, "A quantum-inspired bacterial swarming optimization algorithm for discrete optimization problems," in *Advances in Swarm Intelligence*, pp. 29–36, Springer, Berlin, Germany, 2012.

[25] D. Ding, D. Qi, X. Luo, J. Chen, X. Wang, and P. Du, "Convergence analysis and performance of an extended central force optimization algorithm," *Applied Mathematics and Computation*, vol. 219, no. 4, pp. 2246–2259, 2012.

[26] R. C. Green II, L. Wang, and M. Alam, "Training neural networks using central force optimization and particle swarm optimization: insights and comparisons," *Expert Systems with Applications*, vol. 39, no. 1, pp. 555–563, 2012.

[27] R. A. Formato, "Central force optimization applied to the PBM suite of antenna benchmarks," 2010, http://arxiv.org/abs/1003.0221.

[28] G. M. Qubati, R. A. Formato, and N. I. Dib, "Antenna benchmark performance and array synthesis using central force optimisation," *IET Microwaves, Antennas and Propagation*, vol. 4, no. 5, pp. 583–592, 2010.

[29] D. F. Spears, W. Kerr, W. Kerr, and S. Hettiarachchi, "An overview of physicomimetics," in *Swarm Robotics*, vol. 3324 of *Lecture Notes in Computer Science: State of the Art*, pp. 84–97, Springer, Berlin, Germany, 2005.

[30] L. Xie and J. Zeng, "An extended artificial physics optimization algorithm for global optimization problems," in *Proceedings of the 4th International Conference on Innovative Computing, Information and Control (ICICIC '09)*, pp. 881–884, December 2009.

[31] L. Xie, J. Zeng, and Z. Cui, "The vector model of artificial physics optimization algorithm for global optimization problems," in *Intelligent Data Engineering and Automated Learning—IDEAL 2009*, pp. 610–617, Springer, Berlin, Germany, 2009.

[32] E. Rashedi, H. Nezamabadi-Pour, and S. Saryazdi, "BGSA: binary gravitational search algorithm," *Natural Computing*, vol. 9, no. 3, pp. 727–745, 2010.

[33] H. R. Hassanzadeh and M. Rouhani, "A multi-objective gravitational search algorithm," in *Proceedings of the 2nd International Conference on Computational Intelligence, Communication Systems and Networks (CICSyN '10)*, pp. 7–12, July 2010.

[34] S. Mirjalili and S. Z. M. Hashim, "A new hybrid PSOGSA algorithm for function optimization," in *Proceedings of the International Conference on Computer and Information Application (ICCIA '10)*, pp. 374–377, December 2010.

[35] E. Rashedi, H. Nezamabadi-Pour, S. Saryazdi, and M. Farsangi, "Allocation of static var compensator using gravitational search algorithm," in *Proceedings of the 1st Joint Congress on Fuzzy and Intelligent Systems*, pp. 29–31, 2007.

[36] B. Shaw, V. Mukherjee, and S. P. Ghoshal, "A novel opposition-based gravitational search algorithm for combined economic and emission dispatch problems of power systems," *International Journal of Electrical Power and Energy Systems*, vol. 35, no. 1, pp. 21–33, 2012.

[37] S. Duman, U. Guvenc, Y. Sonmez, and N. Yorukeren, "Optimal power flow using gravitational search algorithm," *Energy Conversion and Management*, vol. 59, pp. 86–95, 2012.

[38] P. Purwoharjono, M. Abdillah, O. Penangsang, and A. Soeprijanto, "Voltage control on 500 kV Java-Bali electrical power system for power losses minimization using gravitational search

algorithm," in *Proceedings of the 1st International Conference on Informatics and Computational Intelligence (ICI '11)*, pp. 11–17, December 2011.

[39] S. Duman, Y. Soonmez, U. Guvenc, and N. Yorukeren, "Optimal reactive power dispatch using a gravitational search algorithm," *IET Generation, Transmission & Distribution*, vol. 6, no. 6, pp. 563–576, 2012.

[40] S. Mondal, A. Bhattacharya, and S. Halder, "Solution of cost constrained emission dispatch problems considering wind power generation using gravitational search algorithm," in *Proceedings of the International Conference on Advances in Engineering, Science and Management (ICAESM '12)*, pp. 169–174, IEEE, 2012.

[41] A. Bhattacharya and P. K. Roy, "Solution of multi-objective optimal power flow using gravitational search algorithm," *IET Generation, Transmission & Distribution*, vol. 6, no. 8, pp. 751–763, 2012.

[42] S. Duman, Y. Sonmez, U. Guvenc, and N. Yorukeren, "Application of gravitational search algorithm for optimal reactive power dispatch problem," in *Proceedings of the International Symposium on Innovations in Intelligent Systems and Applications (INISTA '11)*, pp. 1–5, IEEE, June 2011.

[43] S. Duman, U. Guvenc, and N. Yurukeren, "Gravitational search algorithm for economic dispatch with valve-point effects," *International Review of Electrical Engineering*, vol. 5, no. 6, pp. 2890–2895, 2010.

[44] S. Duman, A. B. Arsoy, and N. Yorukeren, "Solution of economic dispatch problem using gravitational search algorithm," in *Proceedings of the 7th International Conference on Electrical and Electronics Engineering (ELECO '11)*, pp. I54–I59, December 2011.

[45] M. Ghalambaz, A. R. Noghrehabadi, M. A. Behrang, E. Assareh, A. Ghanbarzadeh, and N. Hedayat, "A Hybrid Neural Network and Gravitational Search Algorithm (HNNGSA) method to solve well known Wessinger's equation," *World Academy of Science, Engineering and Technology*, vol. 73, pp. 803–807, 2011.

[46] R. Precup, R. David, E. M. Petriu, S. Preitl, and M. Radac, "Gravitational search algorithm-based tuning of fuzzy control systems with a reduced parametric sensitivity," in *Soft Computing in Industrial Applications*, pp. 141–150, Springer, Berlin, Germany, 2011.

[47] R. Precup, R. David, E. M. Petriu, S. Preitl, and M. Radac, "Fuzzy control systems with reduced parametric sensitivity based on simulated annealing," *IEEE Transactions on Industrial Electronics*, vol. 59, no. 8, pp. 3049–3061, 2012.

[48] M. A. Behrang, E. Assareh, M. Ghalambaz, M. R. Assari, and A. R. Noghrehabadi, "Forecasting future oil demand in Iran using GSA (Gravitational Search Algorithm)," *Energy*, vol. 36, no. 9, pp. 5649–5654, 2011.

[49] M. Khajehzadeh, M. R. Taha, A. El-Shafie, and M. Eslami, "A modified gravitational search algorithm for slope stability analysis," *Engineering Applications of Artificial Intelligence*, vol. 25, 8, pp. 1589–1597, 2012.

[50] A. Hatamlou, S. Abdullah, and H. Nezamabadi-Pour, "Application of gravitational search algorithm on data clustering," in *Rough Sets and Knowledge Technology*, pp. 337–346, Springer, Berlin, Germany, 2011.

[51] M. Yin, Y. Hu, F. Yang, X. Li, and W. Gu, "A novel hybrid K-harmonic means and gravitational search algorithm approach for clustering," *Expert Systems with Applications*, vol. 38, no. 8, pp. 9319–9324, 2011.

[52] C. Li, J. Zhou, B. Fu, P. Kou, and J. Xiao, "T-S fuzzy model identification with a gravitational search-based hyperplane clustering algorithm," *IEEE Transactions on Fuzzy Systems*, vol. 20, no. 2, pp. 305–317, 2012.

[53] A. Bahrololoum, H. Nezamabadi-Pour, H. Bahrololoum, and M. Saeed, "A prototype classifier based on gravitational search algorithm," *Applied Soft Computing Journal*, vol. 12, no. 2, pp. 819–825, 2012.

[54] J. P. Papa, A. Pagnin, S. A. Schellini et al., "Feature selection through gravitational search algorithm," in *Proceedings of the 36th IEEE International Conference on Acoustics, Speech, and Signal Processing (ICASSP '11)*, pp. 2052–2055, May 2011.

[55] B. Zibanezhad, K. Zamanifar, N. Nematbakhsh, and F. Mardukhi, "An approach for web services composition based on QoS and gravitational search algorithm," in *Proceedings of the International Conference on Innovations in Information Technology (IIT '09)*, pp. 340–344, IEEE, December 2009.

[56] S. Duman, D. Maden, and U. Guvenc, "Determination of the PID controller parameters for speed and position control of DC motor using gravitational search algorithm," in *Proceedings of the 7th International Conference on Electrical and Electronics Engineering (ELECO '11)*, pp. I225–I229, IEEE, December 2011.

[57] W. X. Gu, X. T. Li, L. Zhu et al., "A gravitational search algorithm for flow shop scheduling," *CAAI Transaction on Intelligent Systems*, vol. 5, no. 5, pp. 411–418, 2010.

[58] D. Hoffman, "A brief overview of the biological immune system," 2011, http://www.healthy.net/ .

[59] M. Cleric and J. Kennedy, "The particle swarm-explosion, stability, and convergence in a multidimensional complex space," *IEEE Transactions on Evolutionary Computation*, vol. 6, no. 1, pp. 58–73, 2002.

[60] M. S. Innocente and J. Sienz, "Particle swarm optimization with inertia weight and constriction factor," in *Proceedings of the International conference on swarm intelligence (ICSI '11)*, 2011.

[61] R. Mendes, J. Kennedy, and J. Neves, "The fully informed particle swarm: simpler, maybe better," *IEEE Transactions on Evolutionary Computation*, vol. 8, no. 3, pp. 204–210, 2004.

[62] M. Udrescu, L. Prodan, and M. Vlăduţiu, "Implementing quantum genetic algorithms: a solution based on Grover's algorithm," in *Proceedings of the 3rd Conference on Computing Frontiers (CF '06)*, pp. 71–81, ACM, May 2006.

[63] B. Li and L. Wang, "A hybrid quantum-inspired genetic algorithm for multiobjective flow shop scheduling," *IEEE Transactions on Systems, Man, and Cybernetics B*, vol. 37, no. 3, pp. 576–591, 2007.

[64] L. Wang, F. Tang, and H. Wu, "Hybrid genetic algorithm based on quantum computing for numerical optimization and parameter estimation," *Applied Mathematics and Computation*, vol. 171, no. 2, pp. 1141–1156, 2005.

[65] A. Malossini and T. Calarco, "Quantum genetic optimization," *IEEE Transactions on Evolutionary Computation*, vol. 12, no. 2, pp. 231–241, 2008.

[66] A. Layeb, S. Meshoul, and M. Batouche, "quantum genetic algorithm for multiple RNA structural alignment," in *Proceedings of the 2nd Asia International Conference on Modelling and Simulation (AIMS '08)*, pp. 873–878, May 2008.

[67] D. Chang and Y. Zhao, "A dynamic niching quantum genetic algorithm for automatic evolution of clusters," in *Proceedings of the 14th International Conference on Computer Analysis of Images and Patterns*, vol. 2, pp. 308–315, 2011.

[68] J. Xiao, Y. Yan, Y. Lin, L. Yuan, and J. Zhang, "A quantum-inspired genetic algorithm for data clustering," in *Proceedings of the IEEE Congress on Evolutionary Computation (CEC '08)*, pp. 1513–1519, June 2008.

[69] H. Talbi, A. Draa, and M. Batouche, "A new quantum-inspired genetic algorithm for solving the travelling salesman problem," in *Proceedings of the IEEE International Conference on Industrial Technology (ICIT '04)*, vol. 3, pp. 1192–1197, December 2004.

[70] K.-H. Han, K.-H. Park, C.-H. Lee, and J.-H. Kim, "Parallel quantum-inspired genetic algorithm for combinatorial optimization problem," in *Proceedings of the 2001 Congress on Evolutionary Computation*, vol. 2, pp. 1422–1429, IEEE, May 2001.

[71] L. Yan, H. Chen, W. Ji, Y. Lu, and J. Li, "Optimal VSM model and multi-object quantum-inspired genetic algorithm for web information retrieval," in *Proceedings of the 1st International Symposium on Computer Network and Multimedia Technology (CNMT '09)*, pp. 1–4, IEEE, December 2009.

[72] Z. Mo, G. Wu, Y. He, and H. Liu, "quantum genetic algorithm for scheduling jobs on computational grids," in *Proceedings of the International Conference on Measuring Technology and Mechatronics Automation (ICMTMA '10)*, pp. 964–967, March 2010.

[73] Y. Zhang, J. Liu, Y. Cui, X. Hei, and M. Zhang, "An improved quantum genetic algorithm for test suite reduction," in *Proceedings of the IEEE International Conference on Computer Science and Automation Engineering (CSAE '11)*, pp. 149–153, June 2011.

[74] J. Lee, W. Lin, G. Liao, and T. Tsao, "quantum genetic algorithm for dynamic economic dispatch with valve-point effects and including wind power system," *International Journal of Electrical Power and Energy Systems*, vol. 33, no. 2, pp. 189–197, 2011.

[75] J. Dai and H. Zhang, "A novel quantum genetic algorithm for area optimization of FPRM circuits," in *Proceedings of the 3rd International Symposium on Intelligent Information Technology Application (IITA 09)*, pp. 408–411, November 2009.

[76] L. Chuang, Y. Chiang, and C. Yang, "A quantum genetic algorithm for operon prediction," in *Proceedings of the IEEE 26th International Conference on Advanced Information Networking and Applications (AINA '12)*, pp. 269–275, March 2012.

[77] H. Xing, X. Liu, X. Jin, L. Bai, and Y. Ji, "A multi-granularity evolution based quantum genetic algorithm for QoS multicast routing problem in WDM networks," *Computer Communications*, vol. 32, no. 2, pp. 386–393, 2009.

[78] W. Luo, "A quantum genetic algorithm based QoS routing protocol for wireless sensor networks," in *Proceedings of the IEEE International Conference on Software Engineering and Service Sciences (ICSESS '10)*, pp. 37–40, IEEE, July 2010.

[79] J. Wang and R. Zhou, "A novel quantum genetic algorithm for PID controller," in *Proceedings of the 6th International Conference on Advanced Intelligent Computing Theories and Applications: Intelligent Computing*, pp. 72–77, 2010.

[80] B. Han, J. Jiang, Y. Gao, and J. Ma, "A quantum genetic algorithm to solve the problem of multivariate," *Communications in Computer and Information Science*, vol. 243, no. 1, pp. 308–314, 2011.

[81] Y. Zheng, J. Liu, W. Geng, and J. Yang, "Quantum-inspired genetic evolutionary algorithm for course timetabling," in *Proceedings of the 3rd International Conference on Genetic and Evolutionary Computing (WGEC '09)*, pp. 750–753, October 2009.

[82] Y. J. Lv and N. X. Liu, "Application of quantum genetic algorithm on finding minimal reduct," in *Proceedings of the IEEE International Conference on Granular Computing (GRC '07)*, pp. 728–733, November 2007.

[83] X. J. Zhang, S. Li, Y. Shen, and S. M. Song, "Evaluation of several quantum genetic algorithms in medical image registration applications," in *Proceedings of the IEEE International Conference on Computer Science and Automation Engineering (CSAE '12)*, vol. 2, pp. 710–713, IEEE, 2012.

[84] H. Talbi, A. Draa, and M. Batouche, "A new quantum-inspired genetic algorithm for solving the travelling salesman problem," in *Proceedings of the IEEE International Conference on Industrial Technology (ICIT '04)*, pp. 1192–1197, December 2004.

[85] S. Bhattacharyya and S. Dey, "An efficient quantum inspired genetic algorithm with chaotic map model based interference and fuzzy objective function for gray level image thresholding," in *Proceedings of the International Conference on Computational Intelligence and Communication Systems (CICN '11)*, pp. 121–125, IEEE, October 2011.

[86] K. Benatchba, M. Koudil, Y. Boukir, and N. Benkhelat, "Image segmentation using quantum genetic algorithms," in *Proceedings of the 32nd Annual Conference on IEEE Industrial Electronics (IECON '06)*, pp. 3556–3562, IEEE, November 2006.

[87] M. Liu, C. Yuan, and T. Huang, "A novel real-coded quantum genetic algorithm in radiation pattern synthesis for smart antenna," in *Proceedings of the IEEE International Conference on Robotics and Biomimetics (ROBIO '07)*, pp. 2023–2026, IEEE, December 2007.

[88] R. Popa, V. Nicolau, and S. Epure, "A new quantum inspired genetic algorithm for evolvable hardware," in *Proceedings of the 3rd International Symposium on Electrical and Electronics Engineering (ISEEE '10)*, pp. 64–69, September 2010.

[89] H. Yu and J. Fan, "Parameter optimization based on quantum genetic algorithm for generalized fuzzy entropy thresholding segmentation method," in *Proceedings of the 5th International Conference on Fuzzy Systems and Knowledge Discovery (FSKD '08)*, vol. 1, pp. 530–534, IEEE, October 2008.

[90] P. C. Shill, M. F. Amin, M. A. H. Akhand, and K. Murase, "Optimization of interval type-2 fuzzy logic controller using quantum genetic algorithms," in *Proceedings of the IEEE International Conference on Fuzzy Systems (FUZZ-IEEE '12)*, pp. 1–8, June 2012.

[91] M. Cao and F. Shang, "Training of process neural networks based on improved quantum genetic algorithm," in *Proceedings of the WRI World Congress on Software Engineering (WCSE '09)*, vol. 2, pp. 160–165, May 2009.

[92] Y. Sun and M. Ding, "quantum genetic algorithm for mobile robot path planning," in *Proceedings of the 4th International Conference on Genetic and Evolutionary Computing (ICGEC '10)*, pp. 206–209, December 2010.

[93] K. Han and J. Kim, "Quantum-inspired evolutionary algorithm for a class of combinatorial optimization," *IEEE Transactions on Evolutionary Computation*, vol. 6, no. 6, pp. 580–593, 2002.

[94] R. Zhang and H. Gao, "Improved quantum evolutionary algorithm for combinatorial optimization problem," in *Proceedings of the 6th International Conference on Machine Learning and Cybernetics (ICMLC '07)*, vol. 6, pp. 3501–3505, August 2007.

[95] M. D. Platel, S. Sehliebs, and N. Kasabov, "A versatile quantum-inspired evolutionary algorithm," in *Proceedings of the IEEE Congress on Evolutionary Computation (CEC '07)*, pp. 423–430, September 2007.

[96] P. Li and S. Li, "Quantum-inspired evolutionary algorithm for continuous space optimization based on Bloch coordinates of qubits," *Neurocomputing*, vol. 72, no. 1–3, pp. 581–591, 2008.

[97] K. Han and J. Kim, "Quantum-inspired evolutionary algorithm for a class of combinatorial optimization," *IEEE Transactions on Evolutionary Computation*, vol. 6, no. 6, pp. 580–593, 2002.

[98] P. Mahdabi, S. Jalili, and M. Abadi, "A multi-start quantum-inspired evolutionary algorithm for solving combinatorial optimization problems," in *Proceedings of the 10th Annual Genetic and Evolutionary Computation Conference (GECCO '08)*, pp. 613–614, ACM, July 2008.

[99] H. Talbi, M. Batouche, and A. Draao, "A quantum-inspired evolutionary algorithm for multiobjective image segmentation," *International Journal of Mathematical, Physical and Engineering Sciences*, vol. 1, no. 2, pp. 109–114, 2007.

[100] Y. Kim, J. Kim, and K. Han, "Quantum-inspired multiobjective evolutionary algorithm for multiobjective 0/1 knapsack problems," in *Proceedings of the IEEE Congress on Evolutionary Computation (CEC '06)*, pp. 2601–2606, July 2006.

[101] A. Narayan and C. Patvardhan, "A novel quantum evolutionary algorithm for quadratic knapsack problem," in *Proceedings of the IEEE International Conference on Systems, Man and Cybernetics (SMC '09)*, pp. 1388–1392, October 2009.

[102] A. R. Hota and A. Pat, "An adaptive quantum-inspired differential evolution algorithm for 0-1 knapsack problem," in *Proceedings of the 2nd World Congress on Nature and Biologically Inspired Computing (NaBIC '10)*, pp. 703–708, December 2010.

[103] Y. Ji and H. Xing, "A memory storable quantum inspired evolutionary algorithm for network coding resource minimization," in *Evolutionary Algorithms*, InTech, Shanghai, China, 2011.

[104] H. Xing, Y. Ji, L. Bai, and Y. Sun, "An improved quantum-inspired evolutionary algorithm for coding resource optimization based network coding multicast scheme," *International Journal of Electronics and Communications*, vol. 64, no. 12, pp. 1105–1113, 2010.

[105] A. da Cruz, M. M. B. R. Vellasco, and M. Pacheco, "Quantum-inspired evolutionary algorithm for numerical optimization," in *Hybrid Evolutionary Algorithms*, pp. 19–37, Springer, Berlin, Germany, 2007.

[106] G. Zhang and H. Rong, "Real-observation quantum-inspired evolutionary algorithm for a class of numerical optimization problems," in *Proceedings of the 7th international conference on Computational Science, Part IV (ICCS '07)*, vol. 4490, pp. 989–996, 2007.

[107] R. Setia and K. H. Raj, "Quantum inspired evolutionary algorithm for optimization of hot extrusion process," *International Journal of Soft Computing and Engineering*, vol. 2, no. 5, p. 29, 2012.

[108] T. Lau, *Application of quantum-inspired evolutionary algorithm in solving the unit commitment problem [dissertation]*, The Hong Kong Polytechnic University, Hong Kong, 2011.

[109] C. Y. Chung, H. Yu, and K. P. Wong, "An advanced quantum-inspired evolutionary algorithm for unit commitment," *IEEE Transactions on Power Systems*, vol. 26, no. 2, pp. 847–854, 2011.

[110] J. G. Vlachogiannis and K. Y. Lee, "Quantum-inspired evolutionary algorithm for real and reactive power dispatch," *IEEE Transactions on Power Systems*, vol. 23, no. 4, pp. 1627–1636, 2008.

[111] U. Pareek, M. Naeem, and D. C. Lee, "Quantum inspired evolutionary algorithm for joint user selection and power allocation for uplink cognitive MIMO systems," in *Proceedings of the IEEE Symposium on Computational Intelligence in Scheduling (SCIS '11)*, pp. 33–38, April 2011.

[112] J. Chen, "Application of quantum-inspired evolutionary algorithm to reduce PAPR of an OFDM signal using partial transmit sequences technique," *IEEE Transactions on Broadcasting*, vol. 56, no. 1, pp. 110–113, 2010.

[113] J. Jang, K. Han, and J. Kim, "Face detection using quantum-inspired evolutionary algorithm," in *Proceedings of the 2004 Congress on Evolutionary Computation (CEC '04)*, vol. 2, pp. 2100–2106, June 2004.

[114] J. Jang, K. Han, and J. Kim, "Quantum-inspired evolutionary algorithm-based face verification," in *Genetic and Evolutionary Computation—GECCO 2003*, pp. 214–214, Springer, Berlin, Germany, 2003.

[115] K. Fan, A. Brabazon, C. O'Sullivan, and M. O'Neill, "Quantum-inspired evolutionary algorithms for financial data analysis," in *Applications of Evolutionary Computing*, pp. 133–143, Springer, Berlin, Germany, 2008.

[116] K. Fan, A. Brabazon, C. O'Sullivan, and M. O'Neill, "Option pricing model calibration using a real-valued quantum-inspired evolutionary algorithm," in *Proceedings of the 9th Annual Genetic and Evolutionary Computation Conference (GECCO '07)*, pp. 1983–1990, ACM, July 2007.

[117] K. Fan, A. Brabazon, C. OSullivan, and M. ONeill, "Quantum-inspired evolutionary algorithms for calibration of the VG option pricing model," in *Applications of Evolutionary Computing*, pp. 189–198, Springer, Berlin, Germany, 2007.

[118] R. A. de Araújo, "A quantum-inspired evolutionary hybrid intelligent approach for stock market prediction," *International Journal of Intelligent Computing and Cybernetics*, vol. 3, no. 1, pp. 24–54, 2010.

[119] Z. Huang, Y. Wang, C. Yang, and C. Wu, "A new improved quantum-behaved particle swarm optimization model," in *Proceedings of the 4th IEEE Conference on Industrial Electronics and Applications (ICIEA '09)*, pp. 1560–1564, May 2009.

[120] J. Chang, F. An, and P. Su, "A quantum-PSO algorithm for no-wait flow shop scheduling problem," in *Proceedings of the Chinese Control and Decision Conference (CCDC '10)*, pp. 179–184, May 2010.

[121] X. Wu, B. Zhang, K. Wang, J. Li, and Y. Duan, "A quantum-inspired Binary PSO algorithm for unit commitment with wind farms considering emission reduction," in *Proceedings of the Innovative Smart Grid Technologies—Asia (ISGT '12)*, pp. 1–6, IEEE, May 2012.

[122] Y. Jeong, J. Park, S. Jang, and K. Y. Lee, "A new quantum-inspired binary PSO for thermal unit commitment problems," in *Proceedings of the 15th International Conference on Intelligent System Applications to Power Systems (ISAP '09)*, pp. 1–6, November 2009.

[123] H. N. A. Hamed, N. Kasabov, and S. M. Shamsuddin, "Integrated feature selection and parameter optimization for evolving spiking neural networks using quantum inspired particle swarm optimization," in *Proceedings of the International Conference on Soft Computing and Pattern Recognition (SoCPaR '09)*, pp. 695–698, December 2009.

[124] A. A. Ibrahim, A. Mohamed, H. Shareef, and S. P. Ghoshal, "An effective power quality monitor placement method utilizing quantum-inspired particle swarm optimization," in *Proceedings of the International Conference on Electrical Engineering and Informatics (ICEEI '11)*, pp. 1–6, July 2011.

[125] F. Yao, Z. Y. Dong, K. Meng, Z. Xu, H. H. Iu, and K. Wong, "Quantum-inspired particle swarm optimization for power system operations considering wind power uncertainty and carbon tax in Australia," *IEEE Transactions on Industrial Informatics*, vol. 8, no. 4, pp. 880–888, 2012.

[126] Z. Zhisheng, "Quantum-behaved particle swarm optimization algorithm for economic load dispatch of power system," *Expert Systems with Applications*, vol. 37, no. 2, pp. 1800–1803, 2010.

[127] A. Chen, G. Yang, and Z. Wu, "Hybrid discrete particle swarm optimization algorithm for capacitated vehicle routing problem," *Journal of Zhejiang University*, vol. 7, no. 4, pp. 607–614, 2006.

[128] T. J. Ai and V. Kachitvichyanukul, "A particle swarm optimization for the vehicle routing problem with simultaneous pickup and delivery," *Computers and Operations Research*, vol. 36, no. 5, pp. 1693–1702, 2009.

[129] Y. Marinakis, M. Marinaki, and G. Dounias, "A hybrid particle swarm optimization algorithm for the vehicle routing problem," *Engineering Applications of Artificial Intelligence*, vol. 23, no. 4, pp. 463–472, 2010.

[130] S. N. Omkar, R. Khandelwal, T. V. S. Ananth, G. Narayana Naik, and S. Gopalakrishnan, "Quantum behaved Particle Swarm Optimization (QPSO) for multi-objective design optimization of composite structures," *Expert Systems with Applications*, vol. 36, no. 8, pp. 11312–11322, 2009.

[131] L. D. S. Coelho, "Gaussian quantum-behaved particle swarm optimization approaches for constrained engineering design problems," *Expert Systems with Applications*, vol. 37, no. 2, pp. 1676–1683, 2010.

[132] M. Ykhlef, "A quantum swarm evolutionary algorithm for mining association rules in large databases," *Journal of King Saud University*, vol. 23, no. 1, pp. 1–6, 2011.

[133] M. Dorigo and T. Stiitzle, *Ant Colony Optimization*, pp. 153–222, chapter 4, MIT Press, Cambridge, Mass, USA, 1st edition, 2004.

[134] L. Wang, Q. Niu, and M. Fei, "A novel quantum ant colony optimization algorithm," in *Bio-Inspired Computational Intelligence and Applications*, pp. 277–286, Springer, Berlin, Germany, 2007.

[135] X. You, S. Liu, and Y. Wang, "Quantum dynamic mechanism-based parallel ant colony optimization algorithm," *International Journal of Computational Intelligence Systems*, vol. 3, no. 1, pp. 101–113, 2010.

[136] L. Wang, Q. Niu, and M. Fei, "A novel quantum ant colony optimization algorithm and its application to fault diagnosis," *Transactions of the Institute of Measurement and Control*, vol. 30, no. 3-4, pp. 313–329, 2008.

[137] Z. Yu, L. Shuhua, F. Shuai, and W. Di, "A quantum-inspired ant colony optimization for robot coalition formation," in *Chinese Control and Decision Conference (CCDC '09)*, pp. 626–631, June 2009.

[138] K. M. Passino, "Biomimicry of bacterial foraging for distributed optimization and control," *IEEE Control Systems Magazine*, vol. 22, no. 3, pp. 52–67, 2002.

[139] T. Kumbasar, I. Eksin, M. Güzelkaya, and E. Yeşil, "Big bang big crunch optimization method based fuzzy model inversion," in *MICAI 2008: Advances in Artificial Intelligence*, pp. 732–740, Springer, Berlin, Germany, 2008.

[140] T. Kumbasar, E. Yeşil, I. Eksin, and M. Güzelkaya, "Inverse fuzzy model control with online adaptation via big bang-big crunch optimization," in *2008 3rd International Symposium on Communications, Control, and Signal Processing (ISCCSP '08)*, pp. 697–702, March 2008.

[141] M. Aliasghary, I. Eksin, and M. Guzelkaya, "Fuzzy-sliding model reference learning control of inverted pendulum with Big Bang-Big Crunch optimization method," in *Proceedings of the 11th International Conference on Intelligent Systems Design and Applications (ISDA '11)*, pp. 380–384, November 2011.

[142] H. M. Genç, I. Eksin, and O. K. Erol, "Big Bang-Big Crunch optimization algorithm hybridized with local directional moves and application to target motion analysis problem," in *Proceedings of the IEEE International Conference on Systems, Man and Cybernetics (SMC '10)*, pp. 881–887, October 2010.

[143] H. M. Genç and A. K. Hocaoğlu, "Bearing-only target tracking based on Big Bang-Big Crunch algorithm," in *Proceedings of the 3rd International Multi-Conference on Computing in the Global Information Technology (ICCGI '08)*, pp. 229–233, July 2008.

[144] P. Prudhvi, "A complete copper optimization technique using BB-BC in a smart home for a smarter grid and a comparison with GA," in *Proceedings of the 24th Canadian Conference on Electrical and Computer Engineering (CCECE '11)*, pp. 69–72, May 2011.

[145] G. M. Jaradat and M. Ayob, "Big Bang-Big Crunch optimization algorithm to solve the course timetabling problem," in *Proceedings of the 10th International Conference on Intelligent Systems Design and Applications (ISDA '10)*, pp. 1448–1452, December 2010.

[146] D. Debels, B. De Reyck, R. Leus, and M. Vanhoucke, "A hybrid scatter search/electromagnetism meta-heuristic for project scheduling," *European Journal of Operational Research*, vol. 169, no. 2, pp. 638–653, 2006.

[147] P. Chang, S. Chen, and C. Fan, "A hybrid electromagnetism-like algorithm for single machine scheduling problem," *Expert Systems with Applications*, vol. 36, no. 2, pp. 1259–1267, 2009.

[148] A. Jamili, M. A. Shafia, and R. Tavakkoli-Moghaddam, "A hybridization of simulated annealing and electromagnetism-like mechanism for a periodic job shop scheduling problem," *Expert Systems with Applications*, vol. 38, no. 5, pp. 5895–5901, 2011.

[149] M. Mirabi, S. M. T. Fatemi Ghomi, F. Jolai, and M. Zandieh, "Hybrid electromagnetism-like algorithm for the flowshop scheduling with sequence-dependent setup times," *Journal of Applied Sciences*, vol. 8, no. 20, pp. 3621–3629, 2008.

[150] B. Naderi, R. Tavakkoli-Moghaddam, and M. Khalili, "Electromagnetism-like mechanism and simulated annealing algorithms for flowshop scheduling problems minimizing the total weighted tardiness and makespan," *Knowledge-Based Systems*, vol. 23, no. 2, pp. 77–85, 2010.

[151] H. Turabieh, S. Abdullah, and B. McCollum, "Electromagnetism-like mechanism with force decay rate great deluge for the course timetabling problem," in *Rough Sets and Knowl-edge Technology*, pp. 497–504, Springer, Berlin, Germany, 2009.

[152] C. Lee and F. Chang, "Fractional-order PID controller optimization via improved electromagnetism-like algorithm," *Expert Systems with Applications*, vol. 37, no. 12, pp. 8871–8878, 2010.

[153] S. Birbil and O. Feyzioğlu, "A global optimization method for solving fuzzy relation equations," in *Fuzzy Sets and Systems (IFSA '03)*, pp. 47–84, Springer, Berlin, Germany, 2003.

[154] P. Wu, K. Yang, and Y. Hung, "The study of electromagnetism-like mechanism based fuzzy neural network for learning fuzzy if-then rules," in *Knowledge-Based Intelligent Information and Engineering Systems*, pp. 907–907, Springer, Berlin, Germany, 2005.

[155] C. Lee, C. Kuo, H. Chang, J. Chien, and F. Chang, "A hybrid algorithm of electromagnetism-like and genetic for recurrent neural fuzzy controller design," in *Proceedings of the International MultiConference of Engineers and Computer Scientists*, vol. 1, March 2009.

[156] A. Yurtkuran and E. Emel, "A new hybrid electromagnetism-like algorithm for capacitated vehicle routing problems," *Expert Systems with Applications*, vol. 37, no. 4, pp. 3427–3433, 2010.

[157] C. Tsai, H. Hung, and S. Lee, "Electromagnetism-like method based blind multiuser detection for MC-CDMA interference suppression over multipath fading channel," in *2010 International Symposium on Computer, Communication, Control and Automation (3CA '10)*, vol. 2, pp. 470–475, May 2010.

[158] C.-S. Tsou and C.-H. Kao, "Multi-objective inventory control using electromagnetism-like meta-heuristic," *International Journal of Production Research*, vol. 46, no. 14, pp. 3859–3874, 2008.

[159] X. Wang, L. Gao, and C. Zhang, "Electromagnetism-like mechanism based algorithm for neural network training," in *Advanced Intelligent Computing Theories and Applications. With Aspects of Artificial Intelligence*, pp. 40–45, Springer, Berlin, Germany, 2008.

[160] Q. Wu, C. Zhang, L. Gao, and X. Li, "Training neural networks by electromagnetism-like mechanism algorithm for tourism arrivals forecasting," in *Proceedings of the IEEE 5th International Conference on Bio-Inspired Computing: Theories and Applications (BIC-TA '10)*, pp. 679–688, September 2010.

[161] P. Wu and H. Chiang, "The Application of electromagnetism-like mechanism for solving the traveling salesman problems," in *Proceeding of the 2005 Chinese Institute of Industrial Engineers Annual Meeting*, Taichung, Taiwan, December 2005.

[162] P. Wu, K. Yang, and H. Fang, "A revised EM-like algorithm + K-OPT method for solving the traveling salesman problem," in *1st International Conference on Innovative Computing, Information and Control 2006 (ICICIC '06)*, vol. 1, pp. 546–549, August 2006.

[163] C. Su and H. Lin, "Applying electromagnetism-like mechanism for feature selection," *Information Sciences*, vol. 181, no. 5, pp. 972–986, 2011.

[164] K. C. Lee and J. Y. Jhang, "Application of electromagnetism-like algorithm to phase-only syntheses of antenna arrays," *Progress in Electromagnetics Research*, vol. 83, pp. 279–291, 2008.

[165] C. Santos, M. Oliveira, V. Matos, A. Maria, A. C. Rocha, and L. A. Costa, "Combining central pattern generators with the electromagnetism-like algorithm for head motion stabilization during quadruped robot locomotion," in *Proceedings of the 2nd International Workshop on Evolutionary and Reinforcement Learning for Autonomous Robot Systems*, 2009.

[166] X. Guan, X. Dai, and J. Li, "Revised electromagnetism-like mechanism for flow path design of unidirectional AGV systems," *International Journal of Production Research*, vol. 49, no. 2, pp. 401–429, 2011.

[167] A. Yurtkuran and E. Emel, "A new hybrid electromagnetism-like algorithm for capacitated vehicle routing problems," *Expert Systems with Applications*, vol. 37, no. 4, pp. 3427–3433, 2010.

[168] K. F. Pál, "Hysteretic optimization for the traveling salesman problem," *Physica A*, vol. 329, no. 1-2, pp. 287–297, 2003.

[169] B. Gonçalves and S. Boettcher, "Hysteretic optimization for spin glasses," *Journal of Statistical Mechanics*, vol. 2008, no. 1, Article ID P01003, 2008.

[170] X. Yan and W. Wu, "Hysteretic optimization for the capacitated vehicle routing problem," in *Proceedings of the 9th IEEE International Conference on Networking, Sensing and Control (ICNSC '12)*, pp. 12–15, April 2012.

[171] J. Zha, G. Zeng, and Y. Lu, "Hysteretic optimization for protein folding on the lattice," in *Proceedings of the International Conference on Computational Intelligence and Software Engineering (CiSE '10)*, pp. 1–4, December 2010.

[172] A. Kaveh and S. Talatahari, "A charged system search with a fly to boundary method for discrete optimum design of truss structures," *Asian Journal of Civil Engineering*, vol. 11, no. 3, pp. 277–293, 2010.

[173] A. Kaveh and S. Talatahari, "Optimal design of skeletal structures via the charged system search algorithm," *Structural and Multidisciplinary Optimization*, vol. 41, no. 6, pp. 893–911, 2010.

[174] A. Kaveh and S. Talatahari, "Charged system search for optimal design of frame structures," *Applied Soft Computing Journal*, vol. 12, no. 1, pp. 382–393, 2012.

[175] A. Kaveh and S. Talatahari, "Charged system search for optimum grillage system design using the LRFD-AISC code," *Journal of Constructional Steel Research*, vol. 66, no. 6, pp. 767–771, 2010.

[176] A. Kaveh and S. Talatahari, "Geometry and topology optimization of geodesic domes using charged system search," *Structural and Multidisciplinary Optimization*, vol. 43, no. 2, pp. 215–229, 2011.

3

Optimizing the Two-Stage Supply Chain Inventory Model with Full Information Sharing and Two Backorders Costs Using Hybrid Geometric-Algebraic Method

Mohamed E. Seliaman

Department of Information Systems, King Faisal University, P.O. Box 78, Al-Hufof 31982, Saudi Arabia

Correspondence should be addressed to Mohamed E. Seliaman; mseliaman@gmail.com

Academic Editor: Cedric Yiu

We consider the case of a two-stage serial supply chain system. This supply chain system involves a single vendor who supplies a single buyer with a single product. The vendor's production rate is assumed finite. In addition, the demand at the buyer is assumed deterministic. In order to coordinate their replenishment policies and jointly optimize their operational costs, the two supply chain partners fully share their relevant information. For this purpose, we develop an integrated inventory replenishment model assuming linear and fixed backorders costs. Then, we use a hybrid geometric-algebraic method to drive the optimal replenishment policy and the minimum supply chain total cost in a closed form.

1. Introduction

Supply chain integration is concerned with functional integration and coordination among the supply chain partners. In an independently managed supply chain, each member in each stage will optimize his own operational costs in a decentralized fashion. Generally, it has been realized that such managerial independence of the supply chain partners may increase the imbalance between demand and supply. Such independence also has been recognized as direct cause of increased costs. This pushed firms towards the full integration of the supply chain resources and the proper coordination of decisions. Researchers report that closer collaboration among the chain partners, increased level of information sharing, and high level of coordination of various decision processes lead to improved customers service and reduced costs. The significant advances in information and communication technologies facilitated the provision and sharing of the business information necessary for efficiency improvement. This, in turn, facilitated the development in the integrated supply chain management.

In recent years, numerous articles in supply chain modeling have addressed the issue of inventory coordination. Banerjee [1] introduced the concept of joint economic lot-sizing problem (JELS) for the case of a single vendor and a single purchaser under the assumption of deterministic demand and lot for lot policy. Since then, numerous articles in supply chain modeling focused on the integrated vendor-buyer inventory models and the joint economic lot-sizing problem [2–4]. An extensive review of integrated models which deal with the interaction between a buyer and vendor is presented in [5]. This review classified the literature dealing with the integrated models into four main classes. The first class represents models which deal with joint economic lot sizing policies. The second class characterizes models which deal with the coordination of inventory by simultaneously determining the order quantity for the buyer and the vendor. The third class is a group of models which deal with integrated problem but do not determine simultaneously the order quantity of the buyer and the vendor. The last class represents models which deal with buyer-vendor coordination subject to marketing considerations. In the following, we provide an overview of selected relevant models.

Goyal and Szendrovits [6] presented a constant lot size model where the lot is produced through a fixed sequence of manufacturing stages, with a single setup and without interruption at each stage. A one-vendor multibuyer integrated inventory model was developed in [7] with the objective of minimizing the vendor's total annual cost subject to the maximum cost that the buyer may be prepared to incur. The single-vendor multibuyers integrated inventory model was revisited by [8] where he relaxed the constraint of equal-sized shipments of [9] and suggested that the shipment size should grow geometrically.

The idea of producing a single product in a multistage serial production system was extended by [10] to deal with equal and unequal sized batch shipments between stages.

Several benefits of coordinating inventory decisions and sharing related information have been reported in the supply chain management literature. The investigation by [11] on the bullwhip effect in supply chains reported that lack of information sharing can lead to excessive inventory, poor customer service, lost revenues, unplanned capacities, and ineffective logistics. They recommend avoiding managerial independence by integrating various supply chain functions. They also recommend that firms need to device strategies that lead to smaller batches or frequent replenishments. Exchange of substantial quantities of information among the buyer, supplier, and carrier can increase the efficiency and effectiveness of the supply chain [12]. Coordinated replenishment can significantly reduce inventory. Inventory reductions have a significant impact on supply chain activities. Lower inventory levels increase operating revenues and reduce the need for costly facilities [13].

The centralized, coordinated replenishment policy and the decentralized replenishment policy in a two-echelon, multiitem supply chain were investigated in the model developed by [14]. This model determined the optimal common replenishment cycle for end items and the integer multiples of the common replenishment cycle for raw materials. Analysis showed that a centralized, coordinated replenishment policy was always found to be superior to the decentralized replenishment policy in terms of cost reduction, especially when major setup costs were high.

Most of these reviewed research articles used differential calculus to drive the optimal solutions for the integrated production inventory models. However, a recent line of research focused on easier solution methods for the optimization of these types of systems [15, 16]. For example, [17] introduced the use of algebraic optimization approach to the EOQ model with no backorders. Reference [18] used algebraic procedure to the EPQ formula taking shortages into consideration within the case of only one backlog cost per unit and time unit. Reference [19] considered an integrated three-stage inventory system with backorders. They formulated the problem to derive the replenishment policies with four-decision-variables algebraically. Their model was later extended by [20] to include a fourth stage.

The popularity of the algebraic approach for the optimization of production inventory models could be due to the fact that it requires basic knowledge of simple elementary mathematics [21]. An exhaustive literature review on the use of algebraic optimization methods in the development of production inventory systems is presented in [21]. Interested readers are referred to the more recent relevant models presented in [22, 23].

In this paper, we develop a two-stage supply chain inventory model under two types of backordering costs. This model is an extension of [19], which considered only a single stage inventory system. The hybrid geometric-algebraic solution method described in this paper contributes to the trending research that uses nontraditional derivatives based methods. The main purpose of this direction of research is developing useful supply chain models that can be understood and used without the knowledge of differential calculus [19]. To the best of the authors' knowledge, there is no model in the literature of multistage supply chain modeling that used algebra and analytical geometry to drive optimal production-inventory coordination decisions with linear and fixed backorders costs.

The remainder of this paper is organized as follows. The next section presents the notation and assumptions made for developing the model. Section 3 describes the development of the model. A numerical example illustrating the model application is presented in Section 4. Section 5 contains some concluding remarks.

2. Notations and Assumptions

2.1. Notations. The following notations are used in developing the model:

T: basic cycle time, cycle time at the end buyer,

S_b: setup cost for the buyer,

D: the demand rate for the buyer,

S_v: setup cost for the vendor,

K_v: integer multiplier at vendor,

h_r: inventory holding cost per unit time for the buyer,

h_v: inventory holding cost per unit time for the vendor,

T_S: the stock-out time at the buyer,

$\hat{\pi}$: backordering cost per unit per unit time (linear) for the buyer,

π: per unit backorder cost (fixed) for the buyer.

2.2. Assumptions. The following assumptions are made for the two-stage supply chain model:

(a) a single product is produced by the vendor and transferred to the buyer;

(b) replenishment is instantaneous;

(c) production rate and demand rate are deterministic and uniform;

(d) a lot produced by the vendor and sent in equal batches to the buyer;

(e) complete information sharing policy is adopted;

(f) there are two types of backorders costs.

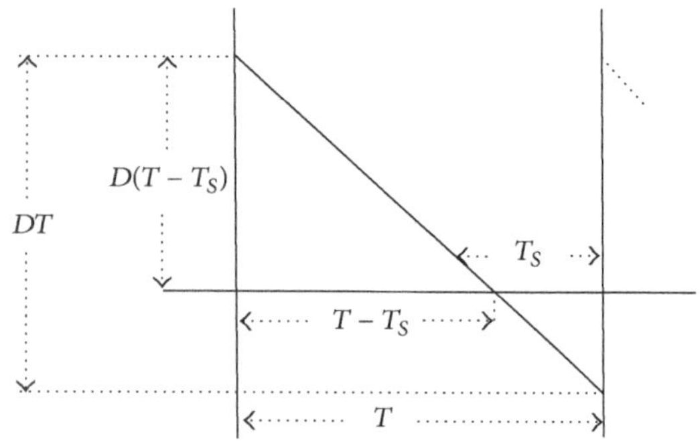

FIGURE 1: The buyer's inventory level behavior.

3. Model Development

We consider a two-stage serial supply chain consisting of a single vendor and a single buyer. This supply chain model is formulated for the integer multiplier inventory replenishment coordination mechanism, where the cycle time at the vendor is an integer multiple of the inventory replenishment cycle time used by the buyer. Shortages are allowed for the end buyer. The buyer's cost consists of the inventory holding cost, the shortage costs, and the setup cost. Figure 1 shows the inventory pattern for the buyer.

As we can see from Figure 1, the total annual cost for the buyer is given by

$$TC_b = h_b\frac{TD}{2} - T_S D h_b + h_b\frac{T_S^2 D}{2T} + \hat{\pi}\frac{T_S^2 D}{2T} + \pi\frac{T_S D}{T} + \frac{S_b}{T}. \quad (1)$$

This total annual cost can be rewritten as

$$TC_b = h_b\frac{TD}{2} + T_S D\left(\frac{\pi}{T} - h_b\right) + \frac{T_S^2 D}{2T}(\hat{\pi} + h_v) + \frac{S_b}{T}. \quad (2)$$

The total costs at the vendor side are made of three parts. The first part is the cost of carrying inventory of raw materials as they are being converted into finished products during the production portion of the cycle. The second part is the cost carrying inventory of the finished products during the nonproduction portion of the cycle [2]. The third part is the setup cost. Therefore, the total cost for the vendor is

$$TC_v = K_v\frac{TD^2}{2P_v}(h_r + h_v) + (K_v - 1)\frac{TD}{2}h_v + \frac{S_v}{K_v T}. \quad (3)$$

The total cost for the entire supply chain is

$$\begin{aligned} TC &= TC_b + TC_v \\ &= \frac{T}{2}\left\{h_b D - h_v D + K_v\left(\frac{D^2}{P_v}(h_r + h_v) + Dh_v\right)\right\} \\ &\quad + \frac{1}{T}\left\{S_b + \frac{S_v}{K_v}\right\} + T_S D\left(\frac{\pi}{T} - h_b\right) \\ &\quad + \frac{T_S^2 D}{2T}(\hat{\pi} + h_v). \end{aligned} \quad (4)$$

This cost model can be rewritten in the following form:

$$TC = TY + \frac{W}{T} + AT_S^2 + BT_S, \quad (5)$$

where

$$\begin{aligned} Y &= \frac{1}{2}\left\{h_b D - h_v D + K_v\left(\frac{D^2}{P_v}(h_r + h_v) + Dh_v\right)\right\}, \\ W &= \left\{S_b + \frac{S_v}{K_v}\right\}, \\ A &= \frac{D(\hat{\pi} + h_v)}{2T}, \\ B &= D\left(\frac{\pi}{T} - h_b\right). \end{aligned} \quad (6)$$

For a fixed replenishment cycle length T, (5) can be optimized with respect T_s using the concept of parabola as in [21]:

$$T_S^* = -\frac{B}{2A} = \frac{Th_b - \pi}{\hat{\pi} + h_v}. \quad (7)$$

Substituting (7) in (5), we get

$$TC = TY + \frac{W}{T} - \frac{D(\pi - Th_b)^2}{2T(\hat{\pi} + h_v)}. \quad (8)$$

This can be further rewritten as

$$TC = TY + \frac{W}{T} - \frac{D\pi^2}{2T(\hat{\pi} + h_v)} + \frac{\pi h_b}{(\hat{\pi} + h_v)} - \frac{Th_b^2}{2(\hat{\pi} + h_v)}. \quad (9)$$

Substituting $\overline{Y} = Y - h_b^2/2(\hat{\pi} + h_v)$ and $\overline{W} = W - D\pi^2/2(\hat{\pi} + h_v)$ in (9), we have

$$TC = T\overline{Y} + \frac{\overline{W}}{T} + \frac{\pi h_b}{(\hat{\pi} + h_v)}. \quad (10)$$

Now, applying the algebraic procedure proposed by [16] and extended by [20], the annual total cost for the entire supply chain in (9) can be represented by factorizing the term $1/T$, and completing the perfect square, one has

$$TC = \frac{1}{T}\left(T^2\overline{Y} - 2T\sqrt{\overline{Y}\,\overline{W}} + \overline{W} + 2T\sqrt{\overline{Y}\,\overline{W}}\right) + \frac{\pi h_b}{(\hat{\pi} + h_v)}. \quad (11)$$

After further factorization, we get

$$TC = \frac{1}{T}\left(T\sqrt{\overline{Y}} - \sqrt{\overline{W}}\right)^2 + 2T\sqrt{\overline{Y}\,\overline{W}} + \frac{\pi h_b}{(\hat{\pi} + h_v)}. \quad (12)$$

It is worthy pointing out that (12) reaches the minimum with respect to T when setting

$$\left(T\sqrt{\overline{Y}} - \sqrt{\overline{W}}\right)^2 = 0. \quad (13)$$

Hence, the optimal batch size T^* is

$$T^* = \sqrt{\frac{\overline{W}}{\overline{Y}}}. \tag{14}$$

Now, for the entire supply chain, the minimum annual total cost is given as

$$TC = 2\sqrt{\overline{Y}\,\overline{W}} + \frac{\pi h_b}{(\hat{\pi} + h_v)}. \tag{15}$$

The optimal batch size T^* is a function of the integer multiplier K_v. We use the method of perfect square to drive the optimal value of this integer multiplier. Substituting for $\overline{Y}$ and $\overline{W}$ into (15), we get

$$TC = \sqrt{2}\left\{(K_v a + b)\left(c + \frac{d}{K_v}\right)\right\}^{1/2} + \frac{\pi h_b}{(\hat{\pi} + h_v)}, \tag{16}$$

where $a = ((D^2/P_v)(h_r + h_v) + Dh_v)$, $b = h_b D - h_v D - h_b^2/2(\hat{\pi} + h_v)$, $c = S_b - D\pi^2/2(\hat{\pi} + h_v)$, and $d = S_v$.

Rewriting (16), we get

$$TC = \sqrt{2}\left\{\frac{1}{K_v}\left[K_v\sqrt{ac} - \sqrt{bd}\right]^2 + \left[\sqrt{ad} + \sqrt{bc}\right]^2\right\}^{1/2} + \frac{\pi h_b}{(\hat{\pi} + h_v)}. \tag{17}$$

From (17), setting

$$\left[K_v\sqrt{ac} - \sqrt{bd}\right]^2 = 0, \tag{18}$$

the optimal value of the integer multiplier K_v^* is derived as follows:

$$K_v^* = \sqrt{\frac{bd}{ac}}. \tag{19}$$

Since the value of K_v^* is a positive integer, the following condition must be satisfied:

$$(K_v^*) \cdot (K_v^* - 1) \le \left(\frac{bd}{ac}\right) \le (K_v^*) \cdot (K_v^* + 1). \tag{20}$$

Now, we can substitute K_v^* from (19) into (14) to find the optimal cycle time size T^*. This optimal value of T^* is used to get T_S^* from (7). Also, substituting K_v^* from (19) into (17) derives the optimal annual total cost in the following closed form:

$$TC^* = \sqrt{2}\left[\sqrt{ad} + \sqrt{bc}\right] + \frac{\pi h_b}{(\hat{\pi} + h_v)}. \tag{21}$$

4. Numerical Example

In this section, we present a numerical example to illustrate the application of developed solution procedure. The relevant data is shown in Table 1. In addition to this data, it is also assumed that holding cost for the raw material at the vendor is $h_r = 0.1$.

By applying the developed hybrid geometric-algebraic method in the previous section, the results of this example are presented in Table 2.

TABLE 1: Data for the example supply chain.

Party	P_i	D	h_i	S_i
Vendor	1500	500	0.2	300
Buyer	—	500	1.5	29

TABLE 2: The example solution.

Parameter	Solution
K_v	3
T^*	0.016
T_S^*	0.113
TC^*	13473.65

5. Conclusion

In this paper, we considered a two-stage supply chain inventory model for coordinating the inventory policies between a vendor and a buyer with full information sharing and planned backorders. We used a hybrid geometric-algebraic method to drive the optimal replenishment policy and the minimum supply chain total cost in a closed form. The described model can be used by the two parties to get benefits of sharing their information and synchronizing their inventory replenishment cycle. The hybrid geometric-algebraic solution method described in this paper contributes to the trending literature that uses nontraditional derivatives based methods.

One clear limitation of the developed model is the assumption of known deterministic demand. Considering a real-world case and relaxing the restrictive model assumptions are good direction for further research. Also, the use of this hybrid geometric-algebraic method for the development of more complex generalized multistage nonserial supply chain inventory models is considered for future research.

Acknowledgment

The author would like to acknowledge the support for this research provided by the Deanship of Scientific Research: King Faisal University under Grant (14045).

References

[1] A. Banerjee, "A joint economic-lot-size model for purchaser and vendor," *Decision Sciences*, vol. 17, no. 3, pp. 292–311, 1986.

[2] M. Khouja, "Optimizing inventory decisions in a multi-stage multi-customer supply chain," *Transportation Research E*, vol. 39, no. 3, pp. 193–208, 2003.

[3] M. Ben-Daya, M. Darwish, and K. Ertogral, "The joint economic lot sizing problem: review and extensions," *European Journal of Operational Research*, vol. 185, no. 2, pp. 726–742, 2008.

[4] M. Khouja and S. K. Goyal, "A review of the joint replenishment problem literature: 1989–2005," *European Journal of Operational Research*, vol. 186, no. 1, pp. 1–16, 2008.

[5] S. K. Goyal and Y. P. Gupta, "Integrated inventory models: the buyer-vendor coordination," *European Journal of Operational Research*, vol. 41, no. 3, pp. 261–269, 1989.

[6] S. K. Goyal and A. Z. Szendrovits, "A constant lot size model with equal and unequal sized batch shipments between production stages," *Engineering Costs and Production Economics*, vol. 10, no. 1, pp. 203–210, 1986.

[7] L. Lu, "A one-vendor multi-buyer integrated inventory model," *European Journal of Operational Research*, vol. 81, no. 2, pp. 312–323, 1995.

[8] S. K. Goyal, "A one-vendor multi-buyer integrated inventory model: a comment," *European Journal of Operational Research*, vol. 82, no. 1, pp. 209–210, 1995.

[9] S. K. Goyal, "A joint economic-lot-size model for a purchaser and vendor: a comment," *Decision Sciences*, vol. 19, no. 1, pp. 236–241, 1988.

[10] M. A. Hoque and S. K. Goyal, "Optimal policy for a single-vendor single-buyer integrated production-inventory system with capacity constraint of the transport equipment," *International Journal of Production Economics*, vol. 65, no. 3, pp. 305–315, 2000.

[11] H. L. Lee, V. Padmanabhan, and S. Whang, "The bullwhip effect in supply chains," *MIT Sloan Management Review*, pp. 93–102, Spring 1997.

[12] J. R. Carter, B. G. Ferrin, and C. R. Carter, "The effect of less-than-truckload rates on the purchase order lot size decision," *Transportation Journal*, vol. 34, no. 3, pp. 35–45, 1995.

[13] T. L. Esper and L. R. Williams, "The value of collaborative transportation management (CTM): its relationship to CPFR and information technology," *Transportation Journal*, vol. 42, no. 4, pp. 55–65, 2003.

[14] J.-M. Chen and T.-H. Chen, "The multi-item replenishment problem in a two-echelon supply chain: the effect of centralization versus decentralization," *Computers and Operations Research*, vol. 32, no. 12, pp. 3191–3207, 2005.

[15] H. M. Wee and C. J. Chung, "A note on the economic lot size of the integrated vendor-buyer inventory system derived without derivatives," *European Journal of Operational Research*, vol. 177, no. 2, pp. 1289–1293, 2007.

[16] L. E. Cárdenas-Barrón, "Optimizing inventory decisions in a multi-stage multi-customer supply chain: a note," *Transportation Research E*, vol. 43, no. 5, pp. 647–654, 2007.

[17] R. W. Grubbström, "Modelling production opportunities—an historical overview," *International Journal of Production Economics*, vol. 41, no. 1–3, pp. 1–14, 1995.

[18] M. E. Seliaman, "Using complete squares method to optimize replenishment policies in a four-stage supply chain with planned backorders," *Advances in Decision Sciences*, vol. 2011, Article ID 745896, 9 pages, 2011.

[19] L. E. Cárdenas-Barrón, "The derivation of EOQ/EPQ inventory models with two backorders costs using analytic geometry and algebra," *Applied Mathematical Modelling*, vol. 35, no. 5, pp. 2394–2407, 2011.

[20] M. E. Seliaman and A. Ahmad, "A generalized algebraic model for optimizing inventory decisions in a multi-stage complex supply chain," *Transportation Research E*, vol. 45, no. 3, pp. 409–418, 2009.

[21] J. T. Teng, L. E. Cárdenas-Barrón, and K. R. Lou, "The economic lot size of the integrated vendor-buyer inventory system derived without derivatives: a simple derivation," *Applied Mathematics and Computation*, vol. 217, no. 12, pp. 5972–5977, 2011.

[22] S. S. Sana, "A production-inventory model of imperfect quality products in a three-layer supply chain," *Decision Support Systems*, vol. 50, no. 2, pp. 539–547, 2011.

[23] K. J. Chung and L. E. Cárdenas-Barrón, "The complete solution procedure for the EOQ and EPQ inventory models with linear and fixed backorder costs," *Mathematical and Computer Modelling*, vol. 55, no. 11-12, pp. 2151–2156, 2012.

4

Designing a Fuzzy Strategic Integrated Multiechelon Agile Supply Chain Network

Morteza Abbasi, Reza Hosnavi, and Mehrdad Mohammadi

Department of Management and Soft Technologies, Malek Ashtar University of Technology, P.O. Box 1774/15875, Tehran, Iran

Correspondence should be addressed to Mehrdad Mohammadi; mehrdadmohamadi@ut.ac.ir

Academic Editor: Zne-Jung Lee

This paper integrates production, distribution and logistics activities at the strategic decision making level, where the objective is to design a multiechelon supply chain network considering agility as a key design criterion. A network with five echelons of supply chains including suppliers, plants, distribution centers, cross-docks, and customer zones is addressed in this paper. The problem has been mathematically formulated as a biobjective optimization model that aims to minimize the cost (fixed and variable) and maximize the plant flexibility and volume flexibility. A novel multiobjective parallel simulating annealing algorithm (MOPSA) is proposed to obtain the Pareto-optimal solutions of the problem. The performance of the proposed solution algorithm is compared with two well-known metaheuristics, namely, nondominated sorting genetic algorithm (NSGA-II) and Pareto archive evolution strategy (PAES). Computational results show that MOPSA outperforms the other metaheuristics.

1. Introduction

The ever-growing global market and impact of competitive environment caused to receive more attention to supply chain for producing and delivering products at low cost, high quality, and short lead time. A supply chain can be defined as a network of suppliers, plants, warehouses, distribution centers, and channels between them which are organized to acquire/achieve raw materials, convert them to finished products, and distribute the finished products to customers; on the other hand, supply chain is a chain that links each element of the manufacturing and supply process from raw materials through to the customer. In most researches, production, distribution and logistics are considered exclusive while considering an integrated approach has more benefit for network. Integrated approach is based on integrating decisions of different functions into a single optimization model. For example, in the supply chain management, demands are satisfied by keeping the service level at a deterministic level then minimizing costs, however; in an integrated approach the costs are minimized and simultaneously the service level is improved by redesigning its supply chain network.

The supply chain network design decisions are classified into three categories based on their importance and the length of the planning horizon. First, strategic (long-range) planning that lasts more than one year; decisions about company selection and facility location are generally placed in this category. Second, tactical (medium) planning that is revised every few months includes decisions about production, inventory, and logistics. Finally, operational or short-range planning involves within-days or shift decisions; raw material, semifinished and finished product flows in the network are operational decisions that are easily classified in this class.

In competitive environment, companies cope with several challenges such as speed of designing, manufacturing, and distributing, while they have to consider simultaneously higher product efficiency and lower operational cost. In this situation, the concept of agile supply chain was proposed to response to these challenges. Agility is used for showing the flexibility and speed in coping with innovative products and unpredictable demands. An agile company should be able to operate in an uncertain and challenging market and use the opportunities that are continuously appearing in competitive environment. Besides, responsiveness is a major factor of an agile supply chain that is as important as a high level of efficiency in cost, quality, and smooth operations flow. In an agile manufacturing when an opportunity appears, the virtual organization can be formed to attain the objective. The virtual organization, according to Goldman et al. [1], is defined as an organizational tool for agile competitors who are integrated by sharing core competencies and resources to produce a particular product that could not be made by each of competitors. Because of short-live durations of virtual organization, strategic and tactical planning should be considered together in the agile manufacturing paradigm.

Uncertainties involved in the supply chain optimization can be categorized into two main groups: (1) environmental and (2) system [2]. Environmental uncertainties are attributed to the performance of each member of supply chain consisting of suppliers, manufacturers, and so forth. System uncertainties are attributed to routine processes in the supply chain such as production and distribution. The above-mentioned uncertainties adversely contribute to the quality of decisions made in strategic, tactical, and operational levels of supply chain. With respect to the alluded points and necessity of considering uncertainties in adopting various decisions in the supply chain networks, researchers, in tune with designing their supply chain networks, have strived to include foregoing two aspects of uncertainties to better represent the practical features of real-world problems. Thus, for this purpose, advanced optimization methodologies are desired. Hence, many researchers have recently made attempts to model the uncertainty in supply chains [3–5].

In the real world, after designing a supply chain network, the costs, demands, distances, times, and other parameters of the problem may change due to uncertain environment. For example, with variability in the transportation time from an origin to a destination, there is a possibility that a flow may not be delivered on time. A failure in on-time delivery may result in a huge and nonmeasureable cost, such as lost-opportunity cost and lost-sale cost due to customer churning, to compensate an unsatisfied customer.

In this paper, we consider a multiechelon supply chain network design problem consisting of five echelon and agility as a design criterion with an integrated approach. The objective of this model is to determine the best suppliers and plants in upstream level and determine the number of distribution and cross-dock that should be opened in downstream level to satisfy demands in lower cost and higher efficiency. Usage of cross-dock caused to decrease the construction and setup costs; this concept is used in this paper at the intermediate level between distribution and customers zones. Also, in order to cope with uncertain environment, some parameters of the model such as demand and cost are considered as fuzzy parameters.

The rest of the paper is organized as follows. Section 2 presents related literature. In Section 3, the fuzzy multiobjective mathematical model for agile supply chain management (ASCM) is presented. Defuzzification of the fuzzy model is proposed in Section 4. Section 5 explains the proposed multiobjective metaheuristic algorithm followed by experimental results in Section 6 followed by conclusion is Section 7.

2. Literature Review

Many studies have been done in traditional supply chains dealing with supplier selection and allocation problem of facilities for satisfying demand. Samadhi and Hoang [6] proposed a three-stage hierarchical procedure for ranking suppliers based on strategic, manufacturing, and logistics factors. De Boer et al. [7] suggested an outranking method for evaluating the supplier by use of the multicriteria factors. Jayaraman [8] considered a classical capacitated warehouse problem. In this paper, the number and location of warehouses are determined and the demand of the customers is satisfied by allocating the customers to warehouses at the minimum cost. Tsiakis et al. [9] addressed a multiproduct, multi-echelon supply chain network under uncertainty. It is assumed that the locations of manufacturing sites are fixed although the number, location, and capacity of warehouses and distribution centers should be determined from a set of potential locations. The objective of this model is to minimize the total cost of the network. This model is solved by using branch-and-bound techniques. Shankar et al. [10] presented a multiobjective optimization of single-product for a multi-echelon supply chain network including suppliers, plants, distribution centers (DCs), and customer zones (CZs). The goal of this model is to find the number and location of plants, the flow of raw materials from suppliers to plants, and the quantity of products to be shipped from plants to DCs, from DCs to CZs with optimizing the objectives simultaneously. The objectives minimize location and shipment costs while the maximum customer demands should be met. They proposed multiobjective hybrid particle swarm optimization (MOHPSO) algorithm for solving their model.

Integrated analysis in supply chain network is an approach based on the integrating the decisions of different functions such as supply, processing, distribution, inventory management, production planning, and facilities location on a single optimization model for a production-distribution system. Sarmiento and Nagi [11] presented a review on integrated model of production-distribution system with considering transportation system. This paper consists of two sections: analyses of production, distribution and inventory planning and the inventory/routing problem. This paper considers decisions at the strategic and tactical levels. Alonso-Ayuso et al. [12] presented a two-stage stochastic supply chain planning. This model determines the size of plant, product selection, assigning product to plants, and vendor selection. The strategic decisions are made in the first stage and tactical decisions are made in second stage. The objective of this model is to maximize the expected profit of the supply chain. Teo and Shu [13] proposed a distribution network design problem with integrating transportation and inventory function. The goal of this model is to determine the number and location of warehouses, assign the retailers to warehouses, and determine the optimal inventory policy. In this model, there is a trade-off between inventory cost, direct shipment cost, and facility location cost, and the objective is to minimize the multi-echelon inventory, transportation, and facility location costs.

Elhedhli and Coffin [14] presented an integrated production-distribution problem involving three echelons: plants, distributions, and customer zones. The objective of this model is to minimize production, distribution, handling, transportation costs, and the fixed costs of opening distribution centers. The solution method for this model is based on two-level Lagrangian relaxation approach and branch-and-price algorithm. Amiri [15] presented a distribution network design problem and considered the multiple levels of capacities for plants and warehouses in supply chain system. In this model, the best strategies for distributing the product from the plants to the warehouses and from the warehouses to the customers are selected. The objective is to minimize total costs of the distribution network by determining the optimum numbers, locations and capacities of plants and warehouses. Van Hoek et al. [16], considered the concept of agility in supply chain. In their model, the agile supply chain consists of customer sensitivity, process integration, and information integration. Dotoli et al. [17] proposed a single and multi-objective optimization model for agile supply chain network. For demonstrating the ability of reconfiguring a flexible system economically, quickly, and environmentally responsible, an Internet-based distributed manufacturing system is considered in this model. Sha and Che [18] offered a way to make manufacturing systems more agile and competitive. They developed a partner selection and production-distribution planning model based on the analytic hierarchy process (AHP), multiattribute utility theory (MAUT), and integer programming (IP), for the virtual integration. They used branch and bound for solving this model. Wu and Barnes [19] presented a review on agile supply chain partner decision making. They concentrated on particular attention given to those methods that are especially relevant for use in agile supply chains.

Pan and Nagi [20] proposed a supply chain design problem with uncertain demand in an agile manufacturing system. They considered the integrated optimization of logistics and production costs. The main objective is to select one company in each echelon. A scenario-based approach is used to handle the uncertainty of demand. The formulation is a robust optimization model with considering the expected total costs, variable cost of uncertain demand, and the penalty of/for unmet demands. The heuristic approach is used for solving this problem. Bachlaus et al. [21] proposed an integrated multi-echelon supply chain network with considering agility. This paper presented a five-echelon supply chain model consisting of suppliers, plants, distribution centers, cross-docks, and customer zones and integrated production, distribution and logistics activities. This model is formulated as a multiobjective optimization model that minimizes the fixed and variable costs and maximizes the plant flexibility and volume flexibility. They used a weighted-sum approach for solving their multiobjective problem while this method always caused suboptimality. They proposed HTPSO algorithm for solving the model.

Cochran and Marquez Uribe [22] presented a modified set covering formulation for capacity planning and equipment selection in supply chains. In this model, use/need of multi-functional equipment to deal with changing demand is considered. Usage of this equipment is costly, but it is necessary to compete with widespread technological developments among competitors and have flexibility in meeting varying supply chain demands. Costantino et al. [23] considered the manufacturing system (MSC) composing of different stages in supply chain planning. They proposed a technique for the strategic management of the supply chain planning and allowed to have an ability of reconfiguration of the chain for the improvement of the MSC agility. They considered the supplier capacity constraints and the level of demands and maximum production capacities with single/multiple sourcing.

Pan and Nagi [24] presented a supply chain network design problem in an agile manufacturing system. They considered a multiple echelons and multiple period model under the situation of multiple customers that have heavy demands. The strategic and tactical decisions are integrated. The main goal of this model is the selection of one or more companies in each echelon, production, inventory, and transportation with minimizing the total operational costs including fixed costs, production, raw material holding, finished products holding, and transportation costs under production and transportation capacity limits. They used Lagrangian heuristic for solving this paper.

The agile supply chain that is presented in this paper consists of cross-dock site at the location between distribution and customer zones. This model integrated a production, distribution, and logistics and formulated as a multiobjective optimization model. Due to the complexity of the proposed integrated model, a well-known multiobjective metaheuristic algorithm, namely, multiobjective parallel simulated annealing (MOPSA) is developed for obtaining Pareto solutions of the model. Two other multiobjective metaheuristic algorithms, called NSGA-II and PAES, are considered in order to show the higher performance of the proposed MOPSA.

This paper has been prepared with the following innovations:

(i) designing a fuzzy integrated agile supply chain at the strategic level by integrating production, distribution, and logistics across the whole chain,

(ii) presenting a multiobjective optimization model with considering objectives of fixed and variable costs and volume flexibility,

(iii) developing a new multiobjective metaheuristic algorithm, namely, MOPSA,

(iv) extracting nondominated Pareto front solutions of the multiobjective model.

3. Fuzzy Multiobjective Mathematical Model

In this section, an integrated agile supply chain network model has been mathematically formulated. The model represents the concept of agility in the supply chain network that decides its ability for rapid responding to the customer demands. The proposed model consists of 5 echelons: suppliers, plants, distribution centers, cross-docks, and customer zones, and these echelons are linked together with a forward flow of material and backward flow of information. The aim of this paper is to integrate the production, distribution and logistics activities at minimum cost, maximum plant flexibility, and volume flexibility. This model optimizes the material flow and determines the optimal number of suppliers, plants, and distribution centers and assignments of cross-docks to customer zones.

The transportation costs and capacities are modeled as fuzzy parameters in the proposed formulation. The following notation is used in the formulation of the agile supply chain network design (ASCND) model.

Indices

$s = 1, 2, \ldots, S$: Set of suppliers

$i = 1, 2, \ldots, I$: Set of plants

$j = 1, 2, \ldots, J$: Set of possible distribution centers

$k = 1, 2, \ldots, K$: Set of possible cross-docks

$d = 1, 2, \ldots, D$: Set of customer zones

$p = 1, 2, \ldots, P$: Product type

$r = 1, 2, \ldots, R$: Raw material type.

Notations

Q_{pi}: Quantity of product p produced at plant i

q_{pij}: Quantity of product p transported from plant i to distribution center j

q_{rsi}: Quantity of raw material r transported from supplier s to plant i

AL_k: Agility level of cross-dock k

AL_j: Agility level of distribution center j

AL_i: Agility level of plant i

AL_s: Agility level of supplier s

Cap_k: Capacity of cross-dock k to handle product families

Cap_j: Capacity of distribution center j to handle product families

CDF: Cross-dock flexibility

C_{mk}: Cost to supply product p from cross-dock k which would be used by the customer zone m

DCF: Distribution center flexibility

D_{mp}: Demand from customer zone m for product p

DCVF_m: Distribution volume flexibility of customer zone m

C_k: Fixed operating cost to open cross-dock k

F_i: Fixed cost for plant i

F_j: Fixed cost for distribution center j

LB_h: Lower bound of high agility performance index (0.8)

UB_h: Upper bound of high agility performance index (1.0)

P: Total number of products to be manufactured

K: Total number of cross-docks

J: Total number of distribution centers

LB_l: Lower bound of low agility performance index (0.4)

UB_l: Upper bound of low agility performance index (0.6)

M: Total number of customer zones

LB_m: Lower bound of medium agility performance index (0.6)

UB_m: Upper bound of medium agility performance index (0.8)

l_{pi}: Minimum production volume for product p at plant i

l_j: Minimum throughput at distribution center j

CD: Maximum number of cross-docks to be opened

DC: Maximum number of distribution centers to be opened

U_{ip}: Maximum production volume for product p at plant i

U_j: Maximum throughput at distribution center j

norm($\cdot$): Represents normalized objective ($\cdot$)

Z: Final integrated objective function

I: Total number of plants

Cap_i: Production capacity for each plant i

Cap_{rs}: Production capacity of supplier s for raw material r

PF: Plant flexibility

PVF_m: Plant volume flexibility

STp_i: Standard units at plant i per unit of product p

R: Total number of raw material to be supplied

S: Total number of suppliers

SF: Supplier flexibility

ST_{jp}: Standard units at distribution center j per unit of product p

TC_m: Total cost of m customer zone

C_{rs}: Unit cost of raw material r for supplier s

C_{pj}: Unit cost of throughput (handling and inventory) for product p at distribution center j

C_{pi}: Unit production cost for product pat plant i

UR_{rp}: Utilization rate for each raw material r per unit of product p

C_{jkp}: Unit transportation cost from distribution center j to cross-dock k for product p

C_{ijp}: Unit transportation cost from plant i to distribution center j for product p

C_{rsi}: Unit transportation cost from supplier s to plant i for raw material r.

Decision Variables

$$X_i = \begin{cases} 1 & \text{if plant } i \text{ is opened} \\ 0 & \text{otherwise,} \end{cases}$$

$$Y_j = \begin{cases} 1 & \text{if distribution } j \text{ is opened} \\ 0 & \text{otherwise,} \end{cases}$$

$$V_k = \begin{cases} 1 & \text{if cross-dock } k \text{ is opened} \\ 0 & \text{otherwise,} \end{cases}$$

$$R_{kjp} = \begin{cases} 1 & \text{if cross-dock } k \text{ is assigned to distribution } j \text{ for product } p \\ 0 & \text{otherwise,} \end{cases} \quad (1)$$

$$A_{mkp} = \begin{cases} 1 & \text{if customer zone } m \text{ is assigned to cross-dock } k \text{ for product } p \\ 0 & \text{otherwise,} \end{cases}$$

$$N_s = \begin{cases} 1 & \text{if supplier } s \text{ is selected} \\ 0 & \text{otherwise,} \end{cases}$$

$$\begin{aligned} \text{TC}_m = & \left[\sum_{i=1}^{I}\sum_{s=1}^{S}\sum_{r=1}^{R}\left(\widetilde{C_{rsi}} + \widetilde{C_{rs}}\right) \cdot q_{rsp}\right] \\ & + \left[\sum_{i}^{I} F_i \cdot x_i + \sum_{p}^{P}\sum_{i}^{I} \widetilde{C_{pi}} \cdot q_{pi} + \sum_{p}^{P}\sum_{i}^{I}\sum_{j}^{J} \widetilde{C_{pij}} \cdot q_{pij}\right] \\ & + \left[\sum_{k}^{K} F_k \cdot y_k + \sum_{p}^{P}\sum_{j}^{J}\sum_{k}^{K} \widetilde{C_{pjk}} \cdot R_{kjp} \right. \\ & \left. + \sum_{m}^{M}\sum_{p}^{P}\sum_{k}^{K}\sum_{j}^{J} \widetilde{C_{pj}} \cdot D_{mp} \cdot R_{kjp} \cdot A_{mkp}\right] \\ & + \left[\sum_{j}^{J} F_k \cdot v_k + \sum_{m}^{M}\sum_{p}^{P}\sum_{k}^{K} \widetilde{C_{mk}} \cdot A_{mkp}\right]. \end{aligned} \quad (2)$$

The objective function (2) minimizes the total cost including fixed and variable costs, and are included four components, that is, (a) purchasing cost of raw material and transportation cost from suppliers to plants, (b) the fixed and variable costs of plant operations and transportation cost, (c) the variable cost of handling and inventory of products at DCs, and transportation cost to ship from plant to DCs, and (d) the transportation cost to ship products from DCs to cross-docks and distribution costs to transfer product families from cross-docks to customer zones:

$$\text{SF} = \sum_{p=1}^{P}\sum_{r=1}^{R}\sum_{s=1}^{S}\left(\widetilde{\text{Cap}}_{rs} \cdot N_s - \widetilde{D}_{mp}\right) \quad \forall m, \quad (3)$$

$$\text{PF} = \sum_{p}^{P}\sum_{i}^{I}\left(\widetilde{\text{Cap}}_i \cdot x_i - \widetilde{D}_{mp}\right) \quad \forall m, \quad (4)$$

$$\text{DCF} = \sum_{p}^{P}\sum_{j}^{J}\left(\widetilde{\text{Cap}}_j \cdot y_j - \widetilde{D}_{mp}\right) \quad \forall m, \quad (5)$$

$$\text{CDF} = \sum_{p}^{P}\sum_{k}^{K}\left(\widetilde{\text{Cap}}_k \cdot V_k - \widetilde{D}_{mp}\right) \quad \forall m. \quad (6)$$

The flexibility objective is divided into two types: distribution volume flexibility and delivery flexibility. Specifically, distribution volume flexibility can be defined as the minimum of the difference between the capacity of suppliers, plants, distribution centers, and cross-docks with total demand of products. In (3), supplier flexibility is equal to the difference between supplier's capacities with total demand of products ordered by customer at zone m. In (4), plant flexibility is the difference between plant's capacities with total demand of products ordered by customer at zone m, and distribution center flexibility and cross-dock flexibility are defined by the same definitions:

$$\text{DVF}_m = \min(\text{SF}, \text{PF}, \text{DCF}, \text{CDF}) \quad \forall m, \quad (7)$$

$$\text{PVF}_m = \sum_{i}^{I}\left(x_i \cdot \widetilde{\text{Cap}}_i - \sum_{p}^{P} q_{pi} \cdot \text{st}_{pi}\right). \quad (8)$$

In (7), distribution volume flexibility is defined as the minimum of supplier flexibility, plant flexibility, distribution center flexibility, and cross-dock flexibility. Equation (8) describes plant volume flexibility as the difference between utilization plant capacities with available plant capacity. Finally, the three objectives of the model are as follows:

$$\begin{aligned} \min Z_1 &= \sum_{m=1}^{M} \text{TC}_m, \\ \max Z_2 &= \sum_{m=1}^{M} \text{DVF}_m, \\ \max Z_3 &= \sum_{m=1}^{M} \text{PVF}_m. \end{aligned} \quad (9)$$

Because of the different dominance of cost and flexibility, the normalized approach is considered to integrate both objectives together. The general objective function consists of normalized cost, distribution volume flexibility, and plant flexibility:

$$\sum_{i}^{I} q_{rsi} \leq \widetilde{\text{Cap}}_{rs} \quad \forall r, s, \tag{10}$$

$$\sum_{s}^{S} \text{UR}_{rp} \cdot q_{rsi} \leq \sum_{s}^{S} \widetilde{\text{Cap}}_{rs} \quad \forall r, p, i, \tag{11}$$

$$\sum_{p}^{P} \text{st}_{pi} \cdot q_{pi} \leq \widetilde{\text{cap}}_{i} \cdot x_{i} \quad \forall i, \tag{12}$$

$$l_{pi} \cdot x_{i} \leq q_{pi} \leq u_{pi} \cdot x_{i} \quad \forall i, p, \tag{13}$$

$$l_{j} \cdot y_{j} \leq \sum_{k} \sum_{p} \text{st}_{pj} \cdot \widetilde{D}_{mp} \cdot R_{kjp} \cdot A_{mkp} \leq u_{j} \cdot y_{j} \quad \forall j, \tag{14}$$

$$\sum_{k}^{K} A_{mkp} = 1 \quad \forall m, p, \tag{15}$$

$$q_{pi} = \sum_{j}^{J} q_{pij} \quad \forall p, i, \tag{16}$$

$$\sum_{i}^{I} \sum_{j}^{J} q_{pij} = \sum_{m}^{M} \widetilde{D}_{mp} \quad \forall p, \tag{17}$$

$$\sum_{i}^{I} q_{pij} = \sum_{m}^{M} \widetilde{D}_{mp} \cdot R_{kjp} \cdot A_{mkp} \quad \forall p, j, \tag{18}$$

$$q_{pij}, q_{pi}, q_{rsi} \geq 0, \quad \forall r, p, i, j, k, \tag{19}$$

$$\sum_{j}^{J} R_{kjp} = V_{k} \quad \forall k, p, \tag{20}$$

$$\sum_{m}^{M} \sum_{p}^{P} \widetilde{d}_{mp} \cdot A_{mkp} \leq \widetilde{\text{cap}}_{k} \quad \forall k, \tag{21}$$

$$\sum_{m}^{M} \sum_{p}^{P} R_{kjp} \cdot \widetilde{\text{cap}}_{k} \leq \widetilde{\text{cap}}_{j} \quad \forall j, \tag{22}$$

$$A_{mkp} \leq V_{k} \quad \forall m, p, k, \tag{23}$$

$$R_{kjp} \leq y_{j} \quad \forall p, k, j, \tag{24}$$

$$\sum_{k}^{K} V_{k} \leq \text{DC}, \tag{25}$$

$$\sum_{j}^{J} y_{j} \leq \text{CD}, \tag{26}$$

$$\text{LB}_{l} \leq \text{AL}_{s} + \text{AL}_{i} + \text{AL}_{j} + \text{AL}_{k} \leq \text{UB}_{l}, \tag{27}$$

$$\text{LB}_{m} \leq \text{AL}_{s} + \text{AL}_{i} + \text{AL}_{j} + \text{AL}_{k} \leq \text{UB}_{m}, \tag{28}$$

$$\text{LB}_{u} \leq \text{AL}_{s} + \text{AL}_{i} + \text{AL}_{j} + \text{AL}_{k} \leq \text{UB}_{u}. \tag{29}$$

Constraint (10) ensures that the total number of raw material shipped to plants by each supplier should not exceed the supplier capacity. Constraint (11) checks the raw material supplied by the supplier meets the production requirements. Constraint (12) checks whether the plant capacity, and the range of the quantity of products to be produced at a plant is considered in Constraint (13). Constraint (14) ensures the minimum and maximum capacities for distribution centers and the customer zones that are assigned to the open DCs in this limited area. Constraint (15) enforces each customer zone to be assigned to exactly one cross-dock. Constraint (16) ensures that the amount of products transfer from plant to distribution centers is equal to the quantity of products that produced at that plant. Constraint (17) ensures that all demands are satisfied and constraint (18) checks the demand requirements at each distribution center. Constraint (19) enforces the nonnegativity for all variables. Constraint (20) ensures that cross-docks should be assigned to only open distribution centers. Constraints (21) and (22) ensure the capacity restriction for distribution centers' cross-docks. Constraint (23) imposes that open cross-dock can satisfy customer demand for products. Constraint (24) ensures that there will be a flow from cross-dock to open distribution centers if the distribution centers are opened. Because of limited capital availability, there is an upper bound for the number of DCs and CDs: a maximum number of cross-docks and distribution centers to be opened. Constrains (25) and (26) ensure that the number of opening cross-docks and distribution centers should not exceed the maximum number of them. Constraints (27), (28), and (29) are modeled to decide the feasibility of agile supply chain. The agility is categorized into three classes: high, medium, and low. Constraint (27) ensures that the agility of selected supplier, plant, distribution centers, and cross-docks should be lesser than the upper bound and greater than its lower bound of low-level agility. Constraints (28) and (29) are similarly defined for medium- and high-level agility performance.

4. Defuzzification Procedure

Jiménez et al. [25] proposed a method to convert the possibilistic mixed integer programming model into an equivalent auxiliary crisp model. Jiménez et al.'s [25] method has some advantages as follows.

(i) This is an efficient method for solving the fuzzy linear problems because it keeps linearity and the number of the objective functions does not increase.

(ii) This is a general ranking method which can be used in ranking method which can be applied to different kinds of membership functions such as triangular, trapezoidal, and nonlinear ones in both symmetric and asymmetric forms.

(iii) This method has strong mathematical concepts such as expected interval and expected value of fuzzy numbers

In the following, some concepts that are used in the Jimenez method are initially defined, and the defuzzification procedure is clearly delineated.

4.1. Concepts Definition. A fuzzy number is a fuzzy set $\tilde{a}$ on the real line R whose membership function $\mu_{\tilde{a}}$ is semicontinuous, such that

$$r = \mu_{\tilde{a}}(x) = \begin{cases} 0 & \forall x \in (-\infty, a_1) \\ f_a(x) & \text{increasing on } [a_1, a_2] \\ 1 & \forall x \in [a_2, a_3] \\ g_a(x) & \text{decreasing on } [a_3, a_4] \\ 0 & \forall x \in (a_4, \infty). \end{cases} \quad (30)$$

An r-cut of a fuzzy number $\tilde{a}$ is defined by $a_r = \{x \in S \mid \mu_{\tilde{a}}(x) \geq r\}$ where S is a universe. Since $\mu_{\tilde{a}}$ is continuous semicontinuous, the r-cuts are closed and bounded intervals, and we represent them by $a_r = [f_a^{-1}(r), g_a^{-1}(r)]$; the expected interval of a fuzzy number $\tilde{a}$, noted EI $(\tilde{a})$, can be defined by following equation:

$$\mathrm{EI}(\tilde{a}) = [E_1^a, E_2^a] = \left[\int_0^1 f_a^{-1}(r), \int_0^1 g_a^{-1}(r)\right]. \quad (31)$$

The expected value of a fuzzy number $\tilde{a}$ noted EV $(\tilde{a})$ is defined as the half-point of its expected interval:

$$\mathrm{EV}(\tilde{a}) = \frac{E_1^a + E_2^a}{2}. \quad (32)$$

Considering $\tilde{a}, \tilde{b}$ as a fuzzy numbers and γ, δ as nonnegative real numbers, the principle to aggregate fuzzy numbers can be computed by the following equation:

$$\begin{aligned} \mathrm{EI}(\delta\tilde{a} + \gamma\tilde{b}) &= \delta \mathrm{EI}(\tilde{a}) + \gamma \mathrm{EI}(\tilde{b}), \\ \mathrm{EV}(\delta\tilde{a} + \gamma\tilde{b}) &= \delta \mathrm{EV}(\tilde{a}) + \gamma \mathrm{EV}(\tilde{b}). \end{aligned} \quad (33)$$

Definition 1. For any pair of fuzzy numbers $\tilde{a}, \tilde{b}$, the degree in which considering $\tilde{a}$ is bigger than $\tilde{b}$ is the following:

$$\mu_M(\tilde{a}, \tilde{b}) = \begin{cases} 0 & \text{if } E_2^a - E_1^b < 0 \\ \dfrac{E_2^a - E_1^b}{E_2^a - E_1^b - (E_1^a - E_2^b)} & \text{if } 0 \in [E_1^a - E_2^b] \\ 1 & \text{if } E_2^a - E_1^b > 0. \end{cases} \quad (34)$$

When $\mu_M(\tilde{a}, \tilde{b}) \geq \propto$, it is said that $\tilde{a}$ is bigger than or equal to $\tilde{b}$ at least in a degree $\propto$ and is represented by $\tilde{a} \geq {}_{\propto}\tilde{b}$.

For fuzzy linear constraint $\tilde{a}_i x \geq \tilde{b}_i$, a decision vector $x \in R$ is said to be feasible if

$$\min \{\mu_M(\tilde{a}_i, \tilde{b}_i)\} = \propto, \quad (35)$$

where $\tilde{a}_i = (\tilde{a}_{i1}, \tilde{a}_{i2}, \ldots, \tilde{a}_{in})$.

That is to say

$$\tilde{a}_i x \geq {}_{\propto}\tilde{b}_i, \quad i = 1, \ldots, m. \quad (36)$$

For fuzzy linear constraint $\tilde{a}_i x = \tilde{b}_i$, a decision vector $x \in R$ is said to be feasible if

$$\tilde{a} \geq {}_{\propto/2}\tilde{b}, \qquad \tilde{a} \leq {}_{\propto/2}\tilde{b}. \quad (37)$$

These equations can be rewritten as follows:

$$\frac{\propto}{2} \leq \mu_M(\tilde{a}, \tilde{b}) \leq 1 - \frac{\propto}{2}. \quad (38)$$

4.2. Defuzzified Model. In this method, each individual fuzzy parameter that exists in each objective function is replaced by its *expected value*, and for defuzzifying constraints, the concept of $\propto$ feasibility is employed:

$$\begin{aligned} \min Z_1 = \sum_{m=1}^{M} \Bigg\{ & \left[\sum_{i=1}^{I}\sum_{s=1}^{S}\sum_{r=1}^{R} (\mathrm{EV}(C_{rsi}) + \mathrm{EV}(C_{rs})) \cdot q_{rsp}\right] \\ & + \left[\sum_{i}^{I} F_i \cdot x_i + \sum_{p}^{P}\sum_{i}^{I} \mathrm{EV}(C_{pi}) \cdot q_{pi} + \sum_{p}^{P}\sum_{i}^{I}\sum_{j}^{J} \mathrm{EV}(C_{pij}) \cdot q_{pij}\right] \\ & + \left[\sum_{k}^{K} F_k \cdot y_k + \sum_{p}^{P}\sum_{j}^{J}\sum_{k}^{K} \mathrm{EV}(C_{pjk}) \cdot R_{kjp} + \sum_{m}^{M}\sum_{p}^{P}\sum_{k}^{K}\sum_{j}^{J} \mathrm{EV}(C_{pj}) \cdot \mathrm{EV}(D_{mp}) \cdot R_{kjp} \cdot A_{mkp}\right] \\ & + \left[\sum_{j}^{J} F_k \cdot v_k + \sum_{m}^{M}\sum_{p}^{P}\sum_{k}^{K} \mathrm{EV}(C_{mk}) \cdot A_{mkp}\right]\Bigg\} \end{aligned}$$

$$\max Z_2 = \sum_{m=1}^{M} \mathrm{DVF}_m$$

$$\max Z_1 = \sum_{m=1}^{M} \mathrm{PVF}_m$$

s.t.:

$$\mathrm{SF} \leq \sum_{p=1}^{P}\sum_{r=1}^{R}\sum_{s=1}^{S} \left(\mathrm{Cap}_{rs} \cdot N_s - \left[\frac{\alpha}{2} \times E_2^{\tilde{D}_{mp}} + \left(1 - \frac{\alpha}{2}\right) \times E_1^{\tilde{D}_{mp}}\right]\right) \quad \forall m,$$

$$\mathrm{SF} \geq \sum_{p=1}^{P}\sum_{r=1}^{R}\sum_{s=1}^{S} \left(\mathrm{Cap}_{rs} \cdot N_s - \left[\left(1 - \frac{\alpha}{2}\right) \times E_2^{\tilde{D}_{mp}} + \frac{\alpha}{2} \times E_1^{\tilde{D}_{mp}}\right]\right) \quad \forall m,$$

$$\mathrm{PF} \le \sum_{p}^{P}\sum_{i}^{I}\left(\mathrm{Cap}_i \cdot x_i - \left[\frac{\alpha}{2} \times E_2^{\tilde{D}_{mp}} + \left(1-\frac{\alpha}{2}\right) \times E_1^{\tilde{D}_{mp}}\right]\right) \quad \forall m,$$

$$\mathrm{PF} \ge \sum_{p}^{P}\sum_{i}^{I}\left(\mathrm{Cap}_i \cdot x_i - \left[\left(1-\frac{\alpha}{2}\right) \times E_2^{\tilde{D}_{mp}} + \frac{\alpha}{2} \times E_1^{\tilde{D}_{mp}}\right]\right) \quad \forall m,$$

$$\mathrm{DCF} \le \sum_{p}^{P}\sum_{j}^{J}\left(\mathrm{Cap}_j \cdot y_j - \left[\frac{\alpha}{2} \times E_2^{\tilde{D}_{mp}} + \left(1-\frac{\alpha}{2}\right) \times E_1^{\tilde{D}_{mp}}\right]\right) \quad \forall m,$$

$$\mathrm{DCF} \ge \sum_{p}^{P}\sum_{j}^{J}\left(\mathrm{Cap}_j \cdot y_j - \left[\left(1-\frac{\alpha}{2}\right) \times E_2^{\tilde{D}_{mp}} + \frac{\alpha}{2} \times E_1^{\tilde{D}_{mp}}\right]\right) \quad \forall m,$$

$$\mathrm{CDF} \le \sum_{p}^{P}\sum_{k}^{K}\left(\mathrm{Cap}_k \cdot V_k - \left[\frac{\alpha}{2} \times E_2^{\tilde{D}_{mp}} + \left(1-\frac{\alpha}{2}\right) \times E_1^{\tilde{D}_{mp}}\right]\right) \quad \forall m,$$

$$\mathrm{CDF} \ge \sum_{p}^{P}\sum_{k}^{K}\left(\mathrm{Cap}_k \cdot V_k - \left[\left(1-\frac{\alpha}{2}\right) \times E_2^{\tilde{D}_{mp}} + \frac{\alpha}{2} \times E_1^{\tilde{D}_{mp}}\right]\right) \quad \forall m,$$

$$\mathrm{DVF}_m = \min(\mathrm{SF}, \mathrm{PF}, \mathrm{DCF}, \mathrm{CDF}) \quad \forall m,$$

$$\mathrm{PVF}_m \le \sum_{i}^{I}\left(x_i \cdot \left[\left(1-\frac{\alpha}{2}\right) \times E_2^{\mathrm{Cap}_i} + \frac{\alpha}{2} \times E_1^{\mathrm{Cap}_i}\right] - \sum_{p}^{P} q_{pi} \cdot st_{pi}\right),$$

$$\mathrm{PVF}_m \ge \sum_{i}^{I}\left(x_i \cdot \left[\frac{\alpha}{2} \times E_2^{\mathrm{Cap}_i} + \left(1-\frac{\alpha}{2}\right) \times E_1^{\mathrm{Cap}_i}\right] - \sum_{p}^{P} \left(q_{pi} \cdot \mathrm{st}_{pi}\right)\right),$$

$$\sum_{i}^{I} q_{rsi} \le \left[(1-\alpha) \times E_2^{\mathrm{Cap}_{rs}} + \alpha \times E_1^{\mathrm{Cap}_{rs}}\right] \quad \forall r, s,$$

$$\sum_{s}^{S} \mathrm{UR}_{rp} \cdot q_{rsi} \le \sum_{s}^{S}\left[(1-\alpha) \times E_2^{\mathrm{Cap}_{rs}} + \alpha \times E_1^{\mathrm{Cap}_{rs}}\right] \quad \forall r, p, i,$$

$$\sum_{p}^{P} \mathrm{st}_{pi} \cdot q_{pi} \le \left[(1-\alpha) \times E_2^{\mathrm{Cap}_{rs}} + \alpha \times E_1^{\mathrm{Cap}_{rs}}\right] \cdot x_i \quad \forall i,$$

$$l_{pi} \cdot x_i \le q_{pi} \le u_{pi} \cdot x_i \quad \forall i, p,$$

$$l_j \cdot y_j \le \sum_{k}^{K}\sum_{p}^{P} \mathrm{st}_{pj} \cdot \left[(1-\alpha) \times E_2^{D_m} + \alpha \times E_1^{D_m}\right] \cdot R_{kjp} \cdot A_{mkp} \quad \forall j,$$

$$\sum_{k}^{K}\sum_{p}^{P} \mathrm{st}_{pj} \cdot \left[\alpha \times E_2^{Dm} + (1-\alpha) \times E_1^{Dm}\right] \cdot R_{kjp} \cdot A_{mkp} \le u_j \cdot y_j \quad \forall j,$$

$$\sum_{k}^{K} A_{mkp} = 1 \quad \forall m, p,$$

$$q_{pi} = \sum_{j}^{J} q_{pij} \quad \forall p, i,$$

$$\sum_{i}^{I}\sum_{j}^{J} q_{pij} \le \sum_{m}^{M}\left[\left(1-\frac{\alpha}{2}\right) \times E_2^{Dm} + \frac{\alpha}{2} \times E_1^{Dm}\right] \quad \forall p,$$

$$\sum_{i}^{I}\sum_{j}^{J} q_{pij} \ge \sum_{m}^{M}\left[\frac{\alpha}{2} \times E_2^{Dm} + \left(1-\frac{\alpha}{2}\right) \times E_1^{Dm}\right] \quad \forall p,$$

$$\sum_{i}^{I} q_{pij} \le \sum_{m}^{M}\left[\left(1-\frac{\alpha}{2}\right) \times E_2^{Dm} + \frac{\alpha}{2} \times E_1^{Dm}\right] \cdot R_{kjp} \cdot A_{mkp} \quad \forall p, j,$$

$$\sum_{i}^{I} q_{pij} \ge \sum_{m}^{M}\left[\frac{\alpha}{2} \times E_2^{Dm} + \left(1-\frac{\alpha}{2}\right) \times E_1^{Dm}\right] \cdot R_{kjp} \cdot A_{mkp} \quad \forall p, j,$$

$$q_{pij}, q_{pi}, q_{rsi} \ge 0, \quad \forall r, p, i, j, k,$$

$$\sum_{j}^{J} R_{kjp} = V_k \quad \forall k, p,$$

$$\sum_{m}^{M}\sum_{p}^{P} d_{mp} \cdot A_{mkp} \le \left[(1-\alpha) \times E_2^{\mathrm{Cap}_k} + \alpha \times E_1^{\mathrm{Cap}_k}\right] \quad \forall k,$$

$$\sum_{m}^{M}\sum_{p}^{P} R_{kjp} \cdot \left[\alpha \times E_2^{\mathrm{Cap}_{rs}} + (1-\alpha) \times E_1^{\mathrm{Cap}_{rs}}\right] \le \left[(1-\alpha) \times E_2^{\mathrm{Cap}_j} + \alpha \times E_1^{\mathrm{Cap}_j}\right] \quad \forall j,$$

$$A_{mkp} \le V_k \quad \forall m, p, k,$$

$$R_{kjp} \le y_j \quad \forall p, k, j,$$

$$\sum_{k}^{K} V_k \leq \text{DC},$$

$$\sum_{j}^{J} y_j \leq \text{CD},$$

$$\text{LB}_l \leq \text{AL}_s + \text{AL}_i + \text{AL}_j + \text{AL}_k \leq \text{UB}_l,$$

$$\text{LB}_m \leq \text{AL}_s + \text{AL}_i + \text{AL}_j + \text{AL}_k \leq \text{UB}_m,$$

$$\text{LB}_u \leq \text{AL}_s + \text{AL}_i + \text{AL}_j + \text{AL}_k \leq \text{UB}_u. \quad (39)$$

5. Proposed Multiobjective Algorithm

In addition to validating the higher performance and justifing the use of MOPSA in under-hand HLP, the proposed MOPSA was compared with several well-known metaheuristic algorithms existing in the literature (such as NSGA-II, NRGA, MOTS, MOSS, MOIWO, and PAES) while the proposed MOPSA had better performance. However, in this paper, the comparison with the best two of them (i.e., NSGA-II and PAES) is reported. For more details about NSGA-II and PAES, the interested readers are referred to [26, 27].

5.1. SA in General. Simulated annealing is a local search-based heuristic that is capable of escaping from being trapped into a local optimum by accepting, with small probability, worse solutions during its iterations. It has been applied successfully to a wide variety of highly complicated combinatorial optimization problems as well as various real-world problems. SA was introduced by Metropolis et al. [28] and used by Kirkpatrick et al. [29] in optimization problems as a first time. The concept of the method is adopted from the "annealing" process used in the metallurgical industry. Annealing is the process by which slow cooling is applied to metals to produce better aligned, low-energy-state crystallization. The optimization procedure of SA searches for a (near) global minimum mimicking the slow cooling procedure in the physical annealing process. It starts from a random initial solution. At each iteration, a new solution is taken from the predefined neighborhood of the current solution.

The objective function value of this new solution is then compared with that of the current best solution in order to determine if an improvement has been achieved. If the objective function value of the new solution is better, that is, being smaller in the case of minimization, the new solution becomes the current solution from which the search continues by proceeding with a new iteration. A new solution with a degraded (larger) objective function value may also be accepted as the new current solution, with a small probability determined by the Boltzmann function, $\exp(-\Delta/kT)$, where Δ is the difference of objective function values between the current solution and the new solution, k is a predetermined constant, and T is the current temperature. The basic idea is not to restrict the search to those solutions that decrease the objective function value but also allow moves that increase the objective function value. This mechanism may avoid the procedure being trapped prematurely in a local minimum.

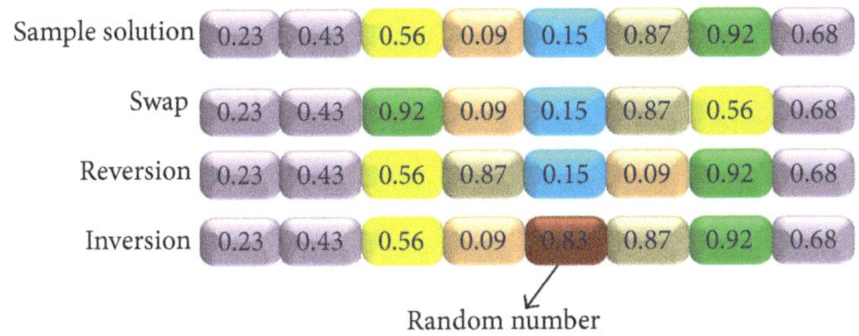

FIGURE 1: Three steps of mutation.

5.2. Proposed MOPSA. In this paper, as a first time in agile supply chain management, we propose a new concept of simulated annealing algorithm adopted in recent studies used in optimization problems, in which, unlike classical SA, more than one solution are used for searching the solution space to obtain Pareto-optimal solutions. So, this kind of SA is called parallel SA (PSA). This PSA is able to search the solution space widely and may also obtain nondominated Pareto solutions in less computational CPU time. Now, the proposed MOPSA will be described in detail.

5.3. Solution Representation. One important decision in designing a metaheuristic method is to decide how to represent and relate solutions in an efficient way to the searching space. Representation should be easy to decode to reduce the cost of the algorithm. The integer coding procedure is adopted to encode the solution, where each solution in the algorithm represents the number of decision variables used in the proposed model. For example, if there are 4 suppliers, 5 plants, 6 DCs, 6 CDs, and 12 CZs for 5 raw materials and 5 products and the objective is to design n agile supply chains, then the length of the solution will be the sum of the decision variables used in the n supply chains to be designed. Each cell of the solution represents the integer value corresponding to the supplier, plant, distribution center, cross-dock, and customer zones for specified amount of raw material and products.

5.4. Initial Solution. The proposed PSA algorithm, unlike classical SA, is started with more than one initial for searching the solution space. The number of initial solutions, namely, nPop, is an input parameter of the PSA algorithm. The initial solutions are created randomly according to Section 5.3.

5.5. Neighborhood Search. We use three consecutive steps for searching the initial solution neighborhood for obtaining the better solution at each iteration. The proposed three steps include mutation, assimilation and crossover that will be described next in detail.

5.5.1. Mutation Step. The first step of the proposed MOPSA is the mutation of the initial solutions to a predefined number *nMutate*. For mutating the initial solution, three different procedures are applied including (a) swap, (b) reversion, and (c) inversion mutation as shown in Figure 1. In the swap mutation, the places of two random selected bits are exchanged. In the reversion mutation, a random part of the

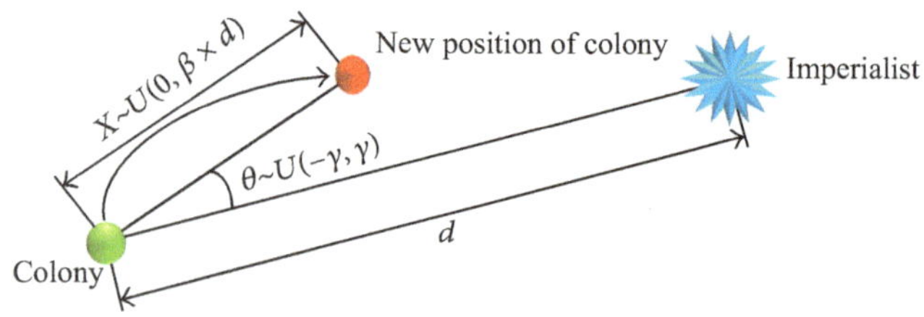

FIGURE 2: Moving colonies toward the imperialist with a random angle θ.

solution is selected and its permutation is reversed. Finally, in the inversion mutation, one bit is chosen randomly and its value is replaced with new random value.

5.5.2. Assimilation Step. At the end of the mutation step, we have *nPop* group of mutated solution with a number of *nMutate*. For example, if *nPop* and *nMutate* are equal to 5 and 10, respectively, then we have totally 50 new mutated solutions which are included in 5 groups with 10 solutions in each group. Now, the mutated solutions, called colonies, start moving toward the best solution in each group, called imperialist, by the assimilation operator of imperialist competitive algorithm (ICA) [24] as illustrated in Figure 2, in which d is the distance between colony and the imperialist and x is the amount of moving of colony toward the imperialist which is uniformly distributed between 0 and $\beta \times d$, in which β is a number greater than 1. For moving colony toward the imperialist in other direction, an θ angle is applied with the uniform distribution between $-\gamma$ and γ. It has been estimated that $\pi/4$ is the most proper value of γ. The best solution in each population group is chosen according to nondominance strategy and crowding distance metric.

5.5.3. Crossover Step. In this step, all the new created solutions in the two previous steps are merged for being combined according to the crossover operator of the genetic algorithm. For applying the crossover on the solutions, three different kinds of crossover are adopted on a pair of solutions, including: (a) one-point crossover, (b) two-point crossover and (c) uniform crossover as they are shown in Figure 3. It must be mentioned that operators a and b are applied for both continous and discrete parts, but operator c is applied for both parts differently. In this step, a binary tournament selection is applied for choosing parents.

5.6. Next Population. At each iteration, the proposed MOPSA needs initial solutions to apply three main neighborhood search strategies again for obtaining nondominated Pareto solutions. In this step, the nondominance strategy and crowding distance metric are applied for choosing the *nPop* better solutions among the whole new created solutions till now to compare them with previous *nPop* initial solutions correspondingly for choosing the next iteration's initial solutions. Three different cases may occur when two corresponding solutions are be compared in order to being chosen for next iteration initial solutions as follows.

(1) The new solution will be chosen for next iteration if dominates the old one.

(2) If two solutions are not dominated with each other, a solution with higher crowding distance metric will be chosen for next iteration.

(3) If the old solution dominates the new one, the probable acceptance function of simulated annealing will be adopted according to the next section.

5.7. Probability Acceptance Function. A new solution with a degraded (larger) objective function value may also be accepted as the new current solution, with a small probability determined by the Boltzmann function, $\exp(-\Delta/kT)$, where Δ is the difference of objective function values between the current solution and the new solution, k is a predetermined constant, and T is the current temperature. The Boltzmann function and Δ in proposed MOPSA are as follows:

$$P = e^{-\Delta f/T}, \tag{40}$$

$$\Delta f = \left| \frac{(f_1(x) - f_1(y))}{f_1(x)} + \frac{(f_2(x) - f_2(y))}{f_2(x)} \right|, \tag{41}$$

where x and y are the new and old solutions, respectively. Also, f_1 and f_2 are the first and the second objective functions, respectively. Variable T, in the Boltzmann function, is the current temperature that is calculated as follows:

$$T_{n+1} = \alpha \cdot T_n, \tag{42}$$

where α is a constant factor between 0 and 1 for decreasing the temperature at each next iteration. According to above description, a worse solution will be accepted for next iteration when a created random value r is less than probability value P calculated according to (40).

5.8. Stopping Criteria. In the proposed MOPSA, the number of function Calls-NFCs is the stopping criteria.

The pseudocode of proposed MOPSA is presented in Pseudocode 1.

6. Experimental Results

This section evaluates the performance of the proposed MOPSA. The performance of the proposed MOPSA is compared with well-known multiobjective evolutionary algorithms (MOEAs), namely, NSGA-II and PAES. At first, a brief discussion about the implementation of NSGA-II and PAES is presented.

6.1. Parameters Setting. It is well known that the quality of an algorithm is significantly influenced by the values of its parameters. In this section, for optimizing the behavior of the proposed algorithms, appropriate tuning of their parameters has been carried out. For this purpose, response surface methodology (RSM) is employed. RSM is defined as a collection of mathematical and statistical method-based experiential, which can be used to optimize processes. Regression equation analysis is used to evaluate the response

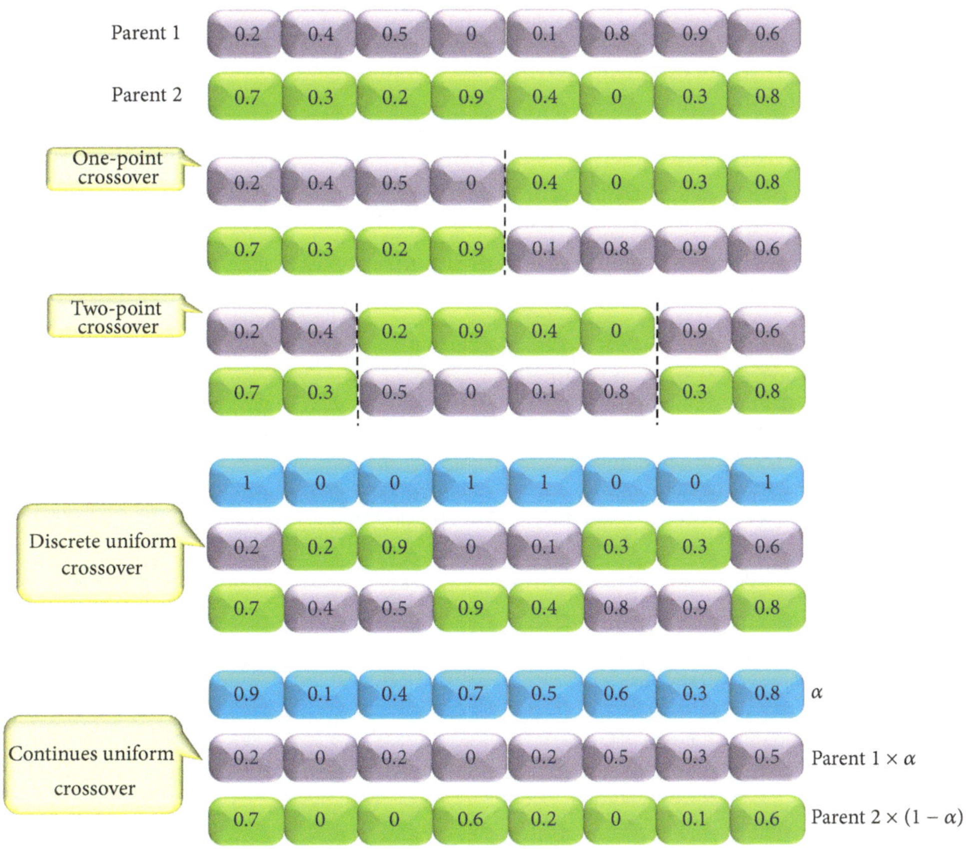

FIGURE 3: Three steps of crossover.

surface model. First of all, parameters of each algorithm that statistically have significant impact on the algorithms are recognized. To select the values that result in solutions with high quality, we consider problems in two different sizes including Small-S and Large-L sizes. To identify significant parameters, two levels for each parameter are considered. Each factor is measured at two levels, which can be coded as −1 when the factor is at its low level (L) and +1 when the factor is at its high level (H). The coded variable can be defined as follows:

$$X_i = \frac{r_i - ((h+l)/2)}{((h-l)/2)}, \tag{43}$$

where X_i and r_i are coded variable and natural variable, respectively. h and l represent high level and low level of factor. Factors and their levels are shown in Table 1.

After developing regression models for each problem size separately, tuned parameters of proposed MOPSA have been shown in Table 2.

For the proposed MOPSA, the NFCs stopping criteria were set on 30000 and 100000 for small- and large-size problems, respectively.

(i) *NSGA-II assumptions and parameters' value.*

(a) The initial population is randomly generated.

(b) One of these operators, namely, one-point crossover, two-point crossover, and uniform crossover, is selected randomly as a crossover operator.

(c) One of these operators, namely, inversion, swap, and reversion, is selected randomly as mutation operator.

(d) The crossover and mutation ratios are set to 0.8 and 0.2, respectively, using RSM method.

(e) The number of the initial population is set to 200 and 300 for small- and large-size problems, respectively.

(f) The NFCs stopping criteria were set on 30000 and 100000 for small- and large-size problems, respectively.

(ii) *PAES assumptions and parameters' value.*

(a) The size of archive is 150.

(b) One of these operators, namely, inversion, swap, and reversion, is selected randomly as revolution.

(c) The NFCs stopping criteria were set on 30000 and 100000 for small- and large-size problems, respectively.

6.2. Comparison Metrics. To validate the reliability of the proposed MOPSA, the following four comparison metrics are taken into account.

```
NFC ← 0
set the parameters of PSA (nPop, nMutate, pCrossover, α, β, T_0)
create initial solution ← nPop
terminate ← false
while (terminate = false) do
        mutate each initial solution ← nMutate
        find the best solution (imperialist) ← non-dominated sorting & crowding distance
        update NFC
        assimilate mutated solutions (colonies) toward imperialist ← β
        merge whole new created solution
        apply crossover ← pCrossover
        update NFC
        find better new solutions ← nPop
        update NFC
        if (new solution dominates old solution) then
            accept the new solution
        else if (no one dominate the other one) then
            calculate crowding distance-CD of each solution
            if (CD of new solution > CD of old solution) then
                accept the new solution
            end if
        else
            apply probable acceptance function ← P = e^(-Δf/T) ← Δf & T
            create random value r
            if (r < P) then
                accept new solution
            else
                accept old solution
            end if
        end if
        update NFC
        T ← α × T
        if (NFC = predefined value) then
            terminate = true
        end if
end while
```

Pseudocode 1

Table 1: Parameters and their levels for small and large sizes.

Factor	T_0		α		*nMutate*		*nPop*		*pCrossover*		β	
Problem size	S	L	S	L	S	L	S	L	S	L	S	L
Lower limit	5	5	0.75	0.75	5	8	3	5	0.4	0.4	1	1
Upper limit	15	15	0.95	0.95	15	20	8	12	0.8	0.8	3	3

Table 2: Tuned parameters of proposed MOPSA.

Factor	T_0		α		*nMutate*		*nPop*		*pCrossover*		β	
Problem size	S	L	S	L	S	L	S	L	S	L	S	L
Tuned value	10	13	0.84	0.91	10	16	5	6	0.5	0.7	1.8	2

Table 3: Data generation and their distribution.

Parameters	AL_k AL_j AL_i AL_s	Cap_k	Cap_j	C_{mk}	D_{mp}	C_k	F_i	F_j
Level	~(0.05, 0.3)	~(100, 500)	~(100, 400)	~(10, 20)	~(20, 50)	~(10^3, 10^4)	~(10^5, 10^6)	~(10^3, 10^4)
Parameters	l_{pi}	l_j	U_{ip}	U_j	Cap_i	Cap_{rs}	ST_{pi}	ST_{jp}	C_{rs}
Level	~(100, 200)	~(100, 300)	~(300, 500)	~(250, 450)	~(300, 500)	~(200, 500)	~(5, 10)	~(5, 10)	~(10, 20)
Parameters	C_{pj}	C_{pi}	UR_{rp}	C_{jkp}	C_{ijp}	C_{rsi}			
Level	~(10, 20)	~(20, 40)	~(0.6, 1)	~(10, 20)	~(10, 20)	~(10, 20)			

TABLE 4: Quality and spacing metrics.

Problem number	Quality metric (QM)			Spacing metric (SM)		
	PAES	NSGA-II	MOPSA	NSGA-II	PAES	MOPSA
1	0	0.08	0.92	0.558	0.879	0.444
2	0.09	0.18	0.73	0.521	0.937	0.315
3	0	0.20	0.80	0.409	0.810	0.385
4	0	0	1	0.395	0.762	0.373
5	0	0.10	0.90	0.608	0.726	0.359
6	0	0.00	1	0.359	0.932	0.282
7	0	0.15	0.85	0.608	0.761	0.274
8	0	0	1	0.595	0.816	0.335
9	0	0.25	0.75	0.870	0.866	0.303
10	0	0.10	0.90	0.672	0.768	0.243
11	0	0	1	0.574	0.893	0.586
12	0	0	1	0.862	0.845	0.303
13	0	0	1	0.674	0.745	0.394
14	0.08	0.17	0.75	0.624	0.935	0.476
15	0.06	0.2	0.74	0.870	0.730	0.314
16	0	0	1	0.517	0.788	0.521
17	0	0	1	0.657	0.771	0.442
18	0	0.12	0.88	0.601	0.859	0.276
19	0	0	1	0.552	0.727	0.486
20	0	0	1	0.562	0.895	0.610
21	0	0	1	0.772	0.819	0.636
22	0	0	1	0.703	0.812	0.481
23	0	0.04	0.96	0.641	0.835	0.423
24	0	0.10	0.90	0.774	0.686	0.359
25	0	0	1	0.531	0.808	0.224
26	0	0	1	0.724	0.767	0.512
27	0	0.05	0.95	0.721	0.866	0.335
28	0	0	1	0.669	0.794	0.572
29	0	0	1	0.555	0.527	0.303
30	0	0	1	0.621	0.862	0.394

(1) Quality metrics: this metric is simply measured by putting together the nondominated solutions found by algorithms, and the ratios between nondominated solutions are reported.

(2) Mean ideal distance (MID): the closeness between Pareto solutions and ideal point $(f_1^{\text{best}}, f_2^{\text{best}})$ is determined by using MID. The equation of MID is defined by

$$\text{MID} = \left(\sum_{i=1}^{n}\sqrt{\left(\frac{f_{1i}-f_1^{\text{best}}}{f_{1,\text{total}}^{\max}-f_{1,\text{total}}^{\min}}\right)^2+\left(\frac{f_{2i}-f_2^{\text{best}}}{f_{2,\text{total}}^{\max}-f_{2,\text{total}}^{\min}}\right)^2}\right)\times(n)^{-1}, \tag{44}$$

where n is the number of nondominated solutions and $f_{i,\text{total}}^{\max}$ and $f_{i,\text{total}}^{\min}$ are the maximum and minimum values of each fitness functions among the all nondominated solutions obtained by the algorithms, respectively. Regarding this definition, an algorithm with a lower value of MID has a better performance.

(3) Diversification metric (DM): this metric measures the spread of the Pareto solutions set and is calculated by

$$\text{DM} = \sqrt{\left(\frac{\max f_{1i}-\min f_{1i}}{f_{1.\text{total}}^{\max}-f_{1.\text{total}}^{\min}}\right)^2+\left(\frac{\max f_{2i}-\min f_{2i}}{f_{2.\text{total}}^{\max}-f_{2.\text{total}}^{\min}}\right)^2}. \tag{45}$$

Regarding this definition, an algorithm with a higher value of DM has a better performance.

(4) Spacing metric (SM): this metric measures the uniformity of the spread of the nondominated set solutions. This metric is defined by

$$\text{SM} = \frac{\sum_{i=1}^{n-1}\left|\bar{d}-d_i\right|}{(n-1)\,\bar{d}}, \tag{46}$$

TABLE 5: Diversification and mean ideal distance metrics.

Problem number	Diversity metric (DM)			Mean ideal distance (MID)		
	PAES	NSGA-II	MOPSA	PAES	NSGA-II	MOPSA
1	0.929	0.872	1.379	0.862	0.569	0.255
2	0.606	1.065	1.143	0.698	0.673	0.258
3	0.617	0.772	1.300	0.795	0.743	0.302
4	0.667	0.805	0.971	0.893	0.621	0.480
5	0.860	0.758	1.111	0.622	0.797	0.317
6	0.893	0.754	1.358	0.776	0.527	0.282
7	0.859	1.048	1.296	0.724	0.596	0.245
8	0.780	0.932	1.380	0.793	0.653	0.519
9	0.819	0.920	1.228	0.679	0.518	0.312
10	0.718	0.758	0.917	0.828	0.718	0.239
11	0.898	1.042	1.325	0.899	0.667	0.430
12	0.675	0.949	1.367	0.656	0.659	0.227
13	0.875	0.840	1.240	0.835	0.749	0.385
14	0.673	0.905	1.279	0.658	0.758	0.203
15	0.771	1.033	1.212	0.898	0.737	0.372
16	0.793	1.022	1.240	0.673	0.606	0.447
17	0.648	0.724	1.098	0.895	0.655	0.282
18	0.836	0.860	1.084	0.638	0.636	0.245
19	0.690	0.911	1.394	0.669	0.481	0.319
20	0.754	0.867	0.918	0.607	0.468	0.312
21	0.833	0.963	1.343	0.782	0.496	0.239
22	0.700	0.951	1.357	0.633	0.649	0.430
23	0.716	0.817	1.298	0.722	0.647	0.227
24	0.847	0.873	0.949	0.866	0.571	0.385
25	0.706	0.706	1.031	0.764	0.572	0.203
26	0.930	1.094	1.068	0.711	0.486	0.372
27	0.994	0.767	1.240	0.662	0.610	0.437
28	0.892	0.742	0.968	0.732	0.639	0.270
29	0.737	0.849	1.261	0.887	0.532	0.334
30	0.834	0.779	0.953	0.637	0.534	0.471

where d_i is the Euclidean distance between consecutive solutions in the obtained nondominated set of solutions and $\overline{d}$ is the average of these distances. Regarding this definition, an algorithm with a lower value of SM has a better performance.

6.3. Data Generation. In order to evaluate the performance of the proposed MOPSA, several test problems are generated in this subsection. The number of potential locations for all five echelons is distributed randomly between 5 and 100 nodes. Other parameters are presented as in Table 3. It should be mentioned that these ranges of data are obtained from real steel industry, and some of the ranges are also generated randomly because of lack of information.

6.4. Computational Results. The proposed MOPSA is applied to a number of test problems, and its performance is compared with NSGA-II and PAES. Table 4 shows the comparison between the proposed MOPSA algorithm and NSGA-II and PAES regarding QM and SM comparison metrics. Besides, Table 5 shows the results of DM and MID comparison metrics. Both Tables 4 to 5 show that the proposed MOPSA is superior to NSGA-II and PAES in each test problem.

(i) The proposed MOPSA can achieve a greater number of Pareto-optimal solutions with higher qualities than both NSGA-II and PAES.

(ii) The proposed MOPSA provides nondominated solutions that have less average values of the spacing metric.

(iii) These data reveal that nondominated solutions obtained by the proposed MOPSA are more uniformly distributed in comparison with NSGA-II.

(iv) The average values of the diversification metric in our proposed PSO are considerably greater than those of NSGA-II (i.e., PSO finds nondominated solutions that diverse more).

(v) In most of the test problems, the values of MID in the proposed MOPSA are smaller than those of NSGA-II and PAES.

(vi) The computational time of the proposed MOPSA was less than 10 minutes even for the largest test problem.

7. Conclusion

This paper integrates production, distribution and logistics activities at the strategic decision making level, where the objective is to design a multi-echelon supply chain network considering agility as a key design criterion. Agility performance index has been conceptualized as a key design criterion which is decided by the manufacturers. A network with five echelons of supply chains including suppliers, plants, distribution centers, cross-docks, and customer zones is addressed in this paper. The problem has been mathematically formulated as a multiobjective optimization model that aims to minimize the cost (fixed and variable) and maximizes the plant flexibility and volume flexibility. Also, in order to cope with uncertain environment, some parameter of the model such as demand and cost are considered as fuzzy parameters. To validate our proposed MOPSA, various test problems were designed to evaluate its performance and reliability in comparison with two conventional multiobjective evolutionary algorithms, known as NSGA-II and PAES. Experimental results showed the higher performance of the proposed MOPSA in all of four comparison metrics. The proposed MOPSA has higher quality in the whole test problems which makes more near optimal solution for the decision makers.

References

[1] S. L. Goldman, R. N. Nagel, and K. Preiss, *Agile Competitors and Virtual Organizations*, Van Nostrand Reinhold, New York, NY, USA, 1995.

[2] C. Ho, "Evaluating the impact of operating environments on MRP system nervousness," *International Journal of Production Research*, vol. 27, no. 7, pp. 1115–1135, 1989.

[3] D. Lee, M. Dong, and W. Bian, "The design of sustainable logistics network under uncertainty," *International Journal of Production Economics*, vol. 128, no. 1, pp. 159–166, 2010.

[4] H. Zerhouni, J. Gayon, and Y. Frein, "Influence of dependency between demands and returns in a reverse logistics system," *International Journal of Production Economics*, vol. 143, no. 1, pp. 62–71, 2010.

[5] J. Kenné, P. Dejax, and A. Gharbi, "Production planning of a hybrid manufacturingremanufacturing system under uncertainty within a closed-loop supply chain," *International Journal of Production Economics*, vol. 135, no. 1, pp. 81–93, 2012.

[6] A. T. M. A. Samadhi and K. Hoang, "Partners selection in a shared-CIM system," *International Journal of Computer Integrated Manufacturing*, vol. 11, no. 2, pp. 173–182, 1998.

[7] L. De Boer, L. Van Der Wegen, and J. Telgen, "Outranking methods in support of supplier selection," *European Journal of Purchasing and Supply Management*, vol. 4, no. 2-3, pp. 109–118, 1998.

[8] V. Jayaraman, "An efficient heuristic procedure for practical-sized capacitated warehouse design and management," *Decision Sciences*, vol. 29, no. 3, pp. 729–744, 1998.

[9] P. Tsiakis, N. Shah, and C. C. Pantelides, "Design of multi-echelon supply chain networks under demand uncertainty," *Industrial and Engineering Chemistry Research*, vol. 40, no. 16, pp. 3585–3604, 2001.

[10] B. L. Shankar, S. Basavarajappa, J. C. H. Chen, and R. S. Kadadevaramath, "Location and allocation decisions for multi-echelon supply chain network-a multi-objective evolutionary approach," *Expert Systems with Applications*, vol. 40, no. 2, pp. 551–562, 2013.

[11] A. M. Sarmiento and R. Nagi, "A review of integrated analysis of production-distribution systems," *IIE Transactions*, vol. 31, no. 11, pp. 1061–1074, 1999.

[12] A. Alonso-Ayuso, L. F. Escudero, A. Garín, M. T. Ortuño, and G. Pérez, "An approach for strategic supply chain planning under uncertainty based on stochastic 0-1 programming," *Journal of Global Optimization*, vol. 26, no. 1, pp. 97–124, 2003.

[13] C. Teo and J. Shu, "Warehouse-retailer network design problem," *Operations Research*, vol. 52, no. 3, pp. 396–497, 2004.

[14] S. Elhedhli and J. Coffin, "Efficient production-distribution system design," *Management Science*, vol. 51, no. 7, pp. 1151–1164, 2005.

[15] A. Amiri, "Designing a distribution network in a supply chain system: formulation and efficient solution procedure," *European Journal of Operational Research*, vol. 171, no. 2, pp. 567–576, 2006.

[16] R. I. Van Hoek, A. Harrison, and M. Christopher, "Measuring agile capabilities in the supply chain," *International Journal of Operations and Production Management*, vol. 21, no. 1-2, pp. 126–147, 2001.

[17] M. Dotoli, M. P. Fanti, C. Meloni, and M. Zhou, "Design and optimization of integrated e-supply chain for agile and environmentally conscious manufacturing," *IEEE Transactions on Systems, Man, and Cybernetics A*, vol. 36, no. 1, pp. 62–75, 2006.

[18] D. Y. Sha and Z. H. Che, "Virtual integration with a multi-criteria partner selection model for the multi-echelon manufacturing system," *International Journal of Advanced Manufacturing Technology*, vol. 25, no. 7-8, pp. 793–802, 2005.

[19] C. Wu and D. Barnes, "A literature review of decision-making models and approaches for partner selection in agile supply chains," *Journal of Purchasing and Supply Management*, vol. 17, no. 4, pp. 256–274, 2011.

[20] F. Pan and R. Nagi, "Robust supply chain design under uncertain demand in agile manufacturing," *Computers and Operations Research*, vol. 37, no. 4, pp. 668–683, 2010.

[21] M. Bachlaus, M. K. Pandey, C. Mahajan, R. Shankar, and M. K. Tiwari, "Designing an integrated multi-echelon agile supply chain network: a hybrid taguchi-particle swarm optimization approach," *Journal of Intelligent Manufacturing*, vol. 19, no. 6, pp. 747–761, 2008.

[22] J. K. Cochran and A. Marquez Uribe, "A set covering formulation for agile capacity planning within supply chains," *International Journal of Production Economics*, vol. 95, no. 2, pp. 139–149, 2005.

[23] N. Costantino, M. Dotoli, M. Falagario, M. P. Fanti, and A. M. Mangini, "A model for supply management of agile manufacturing supply chains," *International Journal of Production Economics*, vol. 135, no. 1, pp. 451–457, 2012.

[24] F. Pan and R. Nagi, "Multi-echelon supply chain network design in agile manufacturing," *Omega*, vol. 41, no. 6, pp. 969–983, 2013.

[25] M. Jiménez, M. Arenas, A. Bilbao, and M. V. Rodríguez, "Linear programming with fuzzy parameters: an interactive method

resolution," *European Journal of Operational Research*, vol. 177, no. 3, pp. 1599–1609, 2007.

[26] K. Deb, A. Pratap, S. Agarwal, and T. Meyarivan, "A fast and elitist multiobjective genetic algorithm: NSGA-II," *IEEE Transactions on Evolutionary Computation*, vol. 6, no. 2, pp. 182–197, 2002.

[27] D. W. Corne, J. D. Knowles, and M. J. Oates, "The Pareto envelope-based selection algorithm for multi objective optimization," in *Parallel Problem Solving from Nature-PPSN VI*, M. Schoenauer, K. Deb, G. Rudolph et al., Eds., pp. 839–848, Springer, Berlin, Germany, 2000.

[28] N. Metropolis, A. W. Rosenbluth, M. N. Rosenbluth, A. H. Teller, and E. Teller, "Equation of state calculations by fast computing machines," *The Journal of Chemical Physics*, vol. 21, no. 6, pp. 1087–1092, 1953.

[29] S. Kirkpatrick, C. D. Gelatt Jr., and M. P. Vecchi, "Optimization by simulated annealing," *Science*, vol. 220, no. 4598, pp. 671–680, 1983.

5

Power Optimization of Tilted Tomlinson-Harashima Precoder in MIMO Channels with Imperfect Channel State Information

Hossein Khaleghi Bizaki,[1] Morteza Khaleghi Hojaghan,[2] and Seyyed Mohammad Razavizadeh[3]

[1] *Malek Ashtar University of Technology, Tehran, Iran*
[2] *Tabriz University, Tabriz, Iran*
[3] *Iran University of Science and Technology (IUST), Tehran, Iran*

Correspondence should be addressed to Hossein Khaleghi Bizaki; bizaki@yahoo.com

Academic Editor: Kai-Kit Wong

This paper concentrates on the designing of a robust Tomlinson-Harashima Precoder (THP) over multiple-input multiple-output (MIMO) channels in wireless communication systems with assumption of imperfect channel state information (CSI) at the transmitter side. With the assumption that the covariance matrix of channel estimation error is available at the transmitter side, we design a THP that presents robustness against channel uncertainties. In the proposed robust THP, the transmit power is further minimized by using the Tilted constellation concept. This power minimization reduces the interchannel Interference (ICI) between subchannels and, furthermore, recovers some part of the THP's power loss. The bit error rate (BER) of the proposed system is further improved by using a power loading technique. Finally, the simulation results compare the performance of our proposed robust THP with a conventional MIMO-THP.

1. Introduction

Recently, due to the impressive capacity of multiple-input multiple-output (MIMO) channel, many companies and researches have been attracted toward designing transceivers for MIMO systems and especially robust transceivers in the case of channel uncertainties. It is well known that inter channel interference (ICI) is one of the important problems of the MIMO systems. It is possible to preeliminate the ICI at the transmitter side by using precoding techniques and consequently reduce transmission power. Due to channel estimation errors, channel variations during time, and channel feedback errors, the assumption of perfect CSI at the transmitter side is not always true and hence the CSI at the transmitter side has uncertainties [1]. Due to these uncertainties, the ICI will be increased and consequently leads to BER growth. Linear precoder in [2] is designed for imperfect CSI, but it could only slightly reduce uncertainties effects by using statistical information of the channel. Nonlinear Tomlinson-Harashima Precoder (THP) in [3] has been considered for imperfect CSI assumption, and we combine it with the tilted constellation method in our work.

In this paper, we propose a design of a robust MIMO-THP that can reduce uncertainties effects more than the proposed methods in [2, 3] and subsequently can further reduce BER. Similar to [3], we utilize second-order statistics of the uncertainties of the MIMO channels at the transmitter side. Moreover, by using tilted constellation concept in [4], we can significantly reduce transmit power. It should be noted that transmit power reduction can decrease ICI between subchannels and in addition recover some parts of the THP's power loss. This ICI reduction between subchannels is one of the main challenges in the designing of our precoder.

Tilted constellation method is introduced in [4] for SISO channel. Based on this idea, we have extended it for the MIMO channel. Here, the power reduction is done by tilting or rotating the ordinary constellation through a proper angle. The desired angle is one that minimizes the transmit power and is selected from a set of possible angles. The results show that the power reduction will be greater at high SNRs values.

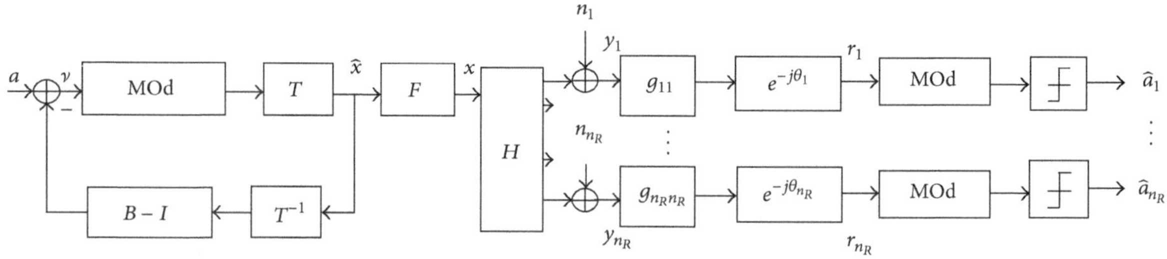

FIGURE 1: The proposed tilted MIMO-THP in decentralized scenario.

To do the tilted constellation method, we use different angles for tilting the constellation in each antenna. Then, we find a set of proper angles, which are selected in a manner to minimize the transmit power in all antennas. The transmit power reduction can reduce the transmission block power, lower than the power of modulated symbols in each block. Finally, we can send proper angles to the receiver within the main data frame. We will show that the proposed method outperforms the conventional MIMO-THP in most SNRs and also in all symbol block lengths.

We can adjust transmit power of each antenna when the CSI (perfect/imperfect) is available at the transmitter side which is known as power loading [5]. It should be noted that in the MIMO systems some of the corresponding parallel subchannels (corresponding to each antenna) might have very low channel condition or might be useless for sometimes [5]. In this situation, the transmitter can adjust transmit power of each antenna by redistributing the available transmit power to get a better average error rate [5]. Moreover, by utilizing power loading, we improve the BER performance of the tilted MIMO-THP. We also study the complexity of our proposed method with respect to some previous methods.

This paper is organized as follows. In Section 2, the system model of the Tilted MIMO-THP in imperfect CSI scenario is introduced. In Section 3, the proposed robust tilted MIMO-THP is described for imperfect CSI. Section 4 presents the power loading technique to improve the performance of the robust precoder. In Section 5, the simulation results are given to show the performance of the proposed method and its comparison with the conventional MIMO-THP. Finally, conclusions are drawn in Section 6.

2. System Model

In this paper, an MIMO communication system with n_T transmit antenna at the base station and n_R users at the receivers side, each equipped with one receive antenna, are considered. The block diagram of the considered system is shown in Figure 1.

The n_T input data symbols and the transmitted signals are denoted by vector $\mathbf{a}$ and vector $\mathbf{x}$, respectively. Both vectors are of size $n_T \times 1$. The channel matrix is modeled as $\mathbf{H} = \mathbf{H_0} + \triangle\mathbf{H}$, where $\mathbf{H_0}$ is the $n_R \times n_T$ estimated channel matrix whose elements are independent and identically distributed (i.i.d.) zero mean complex Gaussian random variables with variance $1 - \rho^2$ (i.e, $\mathbf{H_0} \sim \mathrm{CN}(0, 1 - \rho^2)$), and $\triangle\mathbf{H}$ is an $n_R \times n_T$ channel uncertainties matrix whose elements are i.i.d. zero mean complex Gaussian random variables with variance ρ^2(i.e, $\triangle\mathbf{H} \sim \mathrm{CN}(0, \rho^2)$). Thus, $\mathbf{H}$ can be modeled as $\mathbf{H} \sim \mathrm{CN}(0, 1)$. The matrix $\mathbf{T}$ is the proper rotation angles matrix which is named as tilting matrix, to minimize the transmission power. The received signal $\mathbf{y}$ can be denoted as follows:

$$\mathbf{y} = \mathbf{H}\mathbf{x} + \mathbf{n}, \tag{1}$$

where $\mathbf{n}$ is the noise vector whose elements are assumed to be i.i.d. complex Gaussian random variables with zero mean and variance σ^2 (i.e. $\mathbf{n} \sim \mathrm{CN}(0, \sigma^2\mathbf{I_R})$).

As we see in Figure 1, the n_T dimensional vector $\mathbf{a}$ corresponds to input data, and n_T dimensional vector $\mathbf{x}$ represents transmitted signals. The feedback matrix $\mathbf{B}$ is added to preeliminate the interference from previous users, and the tilting matrix $\mathbf{T}$ is added to reduce the transmit power. Finally, in order to keep the power of transmitted symbols within the original constellation boundary, we employ modulo operator at the forward loop. The boundary region of the constellation is related to the order of modulation constellation, which for rectangular M-ary QAM modulation is $t = 2\sqrt{M}$. Using tilted constellation, the signal constellation is further rotated by a set of appropriate angles, represented by the diagonal elements of the Tilting matrix $\mathbf{T}$, to reduce the ICI and power loss of THP.

The resultant signal $\hat{\mathbf{x}}$ is then passed through a unitary feed-forward filter $\mathbf{F}$, to eliminate the residual interference. Finally, the precoded signal is sent through the MIMO channel. All processes to eliminate the interference are performed at the transmitter side; hence, receivers at the user side are left with some simple operations including power scaling (i.e., elements of the diagonal matrix $\mathbf{G}$), de-modulo operation, constellation tilting with reverse angle, and a single user detection.

By using the second-order statistics of the uncertainties of the MIMO channels at the transmitter side, we design a THP that presents robustness against channel uncertainties. We can describe the difference between impressive vector $\mathbf{v}$ and input vector to the decision block as follows [3]:

$$\mathbf{e} = \mathbf{r} - \mathbf{v} = \left[\mathbf{T}^{-1}\mathbf{G}\left(\mathbf{H_0} + \Delta\mathbf{H}\right)\mathbf{F} - \mathbf{B}\right]\hat{\mathbf{x}} + \tilde{\mathbf{n}}. \tag{2}$$

MMSE solution should minimize the signal error as:

$$\arg\min_{\mathbf{B,F,G}} E\left\|\left[\left[\mathbf{T}^{-1}\mathbf{G}\left(\mathbf{H_0} + \Delta\mathbf{H}\right)\mathbf{F} - \mathbf{B}\right]\hat{\mathbf{x}} + \tilde{\mathbf{n}}\right]\right\|^2 \quad \text{so that } E\|\hat{\mathbf{x}}\|^2 \le P_T, \tag{3}$$

where P_T is the total transmitted power. Since, we assume the matrix **F** is unitary thus, the power constraint is guaranteed.

Instead of analysis of (3) we can use orthogonality property as follow [3]:

$$E\left[\mathbf{er}^H\right] = 0. \tag{4}$$

Hence from (4) we have the following:

$$\mathbf{G}\Phi_{\mathbf{rr}} = \mathbf{B}\Phi_{\mathbf{xr}}. \tag{5}$$

We can calculate $\Phi_{\mathbf{rr}}$ and $\Phi_{\mathbf{xr}}$ using (1) as follows:

$$\begin{aligned}\Phi_{\mathbf{rr}} &= E\left[\mathbf{rr}^H\right] = \sigma_{\mathbf{x}}^2\left(\mathbf{H}_0\mathbf{H}_0{}^H + \zeta\mathbf{I} + \mathbf{C}_{\Delta\mathbf{H}}\right)\\ \Phi_{\mathbf{xr}} &= E\left[\mathbf{xr}^H\right] = \sigma_{\mathbf{x}}^2\left(\mathbf{F}^H\mathbf{H}_0{}^H\right).\end{aligned} \tag{6}$$

By substituting (6) into (5), and after some manipulations, we have the following:

$$\mathbf{F}^H = \mathbf{B}^{-1}\mathbf{G}\left(\mathbf{H}_0\mathbf{H}_0{}^H + \zeta\mathbf{I} + \mathbf{C}_{\Delta\mathbf{H}}\right)\mathbf{H}_0{}^{-H}. \tag{7}$$

Since **F** is unitary, we have the following:

$$\mathbf{RR}^H = \left(\mathbf{H}_0\mathbf{H}_0{}^H + \zeta\mathbf{I} + \mathbf{C}_{\Delta\mathbf{H}}\right)\mathbf{H}_0{}^{-H}\mathbf{H}_0{}^{-1}\left(\mathbf{H}_0\mathbf{H}_0{}^H + \zeta\mathbf{I} + \mathbf{C}_{\Delta\mathbf{H}}\right), \tag{8}$$

where $\mathbf{R} = \mathbf{G}^{-1}\mathbf{B}$. Matrix **R** can be obtained by doing the Cholesky factorization of (8). We can calculate matrixes **G**, **B**, and **F** as follows:

$$\begin{aligned}\mathbf{G} &= \operatorname{diag}\left(\frac{1}{\mathbf{r}_{11}}, \dots, \frac{1}{\mathbf{r}_{n_T n_T}}\right),\\ \mathbf{B} &= \mathbf{G}\left(\mathbf{H}_0\mathbf{H}_0{}^H + \zeta\mathbf{I} + \mathbf{C}_{\Delta\mathbf{H}}\right)\mathbf{H}_0{}^{-H} = \mathbf{GR},\\ \mathbf{F} &= \mathbf{H}_0{}^{-1}\left(\mathbf{H}_0\mathbf{H}_0{}^H + \zeta\mathbf{I} + \mathbf{C}_{\Delta\mathbf{H}}\right)\mathbf{R}^{-H}.\end{aligned} \tag{9}$$

Substituting **G**, **B**, and **F** into (9) the error covariance matrix is equal to the following:

$$\Phi_{\mathbf{ee}} = E\left[\mathbf{ee}^H\right] = \sigma_{\mathbf{x}}^2\mathbf{G}\left(\zeta^2\mathbf{H}_0{}^{-H}\mathbf{H}_0{}^{-1} + \zeta\mathbf{I} + \mathbf{C}_{\Delta\mathbf{H}}\right)\mathbf{G}^H, \tag{10}$$

where the scalar $\boldsymbol{\zeta}$ represents the $1/\text{SNR} = \sigma_n^2/\sigma_x^2$.

It can be seen in (10) that, while the channel has uncertainty, the error covariance matrix has an additional term of $\sigma_{\mathbf{x}}^2\mathbf{G}\mathbf{C}_{\Delta\mathbf{H}}\mathbf{G}^H$ with respect to its counterpart in perfect CSI case.

In this paper, we assume the previous relations and matrixes as the conventional robust MIMO-THP, that is, without using tilted constellations.

3. Robust Tilted MIMO-THP

As mentioned before, in the general model of Figure 1, each diagonal element of the Tilting matrix **T** determines the rotation angle of its corresponding antenna signal.

In conventional THP, the main constellation is only extended by the modulo operator on its boundary regions. However, in the tilted-THP, the signal constellation is further rotated by a set of appropriate angles. To do this, first a block of symbols with length $n_T \times N$ is divided to n_T groups each of length N. Then, for every group of symbols which is transmitted from each antenna, the transmitter chooses an appropriate angle from a set of possible angles in such a way that the transmitted power is minimized based on some predefined criteria. From [4], the ordinary constellation is tilted by Q possible angles as $\theta_q; q = 1, \dots, Q$, then the optimal angle θ_q^* is selected by the following equation:

$$q^* = \arg\min_{q=1,\dots,Q}\left\|\operatorname{Mod}\left[\mathbf{a} - \mathbf{i}e^{-j\theta_q}\right]\right\|, \tag{11}$$

where **a** is the group of symbols in each transmit antenna, and **i** illustrates the interference sequence at each antenna due to the previous antennas and is calculated similar to the conventional THP. The difference between the conventional (i.e. un-tilted) and tilted THP is in rotating of the transmitted signal of the previous antennas by their optimal angles. Transmitted power and resulting optimal angle of each antenna can be calculated as follows [4]:

$$\begin{aligned}P_{\text{Tilted}}(Q) &= E\left\{\min_{q=1,\dots,Q}\mathbf{P}_q\right\}\\ \mathbf{P}_q &= \frac{1}{N}\sum_{i=1}^{N}\left\|\operatorname{Mod}\left[\mathbf{a} - \mathbf{i}e^{-j\theta_q}\right]\right\|^2.\end{aligned} \tag{12}$$

The vector $\mathbf{a}_1$, which is transmitted from the first transmit antenna, is not affected by any interference signal, that is why we transmit $\mathbf{a}_1$ without any interference. The optimal tilted angle θ_1 is resulted by minimizing the transmitted power of the first antenna; that is:

$$\mathbf{x}_1 = \mathbf{a}_1 e^{j\theta_1}. \tag{13}$$

For $\mathbf{a}_2$, which is transmitted from the second antenna, the optimal transmit angle is achieved by computing the minimum transmitted power similar to the $\mathbf{a}_1$, but, here, there is an interference due to the $\mathbf{a}_1$, that is, $\mathbf{b}_{21}$, and hence we have the following:

$$\mathbf{x}_2 = \left(\operatorname{Mod}\left(\mathbf{a}_2 - \mathbf{b}_{21}\mathbf{x}_1 e^{-j\theta_2}\right)\right)\mathbf{e}^{j\theta_2}. \tag{14}$$

Similarly, for the kth antenna, we have the following:

$$\mathbf{x}_k = \left(\operatorname{Mod}\left(\mathbf{a}_k - \sum_{j=1}^{k-1}\mathbf{b}_{kj}\mathbf{x}_j e^{-j\theta_k}\right)\right)e^{j\theta_k}, \tag{15}$$

where θ_k is the optimal tilted angle for the kth antenna. At the receiver side, the received signal for the kth antenna can be shown as follows:

$$\mathbf{r}_k = \left(\mathbf{b}_k\mathbf{x} + \mathbf{n}_k\right)e^{-j\theta_k}, \tag{16}$$

where $\mathbf{b}_k$ denotes the kth row of the matrix $\mathbf{B} = [\mathbf{b}_1, \dots, \mathbf{b}_{k,\dots}]^{\mathrm{T}}$. By using (15), (16) can be written as

$$\mathbf{r}_k = \left(\mathbf{x}_k + \sum_{j=1}^{k-1}\mathbf{b}_{kj}\mathbf{x}_j + \mathbf{n}_k\right)e^{-j\theta_k}. \tag{17}$$

Since $b_{kk} = 1$, after modulo operator we obtain the following:

$$\begin{aligned}\hat{\mathbf{a}}_k &= \text{Mod}(\mathbf{r}_k)\\ &= \text{Mod}\left(\left(\mathbf{a}_k e^{j\theta_k} - \sum_{j=1}^{k-1}\mathbf{b}_{kj}\mathbf{x}_j + \sum_{j=1}^{k-1}\mathbf{b}_{kj}\mathbf{x}_j + \mathbf{n}_k\right)e^{-j\theta_k}\right)\\ &= \text{Mod}\left(\left(\mathbf{a}_k e^{j\theta_k} + \mathbf{n}_k\right)e^{-j\theta_k}\right) = \text{Mod}\left(\mathbf{a}_k + \mathbf{n}_k e^{-j\theta_k}\right).\end{aligned} \tag{18}$$

As shown in [3], in the tilted constellation case, the matrix **B** is similar to the untilted or regular MIMO-THP case. Its difference is only in rotation of the phase of the noise. Since the noise has circularly symmetric complex Gaussian distribution, for the conventional case, all equations and results for the decomposition of matrices in the tilted constellation case are the same.

If we use the tilted constellation for detecting the original symbols directly, the performance will be degraded due to rotation and reduction of minimum distance between symbols. Hence, we should send the optimal angles to the receiver to detect the desired symbols. Upon receipt, first, the symbols are reversely rotated at each antenna and then are detected in a way similar to the conventional THP.

4. Power Loading

The SER of MIMO-THP in imperfect CSI can be approximated as follows [6]:

$$\text{SER} \approx \sum_{k=1}^{K}\left(1-\frac{1}{\sqrt{M_k}}\right)Q\left(\sqrt{\frac{3}{M_k-1}\frac{(P_{\text{Tilted}})_k}{\sigma_e^2}}\right), \tag{19}$$

where $(P_{\text{Tilted}})_k$ is the transmitted power of kth transmit antenna and σ_e^2 is noise power [3]:

$$\sigma_e^2 = [\Phi_{\mathbf{ee}}]_{kk} = \frac{1}{|\mathbf{r}_{kk}|^2}\left(\sigma_n^2 + \sum_{j=1}^{K}(P_{\text{Tilted}})_j|\delta_{kj}|^2 + \beta_k\right). \tag{20}$$

Here, $\delta_{ij} = [\Delta\mathbf{H}]_{ij}$, $\beta_k = \sigma_n^4\sum_{j=1}^{K}|h_{kj}|^2/(P_{\text{Tilted}})_j$, and $h_{ij} = [\mathbf{H}^{-H}\mathbf{H}^{-1}]_{ij}$. With the assumption of small error, that is, $\alpha(P_{\text{Tilted}})_j \le \sigma_n^2$; for all j [3], we can approximate the σ_e^2 value as follows:

$$\sigma_e^2 = \frac{1}{|\mathbf{r}_{kk}|^2}\left(\sigma_n^2 + \alpha(P_{\text{Tilted}})_T + \beta_k\right), \tag{21}$$

where $(P_{\text{Tilted}})_T$ is the total tilted power. At the above equation, we assumed the worst-case (i.e., $\alpha = \max_{\mathbf{B,F,G}} |\delta_{ij}|^2$; for all i, j). We distribute the power to minimize SER in imperfect CSI by using the Lagrange method, so that it can be written as follows:

$$\begin{aligned}L = &\sum_{k=1}^{K}\left(1-\frac{1}{\sqrt{M_k}}\right)Q\left(\sqrt{(P_{\text{Tilted}})_k\beta_k}\right)\\ &- \lambda\left(K - \sum_{k=1}^{K}(P_{\text{Tilted}})_k\right).\end{aligned} \tag{22}$$

$\Lambda = 0.01$; % initial value for Λ
$P_{\text{Total}} = nT$; % Total transmit power
Calculate A_k, β_k, Λ
Loop: Calculate $\hat{\mathbf{P}}$
If $\mathbf{P}_{\text{transmit}} - \hat{\mathbf{P}} \to 0$ then $\mathbf{P}_{\text{transmit}} = \hat{\mathbf{P}}$ end
Else if $\hat{\mathbf{P}} > \mathbf{P}_{\text{transmit}}$, Break;
Else $\Lambda = (P_T/\hat{P}_T) \times \Lambda$, go to loop
end

ALGORITHM 1: The proposed power loading algorithm.

Hence, as in [3], we have the following:

$$\beta_k = \frac{3}{M_k - 1}\frac{1}{\sigma_n^2 + \alpha(P_{\text{Tilted}})_T + \beta_k}. \tag{23}$$

Unfortunately, we did not find any closed form solution for (23), so we should use some suboptimal solution as in [3].

By Solving $\partial L/\partial(P_{\text{Tilted}})_k = 0$ for $(P_{\text{Tilted}})_k$, we have the following:

$$(P_{\text{Tilted}})_k = \frac{1}{\beta_k}W\left(\frac{\beta_k}{A_k}\Lambda\right), \tag{24}$$

where $A_k = (M_k/(\sqrt{M_k}-1)^2)(1/\beta_k)$, $\beta_k = (3/(M_k-1))(|r_{kk}|^2/(\sigma_n^2 + \alpha(P_{\text{Tilted}})_T + \beta))$, $\Lambda = $ cte, and $W(x)$ is real Lambert function [3].

Thus, the power vector $\mathbf{P} = [(P_{\text{tilted}})_1, \ldots, (P_{\text{tilted}})_{n_T}]$ can be obtained by using the following unique repeated solution [3].

(I) A small positive value Λ is chosen so that

$$\sum_{k=1}^{K}\frac{1}{A_k} \le P_T. \tag{25}$$

(II) The value $\hat{P}_T$ is calculated as follows:

$$\hat{P}_T = \sum_{k=1}^{K}\frac{1}{\beta_k}W\left(\frac{\beta_k}{A_k}\Lambda\right). \tag{26}$$

(III) If $\hat{P}_T$ is not close enough to P_T, Λ multiply $P_T/\hat{P}_T$ and be returned to the stage II.

(IV) Based on (24), vector **P** is calculated.

This algorithm is present as pseudocode in Algorithm 1.

Now, we discuss the complexity of the proposed method. We realize that the proposed method adds some parts to the conventional THP method. To compare the complexities of the proposed method and the conventional THP, we only consider those parts that are changed in our method. For example, for the case of $n_R = n_T = 4$, the complexity of the conventional THP is $15n$ at the transmitter side and is $8n$ at the receiver side while for the proposed Tilted-THP method this complexity is $25n$ at the transmitter side and is

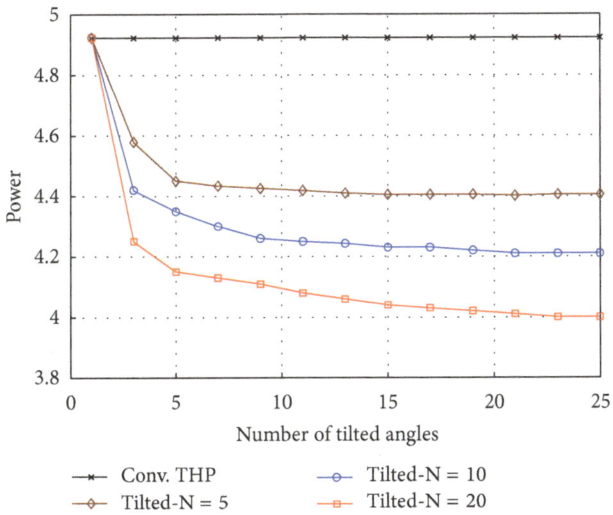

FIGURE 2: Transmitted power versus the number of tilted angles in different block length $N = 5, 10, 20$ (SNR $= 10$ dB, $n_T = n_R = 4$, 4QAM).

12n at the receiver side, where $n = N/n_T$ is the length of the input symbol to each antenna. As it can be seen, the increased complexity is suitable. On the other hand, as we will show in the next section in simulation results, the performance gain that we attain by the proposed method is more considerable with respect to the above complexity increasing.

5. Simulation Results

To evaluate the performance of our proposed method, we use some simulations. To do this we assume an MIMO-THP system with 4 transmit and 4 receive antennas with 4-QAM modulation. We assume that the channel estimation error covariance $\rho = 0.1$ and 0.05. In our figures, the labels "conv," "Rob," and "P.L." are corresponding to conventional, robust, and power loading methods, respectively.

Figure 2 shows the transmitted power of the MMSE-THP system versus the number of tilted angles. In this figure, the SNR is assumed to be 10 dB. It can be seen that by increasing the number of available tilted angles, the transmitted power reduces until it converges to a constant floor. Also, it can be seen that by decreasing the block length, the power reduction will be more and the power converges faster to its floor. The reason of this floor is that we assumed that the matrix $\mathbf{F}$ is unitary. As can be seen, for example, the transmitted power for the conventional MIMO-MMSE-THP for $Q = 25$ is 4.91 but, for the Tilted constellation and with the same parameters, the transmitted power decreases to 4 and 4.24 when the block length increases to 5 and 10, respectively.

Figure 3 shows BER of the MMSE and robust-tilted-MMSE precoders versus the Eb/No for different block lengths and for $\rho = 0.05$ with and without the power loading. As it is seen in this figure, with the Robust-Tilted THP, the BER decreases compared to the conventional one. This BER reduction is more when we use the power loading. Moreover, because of more power reduction in smaller block lengths,

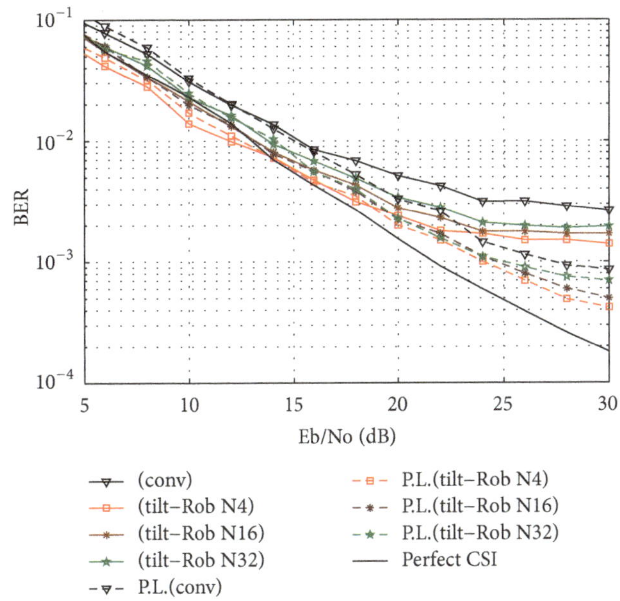

FIGURE 3: BER versus SNR and several block length in MIMO-MMSE-THP with and without power loading ($\rho = 0.05$).

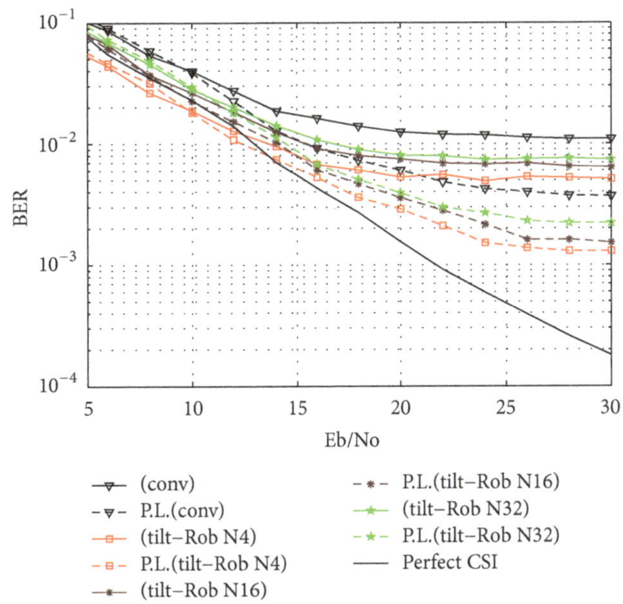

FIGURE 4: BER versus SNR and several block length in MIMO-MMSE-THP with and without power loading ($\rho = 0.1$).

the BER reduction will be more in smaller block length. For example, for SNR $= 30$ dB and $N = 4$ in the conventional MIMO-MMSE-THP, we have BER $= 0.0028$ while in the robust Tilted constellation MIMO-MMSE-THP for with and without power loading we have BER $= 0.0015$ and BER $= 0.0004$, respectively.

Figure 4 shows BER of the MMSE and Robust-Tilted-MMSE precoders versus the Eb/No for different block lengths and for $\rho = 0.1$ with and without the power loading. As it is seen in this figure, with the Robust-Tilted constellation, BER decreased compared to the conventional MIMO-THP which is similar to Figure 3. Again, BER reduction is more when we use the power loading. Due to more power reduction of Tilted constellation for shorter block length, the resultant BER has reduced more for shorter block lengths. More channel uncertainties with respect to Figure 3 lead to BER incensement. For example, at SNR = 30 dB in the conventional MIMO-MMSE-THP, BER = 0.011 while in the robust Tilted constellation MIMO-MMSE-THP for $N = 4$ with and without power loading BER = 0.005 and BER = 0.0012, respectively.

6. Conclusion

In this paper, we proposed a robust-tilted-MIMO-THP scheme for MIMO channels based on MMSE criterion with the assumption that the imperfect CSI is available at the transmitter side. We used second-order terms of the uncertainties of the MIMO channels at the transmitter to design a robust precoder. In the proposed robust method, we minimized the transmit power by using the Tilted constellation. The power minimization will result in ICI reduction between subchannels and in addition recovering some of the THP's power loss. Consequently, this ICI reduction and power recovering result led to BER improvement, especially at high SNRs. The transmission power is more reduced by utilizing more Tilted angles and smaller symbol block length. Also, we demonstrated that the achieved performance gain by using the proposed method is more considerable with respect to its higher complexity. Finally, we showed that we can improve our precoder performance further, by utilizing power loading.

References

[1] X. Geng, L. Jiang, and C. He, "Robust design for generalized vector precoding by minimizing mse with imperfect channel state information," *Chinese Journal of Electronics*, vol. 19, no. 3, pp. 503–506, 2010.

[2] N. Khaled, G. Leus, C. Desset, and H. Man, "A robust joint linear precoder and decoder MMSE design for slowly time-varying MIMO channels," in *Proceedings of the IEEE International Conference on Acoustics, Speech, and Signal Processing (ICASSP '04)*, pp. iv-485–iv-488, May 2004.

[3] H. K. Bizaki and A. Falahati, "Tomlinson-Harashima precoding with imperfect channel state information," *IET Communications*, vol. 2, no. 1, pp. 151–158, 2008.

[4] J. Kang, H. Ku, D. S. Kwon, and C. Lee, "Tomlinson-harashima precoder with tilted constellation for reducing transmission power," *IEEE Transactions on Wireless Communications*, vol. 8, no. 7, pp. 3658–3667, 2009.

[5] H. K. Bizaki and A. Falahati, "Power loading by minimizing the average symbol error rate on MIMO THP systems," in *Proceedings of the 9th International Conference on Advanced Communication Technology (ICACT '07)*, pp. 1323–1326, February 2007.

[6] J. G. Proakis and M. Salehi, *Digital Communications*, McGraw-Hill, 5th edition, 2008.

6

Optimization of Balking and Reneging Queue with Vacation Interruption under N-Policy

P. Vijaya Laxmi,[1] V. Goswami,[2] and K. Jyothsna[1]

[1] *Department of Applied Mathematics, Andhra University, Visakhapatnam 530 003, India*
[2] *School of Computer Application, KIIT University, Bhubaneswar 751 024, India*

Correspondence should be addressed to P. Vijaya Laxmi; vijaya iit2003@yahoo.co.in

Academic Editor: Zhihua Cui

This paper analyzes a finite buffer multiple working vacations queue with balking, reneging, and vacation interruption under N-policy. In the working vacation, a customer is served at a lower rate and at the instants of a service completion; if there are at least N customers in the queue, the vacation is interrupted and the server switches to regular busy period otherwise continues the vacation. Using Markov process and recursive technique, we derive the stationary system length distributions at arbitrary epoch. Various performance measures and some special models of the system are presented. Cost analysis is carried out using particle swarm optimization and quadratic fit search method. Finally, some numerical results showing the effect of model parameters on key performance measures of the system are presented.

1. Introduction

Performance modeling of queueing systems with balking and reneging has attracted many researchers owing to their wide applications in real life congestion problems such as impatient telephone switchboard customers, hospital emergency rooms handling critical patients and perishable goods storage in inventory systems. Balking and reneging are a common phenomena in queues; as a consequence, the customers either decide not to join the queue or depart after joining the queue without getting service due to impatience. Modeling balking and reneging is worthwhile because one obtains new managerial insights. The lost revenues due to balking and reneging in various industries can be enormous. While making decision for the number of servers needed in the service system to meet time-varying demand, the balking and reneging probabilities can be used to estimate the amount of lost business in more practical consideration for the managers as given in Liao [1].

Impatience is the most prominent characteristic as individuals always feel anxious and impatient during waiting for service in real life. The customer's impatient acts should be involved in the study of queueing system to model real situations exactly. Intermittent operation of a service can be economically appealing whenever full time service would result in significant server idle time or would preclude the use of the server in some other productive capacity. On the other hand, having the server inoperative for periods of time may increase the probability of customer losses due to balking and reneging. An $M/M/1$ queue with customers balking and reneging has been discussed in Haight [2] and Haight [3], respectively. The combined impact of balking and reneging with finite capacity in an $M/M/1$ queue has been studied by Ancker and Gafarian [4, 5]. Abou-El-Ata [6] discussed the finite buffer single server queueing system with balking and reneging. Analytical solutions of the single server Markovian overflow queue with balking, reneging and an additional server for longer queue were discussed in Abou-El-Ata and Shawky [7]. A computational algorithm and parameter optimization for a multiserver queue with unreliable server and impatient customers have been discussed by Chia and Jau-Chaun [8]. Choudhury and Medhi [9] presented some aspects of balking and reneging in finite buffer queue. Recently, Rakesh and Sumeet [10] studied an $M/M/1/N$ queueing model with retention of reneged customers and balking.

Queueing systems with server vacations have been studied due to their wide applications in several areas including computer and communication systems, manufacturing and production systems. Queueing systems with server vacations have been studied extensively in Tian and Zhang [11], Ke et al. [12], and so forth. Yue et al. [13] have analyzed an $M/M/1/N$ queue with balking, reneging, and server vacations. Vikas and Deepali [14] studied the state dependent bulk service queue with balking, reneging, and server vacation. However, there are numerous situations where the server remains active during the vacation period. Under working vacation (WV) policy, the server can provide service at a lower speed during vacation period rather than stopping service completely. At the end of a vacation if the queue is nonempty, a service period begins with normal service rate; otherwise the server takes another vacation. This policy is called multiple working vacations ($MWVs$) as introduced by Servi and Finn [15]. Recently, Vijaya Laxmi et al. [16] analyzed an $M/M/1/N$ working vacations queue with balking and reneging.

In order to utilize the server effectively, vacation interruption has become an important aspect, wherein the server interrupts the vacation once some indices of the system, such as the number of waiting customers, achieve a certain value in the vacation period. Li and Tian [17] considered a working vacation $M/M/1$ queue with vacation interruptions. Baba [18] studied an $M/PH/1$ queue with working vacations and vacation interruptions. Zhang and Hou [19] studied the more general $M/G/1$ queue with working vacations and vacation interruptions. An $M/M/1/MWV$ queue with N-policy has been studied in Zhang and Xu [20].

Queueing models with N-policy consider the most common issue of controlling arrivals and reducing the set up costs. The closed-form results in case of impatient customers with working vacations, vacation interruptions under N-policy are still not available. This motivated us to study an N-policy vacation interruptions queue with balking and reneging. This paper also discusses cost optimization problem under a given cost structure using particle swarm optimization (PSO) and quadratic fit search method (QFSM). PSO is a parallel evolutionary computation technique developed by Kennedy and Eberhart [21] based on the social behavior of animals such as bird flocking, fish schooling, and swarm theory. In the basic PSO algorithm the particle velocities build up too fast, and the optimum of the objective function is skipped. Hence, an inertia term introduced by Shi and Eberhart [22] is added to reduce the velocity. PSO can be easily implemented and is computationally inexpensive. An analysis of the PSO algorithm has been given in Qinghai [23] and Rao [24]. QFSM uses a 3-point pattern for fitting a quadratic function that has a unique optimum, see Rardin [25].

This paper deals with a finite buffer single server queue wherein customers may either balk or renege and the server takes multiple working vacations with a provision for vacation interruption. The server interrupts the vacation when the queue size reaches N; otherwise the vacation continues. The steady-state probabilities are obtained at arbitrary epoch through recursive technique which is easy to implement. We formulate a cost model to determine the optimum service rate during normal busy period and during working vacation period using PSO and QFSM.

The rest of the paper is organized as follows. Section 2 presents the description of the model and the steady-state probabilities. Sections 3 and 4 present some special cases, various performance measures, and cost analysis, respectively. Numerical results in the form of tables and graphs are presented in Section 5. Section 6 concludes the paper.

2. Model Description and Steady-State Probabilities

We consider a multiple working vacations $M/M/1/K$ queueing system with vacation interruptions wherein customers either balk or renege. The following notations and assumptions have been used throughout the paper.

(i) Customers arrive one at a time according to Poisson process with arrival rate λ. On arrival a customer either decides to join the queue or balk. Let b_i represent the probability with which the customer either decides to join the queue or balk with probability $1 - b_i$, when there are i customers ahead of him in the system. Furthermore, we assume that $0 \le b_{i+1} \le b_i < 1, 1 \le i \le K - 1$, $b_0 = 1$, and $b_K = 0$.

(ii) After joining the queue each customer will wait a certain length of time, say T, for service to begin before they get impatient and leave the queue without receiving service. This time T is assumed to follow exponential distribution with mean $1/\alpha$. Since the arrival and departure of an impatient customer without receiving service are independent, the average rate of reneging of a customer is given by $(i - 1)\alpha$. Hence, the function of customer's average reneging rate is given by $r_i = (i - 1)\alpha$, $1 \le i \le K$; $r_i = 0$, $i > K$.

(iii) The customers are served on a first-come first-served (FCFS) queue discipline. The service times in normal busy period are assumed to follow exponential distribution with mean $1/\mu$.

(iv) The server takes working vacations whenever the system becomes empty. At a service completion epoch during vacation, if there are at least N customers waiting in the system, the server ends the vacation and switches to normal service; that is, a vacation interruption occurs; otherwise, it continues the vacation. Furthermore, at vacation termination epoch, if there are less than N customers in the queue, the server begins another vacation; else, it switches to the normal busy period.

(v) During any working vacation the server will serve customers at a rate different (generally lower) from the normal service rate μ. The vacation duration and the service times in vacation period are also assumed be exponentially distributed with mean $1/\phi$ and $1/\eta$, respectively.

(vi) The interarrival times, vacation times, service times during normal busy period, and service times during vacation are mutually independent.

2.1. Steady-State Equations. At steady-state, let $P_{0,i}$, $0 \le i \le K$, be the probability that there are i customers in the system when the server is in working vacation and let $P_{1,i}$, $1 \le i \le K$, be the probability that there are i customers in the system when the server is in normal busy period. Using the Markov theory, the following set of steady-state equations may be written

$$\lambda P_{0,0} = \eta P_{0,1} + \mu P_{1,1}, \tag{1}$$

$$(\lambda b_i + \eta + (i-1)\alpha) P_{0,i} = (\eta + i\alpha) P_{0,i+1} + \lambda b_{i-1} P_{0,i-1}, \quad 1 \le i \le N-1, \tag{2}$$

$$(\lambda b_i + \eta + (i-1)\alpha + \phi) P_{0,i} = \lambda b_{i-1} P_{0,i-1} + i\alpha P_{0,i+1}, \quad N \le i \le K-1, \tag{3}$$

$$(\eta + (K-1)\alpha + \phi) P_{0,K} = \lambda b_{K-1} P_{0,K-1}, \tag{4}$$

$$(\mu + \lambda b_i + (i-1)\alpha) P_{1,i} = \lambda b_{i-1} P_{1,i-1} + (\mu + i\alpha) P_{1,i+1}, \quad 1 \le i \le N-1, \tag{5}$$

$$(\mu + \lambda b_i + (i-1)\alpha) P_{1,i} = \lambda b_{i-1} P_{1,i-1} + (\mu + i\alpha) P_{1,i+1} + \eta P_{0,i+1} + \phi P_{0,i}, \quad N \le i \le K-1, \tag{6}$$

$$(\mu + (K-1)\alpha) P_{1,K} = \lambda b_{K-1} P_{1,K-1} + \phi P_{0,K}. \tag{7}$$

The steady-state probabilities $P_{0,i}(P_{1,i})$ can be obtained by solving the previous system of equations recursively that is presented as a theorem later.

Theorem 1. *The steady-state probabilities are given by*

$$P_{0,i} = \frac{\zeta_i}{\sum_{m=0}^{K} \zeta_m + \sum_{m=1}^{K} \xi_m}, \quad 0 \le i \le K, \tag{8}$$

$$P_{1,i} = \frac{\xi_i}{\sum_{m=0}^{K} \zeta_m + \sum_{m=1}^{K} \xi_m}, \quad 1 \le i \le K, \tag{9}$$

where

$$\zeta_K = 1, \qquad \zeta_{K-1} = \frac{\phi + \eta + (K-1)\alpha}{\lambda b_{K-1}},$$

$$\zeta_i = \frac{(\lambda b_{i+1} + \eta + \phi + i\alpha)\zeta_{i+1} - (i+1)\alpha\zeta_{i+2}}{\lambda b_i}, \quad i = K-2, K-3, \ldots, N-1,$$

$$\zeta_i = \frac{(\lambda b_{i+1} + \eta + i\alpha)\zeta_{i+1} - (\eta + (i+1)\alpha)\zeta_{i+2}}{\lambda b_i}, \quad i = N-2, N-3, \ldots, 0,$$

$$\xi_0 = 0, \qquad \xi_1 = \frac{\lambda\zeta_0 - \eta\zeta_1}{\mu},$$

$$\xi_i = \frac{(\mu + \lambda b_{i-1} + (i-2)\alpha)\xi_{i-1} - \lambda b_{i-2}\xi_{i-2}}{\mu + (i-1)\alpha}, \quad i = 2, 3, \ldots, N,$$

$$\xi_i = \frac{(\mu + \lambda b_{i-1} + (i-2)\alpha)\xi_{i-1} - \lambda b_{i-2}\xi_{i-2} - \eta\zeta_i - \phi\zeta_{i-1}}{\mu + (i-1)\alpha}, \quad i = N+1, N+2, \ldots, K. \tag{10}$$

Proof. Solving the system of (1) to (6) recursively and using the normalization condition $\sum_{i=0}^{K} P_{0,i} + \sum_{i=1}^{K} P_{1,i} = 1$, we obtain the results of the theorem. □

3. Special Cases

In this section, results pertaining to some models have been deduced from our model by taking specific values for the parameters α, η, b_i, and N.

Case 1 ($\alpha \to 0$, $b_i = 1$, $0 \le i \le K$, $N = 1$). The model reduces to $M/M/1/K$ queue with multiple working vacations and vacation interruption. We can obtain the steady-state probabilities $P_{0,i}$ $(0 \le i \le K)$ and $P_{1,i}$ $(1 \le i \le K)$ from (8)-(9), where

$$\zeta_K = 1, \qquad \zeta_{K-1} = \frac{\phi + \eta}{\lambda}, \qquad \zeta_i = \frac{(\lambda + \eta + \phi)}{\lambda}\zeta_{i+1}, \quad i = K-2, K-3, \ldots, 0,$$

$$\xi_0 = 0, \qquad \xi_1 = \frac{\lambda\zeta_0 - \eta\zeta_1}{\mu}, \tag{11}$$

$$\xi_i = \frac{(\mu + \lambda)\xi_{i-1} - \lambda\xi_{i-2} - \eta\zeta_i - \phi\zeta_{i-1}}{\mu}, \quad i = 2, 3, \ldots, K,$$

which match with the results of $M/M/1/\infty$ queue with multiple working vacations and vacation interruption Li and Tian [17] for large K values.

Case 2 ($N = 1$ and without vacation interruptions). The model reduces to $M/M/1/K$ queue with balking, reneging and multiple working vacations and results match numerically with the results available in Vijaya Laxmi et al. [16].

Case 3 ($N = 1$, $\eta \to 0$). The model reduces to $M/M/1/K$ queue with balking, reneging and multiple vacations. Taking the average reneging rate as $r_i = (i - n)\alpha$, $n = 0$ in vacation period and $n = 1$ in normal busy period, we can obtain the steady-state probabilities $P_{0,i}$ and $P_{1,i}$ from (8)-(9), where

$$\zeta_K = 1, \qquad \zeta_{K-1} = \frac{K\alpha + \phi}{\lambda b_{K-1}},$$

$$\xi_0 = 0, \qquad \xi_1 = \frac{\lambda\zeta_0 - \alpha\zeta_1}{\mu},$$

$$\zeta_i = \frac{(\lambda b_{i+1} + \phi + (i+1)\alpha)\zeta_{i+1} - (i+2)\alpha\zeta_{i+2}}{\lambda b_i}, \quad i = K-2, K-3, \ldots, 0, \tag{12}$$

$$\xi_i = \frac{(\mu + \lambda b_{i-1} + (i-2)\alpha)\xi_{i-1} - \lambda b_{i-2}\xi_{i-2} - \phi\zeta_{i-1}}{\mu + (i-1)\alpha}, \quad i = 2, 3, \ldots, K.$$

The previous results match numerically with the results of Yue et al. [13].

Case 4 ($\eta \to 0$ and $\phi \to \infty$). The model reduces to $M/M/1/K$ queue with balking and reneging. These results match with the results available in the literature.

4. Performance Measures

Once the steady-state probabilities are obtained, one can evaluate various performance measures of the model like probability that the server is in normal busy period (P_b), probability that the server is in working vacation (P_{wv}), average queue length (L_q), average system length (L_s), and they are given by

$$P_b = \sum_{i=1}^{K} P_{1,i}, \qquad P_{wv} = \sum_{i=0}^{K} P_{0,i},$$

$$L_q = \sum_{i=1}^{K} (i-1)(P_{0,i} + P_{1,i}), \qquad L_s = \sum_{i=1}^{K} i(P_{0,i} + P_{1,i}). \tag{13}$$

When a customer finds i customers in the system upon his arrival, then the instantaneous balking rate is $\lambda(1 - b_i)$, as the probability that a customer balks in the system is $1 - b_i$. Using the concepts of Ancker and Gafarian [4, 5], we can obtain the average balking rate ($B.R.$) as

$$B.R. = \sum_{i=1}^{K} \lambda(1 - b_i) P_{0,i} + \sum_{i=1}^{K} \lambda(1 - b_i) P_{1,i}. \tag{14}$$

If there are i customers in the system and server is available, then there are $(i - 1)$ waiting customers in the queue. Since any one of the $(i - 1)$ customers in the queue may renege, the instantaneous reneging rate is $(i - 1)\alpha$. Again, following the models of Ancker and Gafarian [4, 5], the average reneging rate ($R.R.$) is given by

$$R.R. = \sum_{i=1}^{K} (i-1)\alpha P_{0,i} + \sum_{i=1}^{K} (i-1)\alpha P_{1,i}. \tag{15}$$

The average rate of customer loss ($L.R.$) is the sum of the average balking rate and the average reneging rate. Thus, we have

$$L.R. = B.R. + R.R. \tag{16}$$

4.1. Cost Model. In practice, queueing managers are always interested in minimizing operating cost of unit time. In this section, we first develop a steady-state expected cost function per unit time, in which either service rate during normal busy period or service rate during working vacation period is the decision variable. Our objective is to determine the optimum value of μ or η to minimize the cost function.

Let us define the following cost elements:

$C_\mu \equiv$ service cost per unit time during normal busy period,

$C_\eta \equiv$ service cost per unit time during working vacation period,

$C_{lq} \equiv$ cost per unit time when a customer joins the queue and waits for service,

$C_{lr} \equiv$ cost per unit time when a customer balks or reneges.

The cost parameters are assumed to be linear in the average number of the indicated quantities such as the average queue length (L_q) and the average rate of customer loss ($L.R.$). Based on the definitions of each cost element listed previously and its corresponding system performance measures, the cost minimization problem can be illustrated mathematically as:

$$\text{Minimize}: f(x) = C_\mu \mu + C_\eta \eta + C_{lq} L_q + C_{lr}\, L.R., \tag{17}$$

where x can be either service rate during normal busy period (μ) or service rate during working vacation period (η). The objective is to determine the optimal service rate μ^* or η^* to minimize the cost function f. It is a difficult task to develop analytic results for the optimum value of μ or η, because the expected cost function is highly complex. We note that the derivatives of the operating cost function per unit time are not easily available. We solve the above stated optimization problem using PSO and QFSM. Both methods give identical results as shown in the next section.

4.1.1. Algorithm for PSO

Step 1. Initialize a population of particles of size m with random positions in the problem space. For each particle, evaluate the objective function value.

Step 2. As all the particles will be moving to the optimal point with a velocity, find the velocities of particles. Initially, all particle velocities are assumed to be zero.

Step 3. Store each particle's coordinates associated with the best solution in *p*best and the coordinates of the particle associated with the overall best solution in *g*best.

Table 1: Optimum values for various values of η and λ.

η	$\lambda = 2.0$				$\lambda = 2.2$			
	μ^*	f^*	L_q^*	$L.R.^*$	μ^*	f^*	L_q^*	$L.R.^*$
1.5	2.01604	104.092	1.59964	0.53122	2.23676	112.397	1.75077	0.58658
1.7	2.00395	104.058	1.50002	0.49868	2.23122	112.527	1.65287	0.55431
1.9	1.98578	104.060	1.40553	0.46780	2.21962	112.663	1.55904	0.52338
2.1	1.96241	104.125	1.31656	0.43871	2.20261	112.828	1.46968	0.49392
2.3	1.93473	104.274	1.23334	0.41148	2.18085	113.043	1.38511	0.46603

Table 2: η^* from PSO and QFSM.

μ	PSO		QFSM		
	η^*	f^*	Initial 3-point pattern	η^*	f^*
2.3	1.51326	97.8642	(1.4, 1.5, 1.6)	1.51326	97.8642
2.4	1.45479	98.7762	(1.4, 1.5, 1.6)	1.45479	98.7762
2.5	1.40225	99.7727	(1.3, 1.4, 1.5)	1.40225	99.7727
2.6	1.35504	100.843	(1.3, 1.4, 1.5)	1.35504	100.843
2.7	1.31262	101.979	(1.2, 1.3, 1.4)	1.31262	101.979

Step 4. Compare each particle's fitness with particle's *p*best. If current value is better, then set *p*best value to the current value. Compare each particle's fitness with the population's overall previous best (*g*best), and if any particle's current value is better than *g*best, then set *g*best to the current value.

Step 5. Change the velocity $v_j(i)$ and position $x_j(i)$ of the *j*th particle in the *i*th iteration according to the following equations, respectively,

$$v_j(i) = \varphi(i)\, v_j(i-1) + c_1 \text{rand}(\cdot) \left(p\text{best}(j) - x_j(i-1)\right) + c_2 \text{rand}(\cdot) \left(g\text{best} - x_j(i-1)\right), \quad 1 \le j \le m,$$
$$x_j(i) = x_j(i-1) + v_j(i), \quad 1 \le j \le m, \tag{18}$$

where rand$(\cdot)$ is a random number between $(0, 1)$, c_1 and c_2 are learning factors, and $\varphi(i)$ is the inertia weight given by

$$\varphi(i) = \varphi_{\max} - \left(\frac{\varphi_{\max} - \varphi_{\min}}{i_{\max}}\right) i \tag{19}$$

with $\varphi_{\max}$, $\varphi_{\min}$ being the maximum and minimum values of the inertia weight ($\varphi_{\max} = 0.9$, $\varphi_{\min} = 0.4$ are commonly used) and $i_{\max}$ the maximum number of iterations.

Step 6. Loop to Step 3 until the stopping criterion is met. The stopping criterion is that the number of iterations is completed or the difference between two successive *g*best values is less than a predefined error tolerance.

4.1.2. Algorithm for QFSM. Given a 3-point pattern, we can fit a quadratic function through corresponding functional values that have a unique minimum, x^q, for the given objective function $f(x)$. Quadratic fit uses this approximation to improve the current 3-point pattern by replacing one of its points with approximate optimum x^q. The unique optimum x^q of the quadratic function agreeing with $f(x)$ at 3-point operation (x^l, x^m, x^h) given by Rardin [25] occurs at

$$x^q \cong \frac{1}{2}\Big[\Big(f(x^l)\left[s^m - s^h\right] + f(x^m)\left[s^h - s^l\right] + f(x^h)\left[s^l - s^m\right]\Big) \times \Big(f(x^l)\left[x^m - x^h\right] + f(x^m)\left[x^h - x^l\right] + f(x^h)\left[x^l - x^m\right]\Big)^{-1}\Big], \tag{20}$$

where $s^l = (x^l)^2$, $s^m = (x^m)^2$, $s^h = (x^h)^2$.

5. Numerical Discussions

In this section, we illustrate the results obtained in the previous sections numerically and discuss the effect of system parameters on the performance indices. The differences of parameters, such as the lower service rate (η), vacation rate (ϕ), vacation interruption, and N policy, also influence the various performance measures of the model. We fix the capacity of the system as $K = 10$ and $N = 4$, and the parameters of the model are assumed as $\lambda = 1.8$, $\mu = 2.5$, $\eta = 1.9$, $\phi = 1.2$, and $\alpha = 0.3$, unless they are considered as variables. The balking function is taken as $b_i = 1 - (i/K^2)$, $1 \le i \le K-1$, $b_0 = 1$, $b_K = 0$, and the various cost elements are chosen as $C_\mu = 18$, $C_\eta = 15$, $C_{lq} = 25$, $C_{lr} = 10$. The swarm size, maximum number of iterations, and learning factors of PSO algorithm are taken as 20, 50 and $c_1 = c_2 = 2$, respectively, and the tolerance of QFSM is $\kappa = 1 \times 10^{-6}$.

Table 1 presents the optimum values of μ along with L_q^* and $L.R.^*$ for various η and λ, using particle swarm optimization. For a fixed λ, decreasing trend is observed in μ^*, L_q^*, and $L.R.^*$ with the increase of η. For a fixed η, increasing trend is observed in μ^*, L_q^*, and $L.R.^*$ with

Table 3: Various performance measures for different balking functions, μ, α, and ϕ.

		P_b	P_{wv}	L_q	$L.R.$	f
	$b_i = 1 - (i/K^2)$	0.207206	0.792794	0.943660	0.313380	100.225
$\mu = 2.5$	$b_i = 1/(i+1)$	0.009290	0.990710	0.252221	0.674025	86.5458
	$b_i = e^{-i}$	0.000502	0.999498	0.151711	0.740087	84.6936
	$b_i = 1 - (i/K^2)$	0.137534	0.862466	0.793431	0.264904	113.985
$\mu = 3.5$	$b_i = 1/(i+1)$	0.006559	0.993441	0.248745	0.671253	104.431
	$b_i = e^{-i}$	0.000362	0.999637	0.151550	0.739925	102.688
	$\alpha = 0.2$	0.252143	0.747857	1.101570	0.254461	103.584
$\mu = 2.5$	$\alpha = 0.5$	0.144430	0.855570	0.746728	0.399321	96.1614
	$\alpha = 0.8$	0.088249	0.911750	0.577965	0.484742	92.7966
	$\alpha = 0.2$	0.163313	0.836687	0.878758	0.204519	115.514
$\mu = 3.5$	$\alpha = 0.5$	0.099092	0.900908	0.668530	0.358484	111.798
	$\alpha = 0.8$	0.062491	0.937509	0.542701	0.455723	109.625
	$\phi = 0.9$	0.199081	0.800919	0.951578	0.315933	100.449
$\mu = 2.5$	$\phi = 1.1$	0.204683	0.795317	0.946113	0.314170	100.295
	$\phi = 1.4$	0.211783	0.788217	0.939225	0.31195	100.100
	$\phi = 0.9$	0.131773	0.868227	0.807812	0.269538	114.391
$\mu = 3.5$	$\phi = 1.1$	0.135742	0.864258	0.797890	0.266340	114.111
	$\phi = 1.4$	0.140790	0.859209	0.785357	0.262303	113.757

Table 4: Various performance measures for different balking functions, η, α, and ϕ.

		P_b	P_{wv}	L_q	$L.R.$	f
	$b_i = 1 - (i/K^2)$	0.184638	0.815362	0.854715	0.284490	100.713
$\eta = 2.1$	$b_i = 1/(i+1)$	0.007447	0.992553	0.219831	0.626807	88.2638
	$b_i = e^{-i}$	0.000393	0.999607	0.132791	0.692532	86.7451
	$b_i = 1 - (i/K^2)$	0.154568	0.845432	0.734459	0.245367	101.815
$\eta = 2.4$	$b_i = 1/(i+1)$	0.005443	0.994557	0.181192	0.566167	91.1915
	$b_i = e^{-i}$	0.000279	0.999720	0.110306	0.631206	90.0697
	$\alpha = 0.2$	0.226048	0.773952	1.00062	0.231765	103.833
$\eta = 2.1$	$\alpha = 0.5$	0.127751	0.872249	0.674428	0.361304	96.9737
	$\alpha = 0.8$	0.077759	0.922240	0.521827	0.438267	93.9284
	$\alpha = 0.2$	0.190649	0.809351	0.862265	0.200605	104.563
$\eta = 2.4$	$\alpha = 0.5$	0.106066	0.893934	0.578489	0.310791	98.5701
	$\alpha = 0.8$	0.064377	0.935623	0.448454	0.377461	95.9859
	$\phi = 0.9$	0.176926	0.823074	0.859298	0.285968	100.842
$\eta = 2.1$	$\phi = 1.1$	0.182233	0.817767	0.856141	0.284950	100.753
	$\phi = 1.4$	0.189022	0.810978	0.852124	0.283654	100.640
	$\phi = 0.9$	0.147518	0.852482	0.735379	0.245664	101.841
$\eta = 2.4$	$\phi = 1.1$	0.152358	0.847642	0.734747	0.245460	101.823
	$\phi = 1.4$	0.158625	0.841375	0.733932	0.245197	101.800

the increase of λ. But the minimum cost increases with the increase of η and λ.

To ensure the reliability of PSO algorithm, we compared our results with QFSM, and the results are shown in Table 2. Though both these methods give identical results, the advantage of using PSO algorithm lies in the ease with which it can be tuned and implemented using only a velocity operator to drive the search throughout the search space, whereas the QFSM is dependent on the proper choice of the initial 3-point approximation. It is also observed that convergence is faster in PSO algorithm.

The effect of μ and η on various performance measures of the model, for various balking functions, α and ϕ, is presented in Tables 3 and 4, respectively. Note that for any $\mu(\eta)$, the balking function $b_i = 1 - (i/K^2)$ yields the lowest $L.R.$ which supports our choice of the balking function as $b_i = 1 - (i/K^2)$. On the other hand, for any choice of the balking function P_b, L_q, and $L.R.$ decrease with the increase of $\mu(\eta)$, while P_{wv}

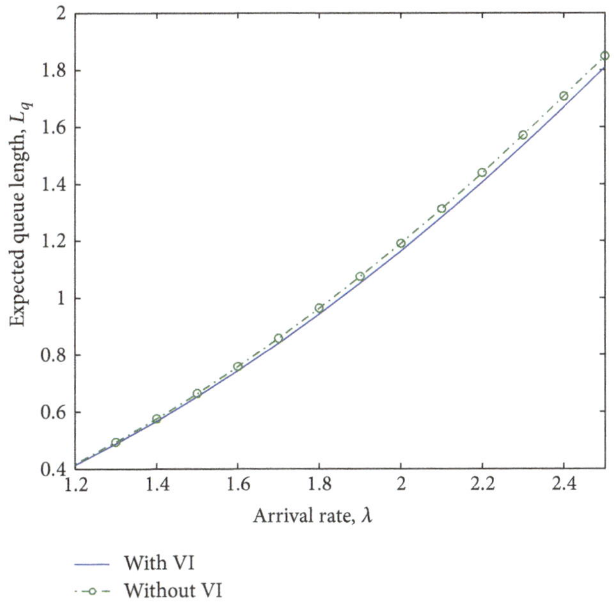

FIGURE 1: Impact of λ on L_q with and without vacation interruption.

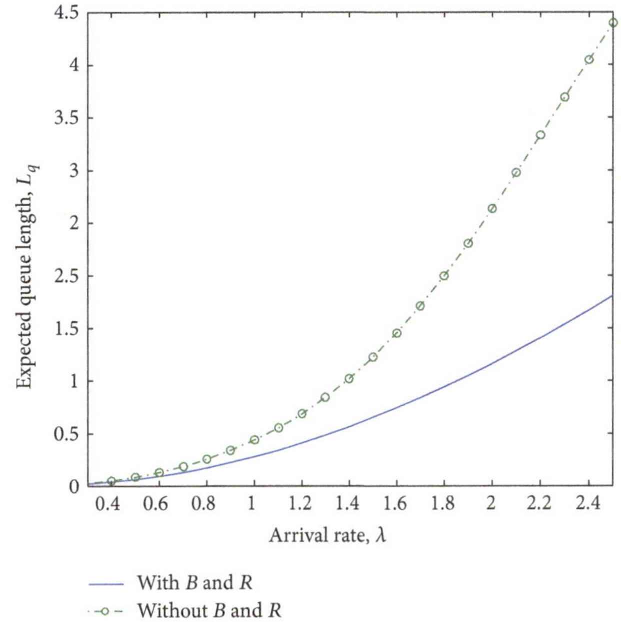

FIGURE 2: λ versus L_q with and without balking and reneging.

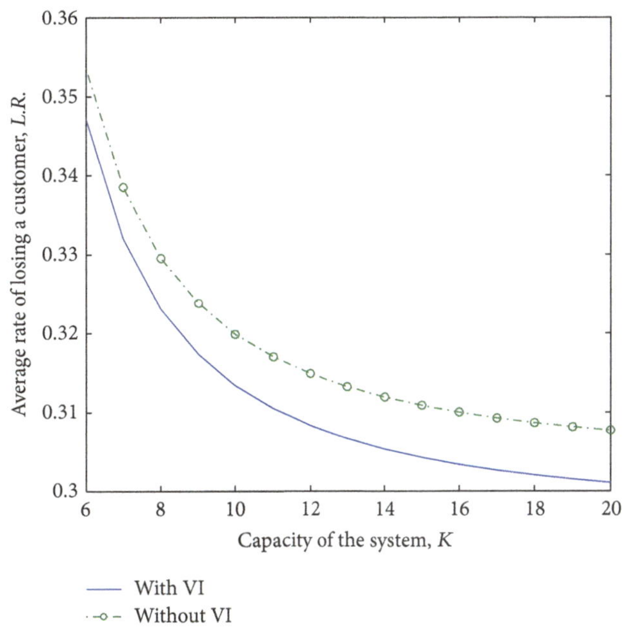

FIGURE 3: Effect of K on $L.R.$

and cost increase with $\mu(\eta)$, as intuitively expected. From the tables it can also be observed that for fixed $\mu(\eta)$, increase in α results in the decrease of P_b, L_q, and cost, whereas the other performance measures P_{wv} and $L.R.$ increase with α. Further, for fixed α, P_b, L_q, and $L.R.$ decrease with the increase of $\mu(\eta)$. It can also be observed that as $\mu(\eta)$ increases, P_b, L_q, and $L.R.$ decrease for any ϕ. Moreover, for fixed $\mu(\eta)$, except P_b all the other performance measures decrease with the increase of ϕ. This is because as ϕ increases, the vacation duration decreases and the server switches to normal busy period during which service is provided with rate μ.

Figure 1 shows the effect of the arrival rate (λ) on the expected number of customers in the queue (L_q) with and without vacation interruptions (VI). It can be observed that with the increase of λ, the expected queue length increases as it should be. Further, one may also observe that for higher values of λ, L_q is less in the vacation interruptions model as compared to the model without vacation interruptions. This is due to the fact that in vacation interruptions model, the server tends to interrupt the vacation and switches to normal service rate when more customers arrive during vacation. The impact of arrival rate (λ) on the expected number of customers in the queue (L_q) with and without balking and reneging (B & R) is shown in Figure 2. It can be seen that as arrival rate increases L_q monotonically increases. Moreover, the model with balking and reneging outperforms the model without balking and reneging.

Figures 3 and 4 plot the effect of the average rate of customer loss ($L.R.$) for different capacity of the system (K) and threshold value (N), respectively, with and without vacation interruptions. From the figures it can be seen that as the buffer size (K) increases the customer loss decreases, whereas it increases with the increase of N, as intuitively expected. Further, in both the figures $L.R.$ is higher in the case of without vacation interruptions as compared to vacation interruptions model implying that a job has to wait more in models without vacation interruptions.

Figure 5 presents the effect of service rate during vacation (η) on the average number of customers in the queue (L_q) for various vacation rates (ϕ). As expected, L_q monotonically decreases as η increases for any ϕ. Further, we observe that for $\eta \leq \mu$, the expected queue lengths decrease with the increase of ϕ, and when η crosses μ, this trend is reversed. This shows that for better performance of the model η should be chosen less than μ. Figure 6 depicts the minimum and

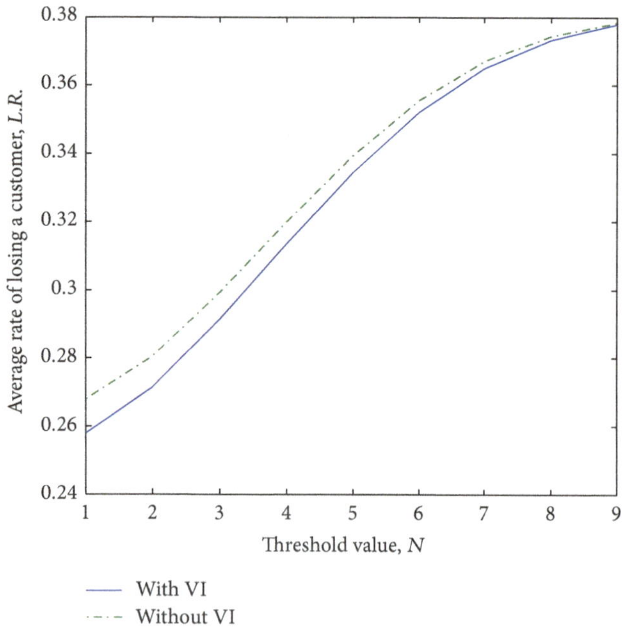

FIGURE 4: Effect of N on $L.R.$

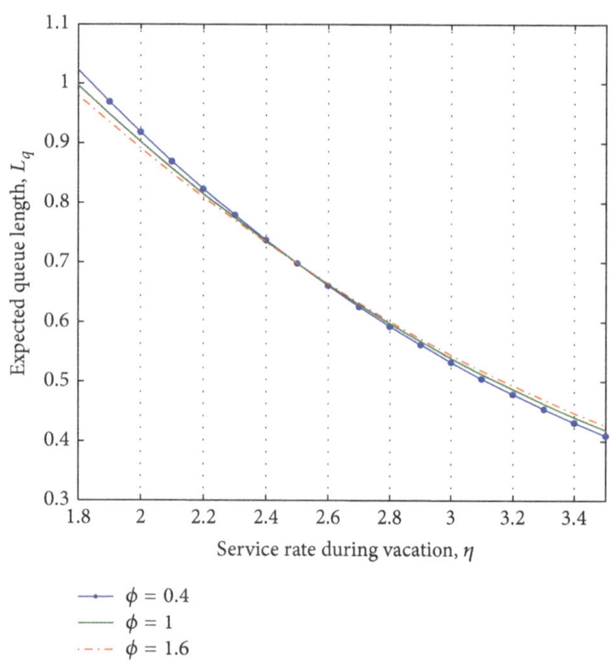

FIGURE 5: Impact of η on L_q.

average costs for number of iterations considered in PSO, when η is a decision variable. From the figure, it is observed that though initially the average cost varies widely from iteration to iteration, stability is attained after some iterations.

6. Conclusions

In this paper, we have carried out an analysis of a finite-buffer balking and reneging Markovian multiple working vacations queue with vacation interruptions under N-policy that have

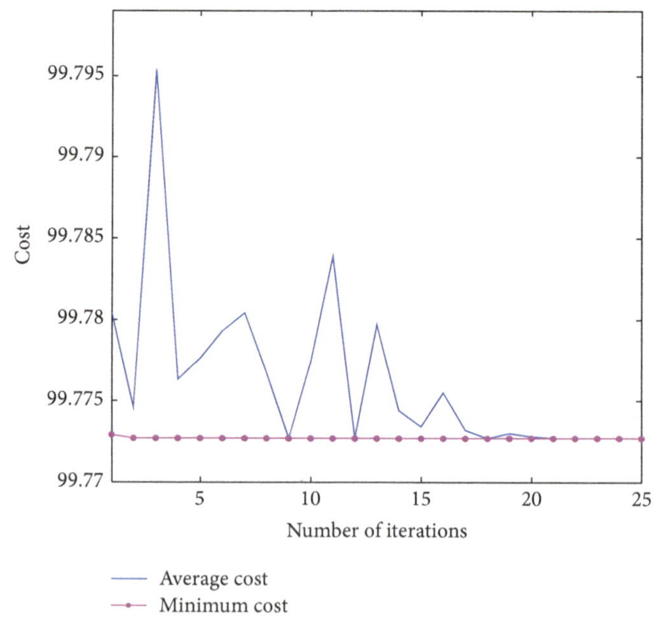

FIGURE 6: Iterations in PSO versus cost.

potential applications in modeling computer and telecommunication systems, computer networks, and so forth. We have obtained the closed-form expressions for the steady-state probabilities using recursive technique. Various performance measures such as expected number of customers in the system, average balking rate, average reneging rate, and the average rate of customer loss due to impatience are evaluated. We have also considered a cost optimization problem using particle swarm optimization and quadratic fit search method. Numerical results have been discussed which may be helpful to explore the effect of parameters on performance measures. The technique used in this paper can be applied to analyze more complex models such as $MAP/M/1/N$ and $M/M/c/N$ multiple working vacations queues with balking and reneging which are left for future investigations.

References

[1] P. Liao, "Optimal staffing policy for queueing systems with cyclic demands: waiting cost approach," in *Proceedings of the 18th Annual Conference of the Production and Operations Management Society (POMS '07)*, Dallas, Tex, USA, 2007.

[2] F. A. Haight, "Queueing with balking," *Biometrika*, vol. 44, pp. 360–369, 1957.

[3] F. A. Haight, "Queueing with reneging," *Metrika*, vol. 2, no. 1, pp. 186–197, 1959.

[4] C. J. Ancker Jr. and A. V. Gafarian, "Some queueing problems with balking and reneging: I," *Operations Research*, vol. 11, no. 1, pp. 88–100, 1963.

[5] C. J. Ancker Jr. and A. V. Gafarian, "Some queueing problems with balking and reneging: II," *Operations Research*, vol. 11, no. 6, pp. 928–937, 1963.

[6] M. O. Abou-El-Ata, "The state-dependent queue: M/M/1/N with reneging and general balk functions," *Microelectronics Reliability*, vol. 31, no. 5, pp. 1001–1007, 1991.

[7] M. O. Abou-El-Ata and A. I. Shawky, "The single-server Markovian overflow queue with balking, reneging and an additional server for longer queues," *Microelectronics Reliability*, vol. 32, no. 10, pp. 1389–1394, 1992.

[8] H. W. Chia and K. Jau-Chaun, "Computational algorithm and parameter optimization for a multi-server system with unreliable servers and impatient customers," *Journal of Computational and Applied Mathematics*, vol. 235, no. 3, pp. 547–562, 2010.

[9] A. Choudhury and P. Medhi, "Some aspects of balking and reneging in finite buffer queues," *RAIRO-Operations Research*, vol. 45, no. 3, pp. 223–240, 2011.

[10] K. Rakesh and K. S. Sumeet, "An M/M/1/N queueing model with retention of reneged customers and balking," *American Journal of Operational Research*, vol. 2, no. 1, pp. 1–5, 2012.

[11] N. Tian and Z. G. Zhang, *Vacation Queueing Models: Theory and Applications*, Springer, New York, NY, USA, 2006.

[12] J. C. Ke, C. H. Wu, and Z. G. Zhang, "Recent developments in vacation queueing models: a short survey," *International Journal of Operations Research*, vol. 7, pp. 3–8, 2010.

[13] D. Yue, Y. Zhang, and W. Yue, "Optimal performance analysis of an M/M/1/N queue system with balking, reneging and server vacation," *International Journal of Pure and Applied Mathematics*, vol. 28, pp. 101–115, 2006.

[14] S. Vikas and P. Deepali, "Performance analysis of state dependent bulk service queue with balking, reneging and server vacation," *International Journal of Operational Research Nepal*, vol. 1, pp. 61–69, 2012.

[15] L. D. Servi and S. G. Finn, "M/M/1 queues with working vacations (M/M/1/WV)," *Performance Evaluation*, vol. 50, no. 1, pp. 41–52, 2002.

[16] P. Vijaya Laxmi, V. Goswami, and K. Jyothsna, "Analysis of finite buffer Markovian queue with balking, reneging and working vacations," *International Journal of Strategic Decision Sciences*, vol. 4, no. 1, pp. 1–24, 2013.

[17] J. Li and N. Tian, "The M/M/1 queue with working vacations and vacation interruptions," *Journal of Systems Science and Systems Engineering*, vol. 16, no. 1, pp. 121–127, 2007.

[18] Y. Baba, "The M/PH/1 queue with working vacations and vacation interruption," *Journal of Systems Science and Systems Engineering*, vol. 19, no. 4, pp. 496–503, 2010.

[19] M. Zhang and Z. Hou, "Performance analysis of M/G/1 queue with working vacations and vacation interruption," *Journal of Computational and Applied Mathematics*, vol. 234, no. 10, pp. 2977–2985, 2010.

[20] Z. Zhang and X. Xu, "Analysis for the M/M/1 queue with multiple working vacations and N-policy," *International Journal of Information and Management Sciences*, vol. 19, no. 3, pp. 495–506, 2008.

[21] J. Kennedy and R. Eberhart, "Particle swarm optimization," in *Proceedings of the IEEE International Conference on Neural Networks*, vol. 4, pp. 1942–1948, Canberra, Australia, December 1995.

[22] Y. Shi and R. C. Eberhart, "Parameter selection in particle swarm optimization," in *Proceedings of the 7th Annual Conference on Evolutionary Programming*, pp. 591–600, Springer, 1998.

[23] B. Qinghai, "Analysis of particle swarm optimization algorithm," *Computer and Information Science*, vol. 3, pp. 180–184, 2010.

[24] S. S. Rao, *Engineering Optimization: Theory and Practice*, John Wiley & Sons, Hoboken, NJ, USA, 2009.

[25] R. L. Rardin, *Optimization in Operations Research*, Prentice Hall, Upper Saddle River, NJ, USA, 1997.

7

Optimal Multiplicative Generalized Linear Search Plan for a Discrete Random Walker

Abd-Elmoneim Anwar Mohamed and Mohamed Abd Allah El-Hadidy

Department of Mathematics, Faculty of Science, Tanta University, Tanta 31527, Egypt

Correspondence should be addressed to Mohamed Abd Allah El-Hadidy; melhadidi@eun.eg

Academic Editor: Bijaya Panigrahi

This paper formulates a search model that gives the optimal search plan for the problem of finding a discrete random walk target in minimum time. The target moves through one of n-disjoint real lines in $\mathbb{R}^n$: we have n-searchers starting the searching process for the target from any point rather than the origin. We find the conditions that make the expected value of the first meeting time between one of the searchers and the target finite. Furthermore, we show the existence of the optimal search plan that minimizes the expected value of the first meeting time and find it. The effectiveness of this model is illustrated using numerical example.

1. Introduction

The search problem for a randomly moving target has a remarkable importance in our life due to its great applicability. This problem is very interesting because it may arise in many real world situations such as searching for randomly moving persons or targets on roads. That mathematical analysis collected all the researches derived from searching applications of the II World War. They solved with beauty two complementary objectives of the search: find the target with (1) the smallest cost and (2) in minimum time. Readers are referred to Koopman [1] for the early works and to Benkoski et al. [2] and Frost and Stone [3] for a more recent survey.

In the linear search problem, the target moves on the real line according to a known random process, and its initial position is given by the value of a random variable X_0 which has known probability distribution function. A searcher starts looking for the target at a point H_0 ($|H_0| < \infty$). The searcher moves continuously along the line in both directions of the starting point H_0 until the target is met. The searcher would change its direction at suitable points many times before meeting its goal. Thus, we consider that the path length which we represented is the cost of the search. The aim of the searcher is to minimize E_{τ_ϕ}, that is, the expected value of the first meeting time τ_ϕ between the searcher and the target. The problem is to find a search plan $\phi(t)$, such that $E_{\tau_\phi} < \infty$; in this case, we call that $\phi(t)$ is a finite search plan and if $E_{\tau_{\phi^*}} < E_{\tau_\phi}$ for all $\phi(t) \in \Phi(t)$, where $\Phi(t)$ terms to the class of all search plans then we call that $\phi^*(t)$ is an optimal search plan.

MC Cabe [4] found a finite search plan for a one-dimensional random walk target when the searcher starts the search from the origin, and the initial position of the target has a standard normal distribution. Mohamed [5] discussed the existence of a finite search plan for a one-dimensional random walk target in general case which means that the search may start from any point on the real line, and the initial position of the target has any distribution. El-Rayes and Mohamed [6] have shown the existence of a search plan which minimizes the expected value of the first meeting time between the searcher and the randomly moving target with imposed conditions. Fristedt and Heath [7] derived the conditions for optimal search path which minimizes the cost of effort of finding a randomly moving target on the real line. Ohsumi [8] presented an optimal search plan for a target moving with Markov process along one of K-nonintersecting arcs to a safe destination within a time limit, where the target starts at a safe base and tries to pass. El-Rayes et al. [9] illustrated this problem when the target moves on the real line with a Brownian motion, and the searcher starts the search from the origin. Recently, Mohamed et al. [10] discussed this problem for a Brownian target motion on one of n-intersected real lines in which any information of the target position is not available to the searchers all the time. Mohamed et al. formulate a search model and find the conditions under which the expected value of the first meeting time between one of the searchers and the target is finite. Furthermore, they showed the existence of the optimal search plan that minimizes the expected value of the first meeting time and found it.

On the other hand, when the target is located somewhere on the real line according to a known probability distribution, the searcher searches for it with known velocity and tries to find it in minimal expected time. It is assumed that the searcher can change the direction of its motion without any loss of time. The target can be detected only if the searcher reaches the target. In an earlier work, this problem has been studied extensively in many variations, mostly by Beck et al. [11–17], Franck [18], Rousseeuw [19], Reyniers [20, 21], and Balkhi [22, 23].

Furthermore, Mohamed et al. [24, 25] have got more interesting results when they studied this problem to find a randomly located target in the plane. The target has symmetric or asymmetric distribution and with less information about this target available to the searchers. More recently, Mohamed and El-Hadidy [26] studied this problem when the target moves with parabolic spiral in the plane and starts its motion from a random point. Also, Mohamed and El-Hadidy [27] disscussed this problem in the plane when the target moves randomly with conditionally deterministic motion.

Some problems of search may impose using more than one searcher such as when we search for a valuable target (e.g., person lost on one of n-disjoint roads) or search for a serious target (e.g., a car filled with explosives which moves randomly in one of n-disjoint roads). Thus, the main contributions of this paper center around studying the problem of searching for a one-dimensional random walker target that is moving on one of a system of n-disjoint continuous real lines in $\mathbb{R}^n$ (i.e., not intersected continuous real lines in the n-space). The problem that is studied here is very interesting where we argue to give the conditions on a strategy (or trajectories) of n-searchers, one on each line, that make the expected value of the first meeting time between one of the searchers and the target be finite and minimum. In this problem there exists a complication of such analysis. This complication is due to the fact that the searchers do not know the initial position of the target but only know its probability distribution. Otherwise, the problem would be reduced to determine the strategy of just one searcher on the same target's line. This problem is already tackled in [6, 7]. This work focuses on the necessary conditions for the existence of finite and optimal search plan that finds a random walker target.

The optimal search plan that is proposed here shows that the special structure of the search problem can be exploited

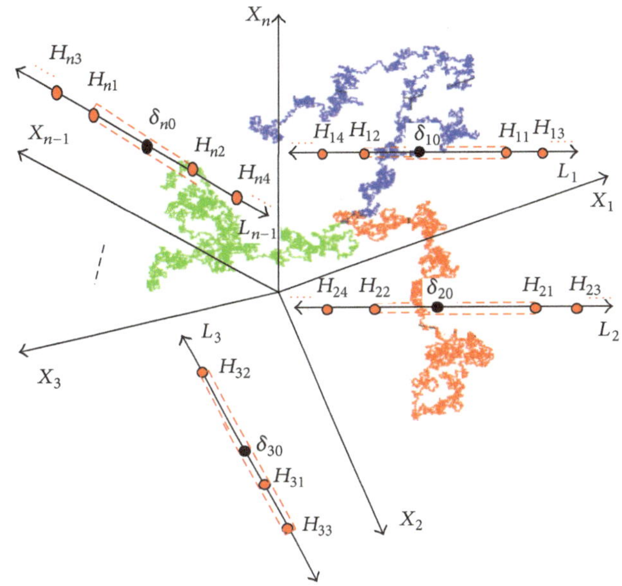

FIGURE 1: The search plan $\phi(t) = (\phi_1(t), \phi_2(t), \ldots, \phi_n(t))$ of the searchers S_i, $i = 1, 2, \ldots, n$.

to obtain the efficient solution. For example, the search plan for a criminal drunk leaves its cache and walks up and down through one of n-disjoint streets, totally disoriented.

This paper is organized as follows. In Section 2 we formulate the problem. We display some properties that the search model should satisfy in Section 3. The search plan and the conditions that make the expected value of the first meeting time between one of the searchers and the target be finite is discussed in Section 4. The existence of optimal search plan that minimizes the expected value of the first meeting time is presented in Section 5. The optimal search plan is studied in Section 6. In Section 7 we illustrate the effectiveness of this model using numerical example. Finally, the paper concludes with a discussion of the results and directions for future research.

2. Problem Description

The problem under study can be formally described as follows. We have n-searchers S_i, $i = 1, 2, \ldots, n$, that start the searching process from any point rather than the origin of the line L_i, $i = 1, 2, \ldots, n$, respectively, as in Figure 1. Each of the searchers moves continuously along its line in both directions of the starting point. The searcher S_i would conduct its search in the following manner. Start at δ_{i0} and go to the left (right) as far as H_{i1}. Then, turn back to explore the right (left) part of δ_{i0} as far as H_{i2}. Retrace the steps again to explore the left (right) part of H_{i1} as far as H_{i3} and so forth. In this paper, we need to determine H_{ij}, $i = 1, 2, \ldots, n$, $j = 1, 2, \ldots$, that minimize the first meeting time between one of the searchers and the moving target.

Let I be a set of integer numbers and I^+ a nonnegative part of I. We also assume that $\{X_{ij}\}_{j\geq 1}$ are sequences of independent identically distributed random variables in L_i, $i = 1, 2, \ldots, n$, respectively. In addition, we let a value of "-1" indicate a step to the left and a value "1" a step to the right, so that, for any $j \geq 1$, we have the n-dimensional probability vectors, $[P(X_{1j} = 1) = p_1, P(X_{2j} = 1) = p_2, \ldots, P(X_{nj} = 1) = p_n]$ and $[P(X_{1j} = -1) = q_1, P(X_{2j} = -1) = q_2, \ldots, P(X_{nj} = -1) = q_n]$, where $[p_1, p_2, \ldots, p_n] + [q_1, q_2, \ldots, q_n] = [1, 1, \ldots, 1]$.

Supposing that, for $t > 0$, $t \in I^+$, $W(t) = \sum_{j=1}^{t} X_{ij}$, $W(0) = 0$ and X_0 is a random variable that represent the target initial position, valued in $2I$ or $(2I + 1)$ and independent of $W(t)$, $t \geq 0$; if $\xi_i > 0$ and X_{ij} such that $0 \leq K_{i1} = (\xi_i + x_i)/2 \leq \xi_i$, where K_{i1} is an integer, then

$$P(W(\xi_i) = K_{i1}) = \begin{cases} \begin{pmatrix} \xi_i \\ K_{i1} \end{pmatrix} p_i^{K_{i1}} q_i^{\xi_i - K_{i1}}, \\ 0, & \text{if } K_{i1} \text{ does not exist.} \end{cases} \tag{1}$$

The target is assumed to move randomly on one of n-disjoint real lines according to the process $\{W(t), t \in I^+\}$, where I^+ is the set of positive integer numbers and $W(t)$ is a one-dimensional random walk motion. The initial position of the target is unknown but the searchers know its probability distribution (i.e., the probability distribution of the target is given at time 0) and the process $\{W(t), t \in I^+\}$ which controls the target's motion.

Figure 1 gives an illustration of the search plan that the n-searchers S_i, $i = 1, 2, \ldots, n$, follow it. Moreover, it is to be noted that this search plan $\phi(t) = (\phi_1(t), \phi_2(t), \ldots, \phi_n(t))$ is a combination of continuous functions $\phi_i(t)$ with speed v_i and given by $\phi_i(t) : I^+ \to I$, $i = 1, 2, \ldots, n$, such that

$$\begin{gathered} |\phi_i(t_1) - \phi_i(t_2)| < v_i |t_1 - t_2| \quad \forall t_1, t_2 \in I^+, \\ \phi_i(0) = 0, \quad i = 1, 2, \ldots, n. \end{gathered} \tag{2}$$

The first meeting time τ_ϕ is a random variable valued in I^+, and it is defined by

$$\tau_\phi = \begin{cases} \inf\{t : \text{either one of } \phi_i(t) = X_0 + W(t), \\ \quad i = 1, 2, \ldots, n\}, \\ \infty, \quad \text{if the set is empty,} \end{cases} \tag{3}$$

where $X_0 \in I$, that is, a random variable independent of $W(t)$, and it represents the initial position of the target. We suppose that $X_0 = X_i$ if the target moves on L_i, $i = 1, 2, \ldots, n$. Also assume that the set of all search plans of the searchers S_i with speeds v_i, $i = 1, 2, \ldots, n$, respectively, satisfying condition (2) be $\Phi_{v_i}(t)$, $i = 1, 2, \ldots, n$. The problem is to find a search plan $\phi(t) = (\phi_1(t), \phi_2(t), \ldots, \phi_n(t)) \in \Phi(t)$ for the n-searchers such that $E_{\tau_\phi} < \infty$, where E_{τ_ϕ} is the expected value of the first meeting time between one of the searchers and the target, and $\Phi(t) = \{(\phi_1(t), \phi_2(t), \ldots, \phi_n(t)); \phi_i(t) \in \Phi_{v_i}(t), i = 1, 2, \ldots, n\}$, is called a class of all sets of the search plans.

Assuming that ϑ_i, λ_i, $i = 1, 2, \ldots, n$, are positive numbers such that $\vartheta_i > 1$, $C_i = (\vartheta_i - 1)/(\vartheta_i + 1) > |p_i - q_i|$, $\lambda_i = b\varpi_i$, $b = 1, 2, \ldots$, and $\varpi_i(1 - C_i)/2$ is positive number also. In this problem, we assume that $v_i = 1$, $i = 1, 2, \ldots, n$. In addition, we define the sequences $\{G_{ij}\}$, $\{d_{ij}\}$, and $\{H_{ij}\}$, $i = 1, 2, \ldots, n$, $j = 1, 2, \ldots$ by $G_{ij} = \lambda_i(\vartheta_i^j - 1)$, $d_{ij} = (-1)^{j+1} C_i [G_{ij} + 1 + (-1)^{j+1}]$, and $H_{ij} = d_{ij} + \delta_{i0}$. We also use the notations $\psi_i(G_{i(2j+1)}) = W(G_{i(2j+1)}) - K_{i1}(G_{i(2j+1)})$, $\widehat{\psi}_i(G_{i(2j)}) = W(G_{i(2j)}) + K_{i2}(G_{i(2j)})$, where $K_{i1}(G_{i(2j+1)}) = C_i G_{i(2j+1)}$, $K_{i2}(G_{i(2j)}) = C_i G_{i(2j)}$ which are positive functions and the following remark.

Remark 1. If $0 < \varkappa_\S < 1$, $\S = 1, 2, \ldots, \kappa$, then $\prod_{\S=1}^{\kappa} \varkappa_\S < \sum_{\S=1}^{\kappa} \varkappa_\S$.

There are known probability measures γ_i, $i = 1, 2, \ldots, n$, such that $\gamma_1 + \gamma_2 + \cdots + \gamma_n = 1$ on $L_1 \cup L_2 \cup \cdots \cup L_n$. And, they describe the location of the target, where γ_i is induced by the position of the target on L_i.

The objective is to obtain the conditions that make the search plan $\phi(t) = (\phi_1(t), \phi_2(t), \ldots, \phi_n(t))$ be finite (i.e., $E_{\tau_\phi} < \infty$). We also need to show the existence of the optimal search plan $\phi^*(t) = (\phi_1^*(t), \phi_2^*(t), \ldots, \phi_n^*(t))$ and find it, that is, to give the minimum expected value of the first meeting time.

3. Properties of the Search Model

In this section we start to discuss some properties that the search model for a one-dimensional random walk target should satisfy.

The searcher S_i would conduct its search as in the above manner that is detailed in Section 2. Consequently, for any line L_i, $i = 1, 2, \ldots, n$, $t \in I^+$, we have the following.

Case 1. If we consider that $\mathbb{Q}$ is a set of positive even numbers such that $\mathbb{Q} = \{2, 4, 6, \ldots, \widehat{n}\}$, for $\mathscr{F} \in \mathbb{Q}$, $j \in I^+$, then we have

$$\begin{gathered} \cdots < H_{i(j+2)} < 0 < H_{ij} < H_{i(j-2)} < \cdots < H_{i2} < \delta_{i0} \\ < H_{i1} < H_{i3} < \cdots; \end{gathered} \tag{4}$$

for any $t \in I^+$, if $G_{i(2j-1)} \leq t \leq G_{i(2j)}$, $1 \leq j \leq \mathscr{F}/2$, then

$$\phi_i(t) = H_{i(2j+1)} - \delta_{i0} - [t - G_{i(2j-1)}]; \tag{5}$$

if $G_{ij} \leq t \leq G_{i(j+1)}$, $j \geq \mathscr{F} + 1$, then

$$\phi_i(t) = H_{ij} + (-1)^j [t - G_{ij}]; \tag{6}$$

if $G_{i(2j)} \leq t \leq G_{i(2j+1)}$, $1 \leq j \leq \mathscr{F}/2$, then

$$\phi_i(t) = \delta_{i0} - H_{i(2j)} + [t - G_{i(2j)}]. \tag{7}$$

Case 2. Also, if O is a set of positive odd numbers such that $O = \{1, 3, 5, \ldots, \widehat{m}\}$, for $\mathscr{F} \in O$, $j \in I^+$, then we have

$$\begin{gathered} \cdots < H_{i2} < \delta_{i0} < H_{i1} < H_{i3} < \cdots < H_{i(j-2)} < H_{ij} \\ < 0 < H_{i(j+2)} < \cdots; \end{gathered} \tag{8}$$

for any $t \in I^+$, if $G_{i(2j-1)} \le t \le G_{i(2j)}$, $1 \le j \le (\mathcal{F}+1)/2$, then

$$\phi_i(t) = \delta_{i0} - H_{i(2j-1)} - \left[t - G_{i(2j-1)}\right]; \quad (9)$$

if $G_{ij} \le t \le G_{i(j+1)}$, $j \ge \mathcal{F}+1$, then

$$\phi_i(t) = H_{ij} + (-1)^j \left[t - G_{ij}\right]; \quad (10)$$

if $G_{i(2j)} \le t \le G_{i(2j+1)}$, $1 \le j \le (\mathcal{F}-1)/2$, then

$$\phi_i(t) = H_{i(2j)} - \delta_{i0} + \left[t - G_{i(2j)}\right]. \quad (11)$$

It is clear that the search path of S_i depends on λ_i, ϑ_i and $j \in I^+$. Since $X_0 \in I$ then the first meeting time is done at $x_i \in I$ where x_i is an integer number. For this reason we will prove the conditions (properties) that make S_i meet the target at x_i as in the following theorems.

Theorem 2. *If C_i is a rational number different from 1 to −1 and, for all $\xi_i \ge 1$, $V(\xi_i) = (W(\xi_i\vartheta_i) - C_i\xi_i\vartheta_i)/2$ then*

(i) *there exists a sequence $\{Y_{ij}\}_{j\ge1}$ of independent identically distributed random variables such that $V(\xi_i) = \sum_{j=1}^{\Gamma} Y_{ij}$, and the distribution of Y_{ij} is concentrated on the integers $E(Y_{ij}) = (\vartheta_i/2)[E(X_{ij}) - C_i]$ and the probability vector $[p(Y_{1j} = \mathcal{F}) > 0, p(Y_{2j} = \mathcal{F}) > 0, \dots, p(Y_{nj} = \mathcal{F}) > 0]$, if and only if $-\vartheta_i(1+C_i)/2 \le \mathcal{F} \le \vartheta_i(1-C_i)/2$.*

(ii) *The probability vector, $[p(V(\xi_1) = x_1) > 0, p(V(\xi_2) = x_2) > 0, \dots, p(V(\xi_n) = x_n) > 0]$, holds if and only if x_i is an integer such that $-\xi_i\vartheta_i(1+C_i)/2 \le x_i \le \xi_i\vartheta_i(1-C_i)/2$.*

(iii) *If $C_i \ne p_i - q_i$ then there exist constants r_{i1} and r_{i2} depending on C_i, p_i such that, for any $x_i \in \mathbb{R}$, where $\mathbb{R}$ is the set of real numbers, if $\xi_i > r_{i1}x_i + r_{i2}$ and if $x_i \ge 0$ we have $[p(0 \le V(\xi_1+1) \le x_1), p(0 \le V(\xi_2+1) \le x_2), \dots, p(0 \le V(\xi_n+1) \le x_n)] \le [p(0 \le V(\xi_1) \le x_1), p(0 \le V(\xi_2) \le x_2), \dots, p(0 \le V(\xi_n) \le x_n)]$, and if $x_i < 0$ we have, $[p(x_1 \le V(\xi_1+1) < 0), p(x_2 \le V(\xi_2+1) < 0), \dots, p(x_n \le V(\xi_n+1) < 0)] \le [p(x_1 \le V(\xi_1) < 0), p(x_2 \le V(\xi_2) < 0), \dots, p(x_n \le V(\xi_n) < 0)]$.*

Proof. (i) Define $Y_{ij} = \sum_{\mathcal{F}=1}^{\vartheta_i}((X_{i(\mathcal{F}+(j-1)\vartheta_i)} - C_i)/2)$, $j \ge 1$, and $[p(Y_{1j} = x_1), p(Y_{2j} = x_2), \dots, p(Y_{nj} = x_n)] = [p(W(\vartheta_1)) = 2x_1 + \vartheta_1C_1, p(W(\vartheta_2)) = 2x_2 + \vartheta_2C_2, \dots, p(W(\vartheta_n)) = 2x_n + \vartheta_nC_n]$.

Consequently, if $[p(Y_{1j} = x_1) > 0, p(Y_{2j} = x_2) > 0, \dots, p(Y_{nj} = x_n) > 0]$, then from (1) we have $x_i + \vartheta_i(1+C_i)/2$ that is an integer, and since x_i is an integer also then we have $0 \le x_i + \vartheta_i(1+C_i)/2 \le \vartheta_i$. Therefore, $-\vartheta_i(1+C_i)/2 \le x_i \le \vartheta_i(1-C_i)/2$. If $x_i = \mathcal{F}$ then we have $[p(Y_{1j} = \mathcal{F}) > 0, p(Y_{2j} = \mathcal{F}) > 0, \dots, p(Y_{nj} = \mathcal{F}) > 0]$, if and only if $-\vartheta_i(1+C_i)/2 \le \mathcal{F} \le \vartheta_i(1-C_i)/2$. In addition, by using (1), $E(\sum_{\mathcal{F}=1}^{\vartheta_i}((X_{i(\mathcal{F}+(j-1)\vartheta_i)} - C_i)/2)) = (\vartheta_i/2)[E(X_{ij}) - C_i]$ is proved.

(ii) Since $V(\xi_i) = (W(\xi_i\vartheta_i) - C_i\xi_i\vartheta_i)/2$ then we have $[p(V(\xi_1) = x_1), p(V(\xi_2) = x_2), \dots, p(V(\xi_n) = x_n)] = [p((W(\xi_1\vartheta_1) - C_1\xi_1\vartheta_1)/2 = x_1), p((W(\xi_2\vartheta_2) - C_2\xi_2\vartheta_2)/2 = x_2), \dots, p((W(\xi_n\vartheta_n) - C_n\xi_n\vartheta_n)/2 = x_n)] = [p(W(\xi_1\vartheta_1)) = 2x_1 + C_1\xi_1\vartheta_1), p(W(\xi_2\vartheta_2)) = 2x_2 + C_2\xi_2\vartheta_2), \dots, p(W(\xi_n\vartheta_n)) = 2x_n + C_n\xi_n\vartheta_n)]$, and using (1) the prove is completed.

(iii) By using (ii), if $x_i \ge 0$ then we have $[p(0 \le V(\xi_1) \le x_1), p(0 \le V(\xi_2) \le x_2), \dots, p(0 \le V(\xi_n) \le x_n)] = [\sum_{j=0}^{[x_1]} p(V(\xi_1) = j), \sum_{j=0}^{[x_2]} p(V(\xi_2) = j), \dots, \sum_{j=0}^{[x_n]} p(V(\xi_n) = j)]$, and if $x_i < 0$ we get $[p(x_1 \le V(\xi_1+1) < 0), p(x_2 \le V(\xi_2+1) < 0), \dots, p(x_n \le V(\xi_n+1) < 0)] = [\sum_{j=[x_1]}^{0} p(V(\xi_1) = j), \sum_{j=[x_2]}^{0} p(V(\xi_2) = j), \dots, \sum_{j=[x_n]}^{0} p(V(\xi_n) = j)]$, where $[x_i]$ means the greatest integer less than or equal to x_i, $i = 1, 2, \dots, n$. It is sufficient to show that

$$\begin{aligned}&[p(V(\xi_1+1) = \mathcal{F}), p(V(\xi_2+1) = \mathcal{F}), \dots,\\ &\quad p(V(\xi_n+1) = \mathcal{F})]\\ &\qquad \le [p(V(\xi_1) = \mathcal{F}), p(V(\xi_2) = \mathcal{F}), \dots,\\ &\qquad\qquad p(V(\xi_n) = \mathcal{F})] \quad \text{if } \xi_i > r_{i1}x_i + r_{i2}.\end{aligned} \quad (12)$$

We have the following cases.

Case (a). If $C_i < -1$, then, from (ii), $[p(V(\xi_1) = \mathcal{F}) > 0, p(V(\xi_2) = \mathcal{F}) > 0, \dots, p(V(\xi_n) = \mathcal{F}) > 0]$, if and only if $2\mathcal{F}/\vartheta_i(1-C_i) \le \xi_i \le -2\mathcal{F}/\vartheta_i(1+C_i)$. We take $r_{i1} = -2/\vartheta_i(1+C_i)$, $r_{i2} = 0$ and then $\xi_i > r_{i1}|\mathcal{F}| \Rightarrow \xi_i > -2\mathcal{F}/\vartheta_i(1+C_i)$ that leads to $[p(V(\xi_1) = \mathcal{F}) = 0, p(V(\xi_2) = \mathcal{F}) = 0, \dots, p(V(\xi_n) = \mathcal{F}) = 0]$. Consequently, (12) holds.

Case (b). If $C_i > 1$, hence $[p(V(\xi_1) = \mathcal{F}) > 0, p(V(\xi_2) = \mathcal{F}) > 0, \dots, p(V(\xi_n) = \mathcal{F}) > 0]$, if and only if $-2\mathcal{F}/\vartheta_i(1+C_i) \le \xi_i \le 2\mathcal{F}/\vartheta_i(1-C_i)$. From (ii) putting $r_{i1} = -2/\vartheta_i(1-C_i)$, $r_{i2} = 0$ then $\xi_i > r_{i1}|\mathcal{F}| \Rightarrow \xi_i > 2\mathcal{F}/\vartheta_i(1-C_i)$ that leads to $[p(V(\xi_1) = \mathcal{F}) = 0, p(V(\xi_2) = \mathcal{F}) = 0, \dots, p(V(\xi_n) = \mathcal{F}) = 0]$.

Then, (12) holds.

Case (c). If $-1 < C_i < 1$, then let $\alpha_i = (1-C_i)/2$, $\beta_i = 1 - \alpha_i$ and $\Delta_i = (\alpha_i/q_i)^{\alpha_i}(\beta_i/p_i)^{\beta_i}$, where $[p_1, p_2, \dots, p_n] + [q_1, q_2, \dots, q_n] = [1, 1, \dots, 1]$. In addition, put $r_{i2} = 1/(\Delta_i - 1)$, $r_{i1} = \Delta_i \max[(1/\alpha_i, 1/\beta_i)\vartheta_i(\Delta_i - 1)]$. Since $E(X_{ij}) \ne C_i$ and $0 < \alpha_i < 1$ then $\alpha_i \ne q_i$ and then $\Delta_i > 1$. Consequently, r_{i1} and r_{i2} are positive and well defined. Assuming that $\xi_i > r_{i1}|\mathcal{F}|$ then $-\beta_i\xi_i\vartheta_i < \mathcal{F} < \alpha_i\xi_i\vartheta_i$; therefore, from (ii), we have $[p(V(\xi_1) = \mathcal{F}) > 0, p(V(\xi_2) = \mathcal{F}) > 0, \dots, p(V(\xi_n) = \mathcal{F}) > 0]$.

It remains to prove that if $[p(V(\xi_1) = \mathcal{F}) > 0, p(V(\xi_2) = \mathcal{F}) > 0, \dots, p(V(\xi_n) = \mathcal{F}) > 0]$ and $\xi_i > r_{i1}|\mathcal{F}| + r_{i2}$ then (2) holds.

Considering that $Z_{ij} = (X_{ij} - C_i)/2$ then $[p(Z_{1j} = \alpha_1) = p_1, p(Z_{2j} = \alpha_2) = p_2, \dots, p(Z_{nj} = \alpha_n) = p_n]$, $[p(Z_{1j} = -\beta_1) = q_1, p(Z_{2j} = -\beta_2) = q_2, \dots, p(Z_{nj} = -\beta_n) = q_n]$, and $[p(V(\xi_1) = \mathscr{F}), p(V(\xi_2) = \mathscr{F}), \dots, p(V(\xi_n) = \mathscr{F})] = [p(\sum_{j=1}^{\xi_1\vartheta_1} Z_{1j} = \mathscr{F}), p(\sum_{j=1}^{\xi_2\vartheta_2} Z_{2j} = \mathscr{F}), \dots, p(\sum_{j=1}^{\xi_n\vartheta_n} Z_{nj} = \mathscr{F})]$. By using (1) we have

$$\left[\frac{p(V(\xi_1+1)=\mathscr{F})}{p(V(\xi_1)=\mathscr{F})}, \frac{p(V(\xi_2+1)=\mathscr{F})}{p(V(\xi_2)=\mathscr{F})}, \dots, \frac{p(V(\xi_n+1)=\mathscr{F})}{p(V(\xi_n)=\mathscr{F})}\right]$$
$$= \left[\prod_{j=1}^{\vartheta_1\beta_1} \frac{\xi_1\vartheta_1 + j}{\Delta_1(\xi_1\vartheta_1 + j + (j\alpha_1 + \mathscr{F})/\beta_1)} \times \prod_{j=1}^{\vartheta_1\alpha_1} \frac{\xi_1\vartheta_1 + j + \vartheta_1\beta_1}{\Delta_1(\xi_1\vartheta_1 + j + (i\beta_1 - \mathscr{F})/\alpha_1)}, \right.$$
$$\prod_{j=1}^{\vartheta_2\beta_2} \frac{\xi_2\vartheta_2 + j}{\Delta_2(\xi_2\vartheta_2 + j + (j\alpha_2 + \mathscr{F})/\beta_2)} \times \prod_{j=1}^{\vartheta_2\alpha_2} \frac{\xi_2\vartheta_2 + j + \vartheta_2\beta_2}{\Delta_2(\xi_2\vartheta_2 + j + (i\beta_2 - \mathscr{F})/\alpha_2)}, \dots,$$
$$\left. \prod_{j=1}^{\vartheta_n\beta_n} \frac{\xi_n\vartheta_n + j}{\Delta_n(\xi_n\vartheta_n + j + (j\alpha_n + \mathscr{F})/\beta_n)} \times \prod_{j=1}^{\vartheta_n\alpha_n} \frac{\xi_n\vartheta_n + j + \vartheta_n\beta_n}{\Delta_n(\xi_n\vartheta_n + j + (i\beta_n - \mathscr{F})/\alpha_n)}\right]. \quad (13)$$

Since $\xi_i > r_{i1}|\mathscr{F}| + r_{i2}$ then every term is strictly less than 1. Consequently, $[p(V(\xi_1 + 1) = \mathscr{F}), p(V(\xi_2 + 1) = \mathscr{F}), \dots, p(V(\xi_n + 1) = \mathscr{F})] < [p(V(\xi_1) = \mathscr{F}), p(V(\xi_2) = \mathscr{F}), \dots, p(V(\xi_n) = \mathscr{F})]$. □

Theorem 3. *If $E(X_{ij}) < C_i$, $C_i \in \mathbb{R}$ (the set of real numbers), then there exists ε_i, $0 < \varepsilon_i < 1$, such that $[p(W(\xi_1)), p(W(\xi_2)), \dots, p(W(\xi_n))] \le [\varepsilon_1^{\xi_1}, \varepsilon_2^{\xi_2}, \dots, \varepsilon_n^{\xi_n}]$, for all ξ_i, $i = 1, 2, \dots, n$.*

Proof. For $\zeta_i > 0$, $[p(W(\xi_1) \ge C_1\xi_1), p(W(\xi_2) \ge C_2\xi_2), \dots, p(W(\xi_n) \ge C_n\xi_n)] = [p(\exp\{\zeta_1(W(\xi_1) - C_1\xi_1)\} \ge 1), p(\exp\{\zeta_2(W(\xi_2) - C_2\xi_2)\} \ge 1), \dots, p(\exp\{\zeta_n(W(\xi_n) - C_n\xi_n)\} \ge 1)] \le [E(\exp\{\zeta_1(W(\xi_1) - C_1\xi_1)\}), E(\exp\{\zeta_2(W(\xi_2) - C_2\xi_2)\}), \dots, E(\exp\{\zeta_n(W(\xi_n) - C_n\xi_n)\})] = [\{f_1(\zeta_1)\}^{\xi_1}, \{f_2(\zeta_2)\}^{\xi_2}, \dots, \{f_n(\zeta_n)\}^{\xi_n}]$, where $[\{f_1(\zeta_1)\}, \{f_2(\zeta_2)\}, \dots, \{f_n(\zeta_n)\}] = [E(\exp\{\zeta_1(X_{1j} - C_1)\}), E(\exp\{\zeta_2(X_{2j} - C_2)\}), \dots, E(\exp\{\zeta_n(X_{nj} - C_n)\})] = [p_1 \exp\{\zeta_1(1 - C_1)\} + q_1 \exp\{\zeta_1(-1 - C_1)\}, p_2 \exp\{\zeta_2(1 - C_2)\} + q_2 \exp\{\zeta_2(-1 - C_2)\}, \dots, p_n \exp\{\zeta_n(1 - C_n)\} + q_n \exp\{\zeta_n(-1 - C_n)\}]$; if $E(X_{ij}) < C_i$, $i = 1, 2, \dots, n$, then $[\{f_1'(0)\}, \{f_2'(0)\}, \dots, \{f_n'(0)\}] < [0, 0, \dots, 0]$, and since $[\{f_1(0)\}, \{f_2(0)\}, \dots, \{f_n(0)\}] < [1, 1, \dots, 1]$, then $[\min_{\zeta_1>0}\{f_1(\zeta_1)\}, \min_{\zeta_2>0}\{f_2(\zeta_2)\}, \dots, \min_{\zeta_n>0}\{f_n(\zeta_n)\}] = [\varepsilon_1, \varepsilon_2, \dots, \varepsilon_n] < [1, 1, \dots, 1]$.

By similar arguments if $E(X_{ij}) > C_i$, $i = 1, 2, \dots, n$, then there exist $\varepsilon_i < 1$ such that $[p(W(\xi_1)), p(W(\xi_2)), \dots, p(W(\xi_n))] \le [\varepsilon_1^{\xi_1}, \varepsilon_2^{\xi_2}, \dots, \varepsilon_n^{\xi_n}]$ for all ξ_i, $i = 1, 2, \dots, n$. □

4. Existence of a Finite Search Plan

In this section, we find the conditions that make the search plan be finite. In addition, we will discuss that under what these conditions are indeed finite. This is a crucial issue related to the existence of a finite search plan. So, we will provide useful theorems that help us to do it.

Theorem 4. *Let γ_i be the measure defined on $\mathbb{R}$ by X_{ij}, $i = 1, 2, \dots, n$, and if $\phi(t) = (\phi_1(t), \phi_2(t), \dots, \phi_n(t))$ is a search plan defined previously, the expectation E_{τ_ϕ} is finite if*

$$\int_{-\infty}^{\delta_{i0}} \left[\sum_{j=1}^{\mathscr{F}/2} \vartheta_i^{2j} p(\psi_i(G_{i(2j)}) < -x_i) + \sum_{j=\mathscr{F}/2+1}^{\infty} \vartheta_i^{2j} p(\tilde{\psi}_i(G_{i(2j)}) \le -x_i)\right] \gamma_i(dx_i), \quad (14)$$

$\int_{\delta_{i0}}^{\infty} \sum_{j=1}^{\infty} \vartheta_i^{2j+1} p(\psi_i(G_{i(2j+1)}) > -x_i)\gamma_i(dx_i)$, are finite when $\delta_{i0} > 0$. And if $\delta_{i0} < 0$, then E_{τ_ϕ} is finite if

$$\int_{\delta_{i0}}^{\infty} \left[\sum_{j=(\mathscr{F}+1)/2}^{\infty} \vartheta_i^{2j+1} p(\psi_i(G_{i(2j+1)}) > -x_i) + \sum_{j=1}^{(\mathscr{F}-1)/2} \vartheta_i^{2j+1} p(\tilde{\psi}_i(G_{i(2j+1)}) \le -x_i)\right] \gamma_i(dx_i), \quad (15)$$

$\int_{-\infty}^{\delta_{i0}} \sum_{j=1}^{\infty} \vartheta_i^{2j} p(\tilde{\psi}_i(G_{i(2j)}) < -x_i)\gamma_i(dx_i)$, are finite.

Proof. The hypotheses X_i and δ_{i0} are valued in $2I$ or $(2I + 1)$ then $X_i + W(t)$ is greater than $\phi_i(t)$ until the first meeting; also if X_i is smaller than δ_{i0} then $X_i + W(t)$ is smaller than $\phi_i(t)$ until the first meeting. Since $\tau_{\phi_i} > t$, $i = 1, 2, \dots, n$, are mutually exclusive events then $p(\tau_{\phi_i} > t) = p(\tau_{\phi_1} > t$ or $\tau_{\phi_2} > t$ or $\cdots$ or $\tau_{\phi_n} > t) = \sum_{i=1}^{n} p(\tau_{\phi_i} > t)$, and, for any $j \ge 0$ we have

$$p(\tau_{\phi_i} > G_{i(2j+1)}) \le \int_{-\infty}^{\delta_{i0}} p(X_i + W(G_{i(2j)}) < H_{i(2j)} \mid X_i = x_i)\gamma_i(dx_i) + \int_{\delta_{i0}}^{\infty} p(X_i + W(G_{i(2j+1)}) > H_{i(2j+1)} \mid X_i = x_i)\gamma_i(dx_i). \quad (16)$$

Using the notation $\tilde{\psi}_i(G_{i(2j)}) = W(G_{i(2j)}) + C_i G_{i(2j)}$, we obtain $W(G_{i(2j)}) - H_{i(2j)} < -X_i = x_i$; then $W(G_{i(2j)}) - (-1)^{2j+1}C_i[G_{i(2j)} + 1 + (-1)^{2j+1}] < -x_i$ leads to $W(G_{i(2j)}) - (-1)C_i[G_{i(2j)} + 1 + (-1)] = W(G_{i(2j)}) + C_i[G_{i(2j)}] =$

$\widetilde{\psi}_i(G_{i(2j)}) < -x_i$. Similarly, by using the notation $\psi_i(G_{i(2j+1)}) = W(G_{i(2j+1)}) - C_i G_{i(2j+1)}$, we get $\psi_i(G_{i(2j+1)}) > -x_i$. Consequently,

$$p\left(\tau_{\phi_i} > G_{i(2j+1)}\right) \leq \int_{-\infty}^{\delta_{i0}} p\left(\widetilde{\psi}_i\left(G_{i(2j)}\right) < -x_i\right)\gamma_i\left(dx_i\right) + \int_{\delta_{i0}}^{\infty} p\left(\psi_i\left(G_{i(2j+1)}\right) > -x_i\right)\gamma_i\left(dx_i\right), \quad (17)$$

also,

$$p\left(\tau_{\phi_i} > G_{i(2j)}\right) \leq \int_{-\infty}^{\delta_{i0}} p\left(\widetilde{\psi}_i\left(G_{i(2j)}\right) < -x_i\right)\gamma_i\left(dx_i\right) + \int_{\delta_{i0}}^{\infty} p\left(\psi_i\left(G_{i(2j-1)}\right) > -x_i\right)\gamma_i\left(dx_i\right). \quad (18)$$

From Remark 1, we obtain

$$E_{\tau_\phi} = \int_0^{\infty} p\left(\tau_\phi > t\right)dt \leq \sum_{j=0}^{\infty}\int_{G_{1j}}^{G_{1(j+1)}} p\left(\tau_{\phi_1} > t\right)dt + \cdots + \sum_{j=0}^{\infty}\int_{G_{nj}}^{G_{n(j+1)}} p\left(\tau_{\phi_n} > t\right)dt, \quad (19)$$

where $G_{i0} = 0$; then

$$\begin{aligned}E_{\tau_\phi} &\leq \sum_{j=0}^{\infty}\int_{G_{1j}}^{G_{1(j+1)}} p\left(\tau_{\phi_1} > G_{1j}\right)dt + \cdots + \sum_{j=0}^{\infty}\int_{G_{nj}}^{G_{n(j+1)}} p\left(\tau_{\phi_n} > t\right)dt \\
&= \sum_{i=1}^{n}\sum_{j=0}^{\infty}\left(G_{i(j+1)} - G_{ij}\right)p\left(\tau_{\phi_i} > G_{ij}\right) \\
&= \sum_{i=1}^{n}\sum_{j=0}^{\infty}\left[\lambda_i\left(\vartheta_i^{j+1} - 1\right) - \lambda_i\left(\vartheta_i^{j} - 1\right)\right]p\left(\tau_{\phi_i} > G_{ij}\right) \\
&= \sum_{i=1}^{n}\sum_{j=0}^{\infty}\lambda_i\vartheta_i^{j}\left(\vartheta_i - 1\right)p\left(\tau_{\phi_i} > G_{ij}\right) \\
&= \sum_{i=1}^{n}\left[\lambda_i\left(\vartheta_i - 1\right)\sum_{j=0}^{\infty}\vartheta_i^{j}p\left(\tau_{\phi_i} > G_{ij}\right)\right] \\
&= \sum_{i=1}^{n}\left[\lambda_i\left(\vartheta_i - 1\right)\times\left(p\left(\tau_{\phi_i} > D\right) + \vartheta_i p\left(\tau_{\phi_i} > G_{i1}\right) + \vartheta_i^2 p\left(\tau_{\phi_i} > G_{i2}\right) + \vartheta_i^3 p\left(\tau_{\phi_i} > G_{i3}\right) + \cdots\right)\right].\end{aligned} \quad (20)$$

If $\delta_{i0} < 0$, then we get

$$\begin{aligned}E_{\tau_\phi} \leq \sum_{i=1}^{n}\Bigg[\lambda_i\left(\vartheta_i - 1\right) \times \Bigg(& p\left(\tau_{\phi_i} > 0\right) + \vartheta_i p\left(\tau_{\phi_i} > G_{i1}\right) \\
&+ \vartheta_i^2\left\{\int_{\delta_{i0}}^{\infty} p\left(\psi_i\left(G_{i1}\right) > -x_i\right)\gamma_i\left(dx_i\right) + \int_{-\infty}^{\delta_{i0}} p\left(\widetilde{\psi}_i\left(G_{i2}\right) < -x_i\right)\gamma_i\left(dx_i\right)\right\} \\
&+ \vartheta_i^3\left\{\int_{\delta_{i0}}^{\infty} p\left(\psi_i\left(G_{i3}\right) > -x_i\right)\gamma_i\left(dx_i\right) + \int_{-\infty}^{\delta_{i0}} p\left(\widetilde{\psi}_i\left(G_{i2}\right) < -x_i\right)\gamma_i\left(dx_i\right)\right\} \\
&+ \vartheta_i^4\left\{\int_{\delta_{i0}}^{\infty} p\left(\psi_i\left(G_{i3}\right) > -x_i\right)\gamma_i\left(dx_i\right) + \int_{-\infty}^{\delta_{i0}} p\left(\widetilde{\psi}_i\left(G_{i4}\right) < -x_i\right)\gamma_i\left(dx_i\right)\right\} \\
&+ \cdots \\
&+ \vartheta_i^{\mathcal{F}}\left\{\int_{\delta_{i0}}^{\infty} p\left(\psi_i\left(G_{i(\mathcal{F}-1)}\right) > -x_i\right)\gamma_i\left(dx_i\right) + \int_{-\infty}^{\delta_{i0}} p\left(\widetilde{\psi}_i\left(G_{i\mathcal{F}}\right) < -x_i\right)\gamma_i\left(dx_i\right)\right\} \\
&+ \vartheta_i^{\mathcal{F}+1}\left\{\int_{\delta_{i0}}^{\infty} p\left(\psi_i\left(G_{i(\mathcal{F}+1)}\right) > -x_i\right)\gamma_i\left(dx_i\right) + \int_{-\infty}^{\delta_{i0}} p\left(\widetilde{\psi}_i\left(G_{i\mathcal{F}}\right) < -x_i\right)\gamma_i\left(dx_i\right)\right\} \\
&+ \vartheta_i^{\mathcal{F}+2}\left\{\int_{\delta_{i0}}^{\infty} p\left(\psi_i\left(G_{i(\mathcal{F}+1)}\right) > -x_i\right)\gamma_i\left(dx_i\right) + \int_{-\infty}^{\delta_{i0}} p\left(\widetilde{\psi}_i\left(G_{i(\mathcal{F}+2)}\right) < -x_i\right)\gamma_i\left(dx_i\right)\right\} \\
&+ \vartheta_i^{\mathcal{F}+3}\left\{\int_{\delta_{i0}}^{\infty} p\left(\psi_i\left(G_{i(\mathcal{F}+3)}\right) > -x_i\right)\gamma_i\left(dx_i\right) + \int_{-\infty}^{\delta_{i0}} p\left(\widetilde{\psi}_i\left(G_{i(\mathcal{F}+2)}\right) < -x_i\right)\gamma_i\left(dx_i\right)\right\} \\
&+ \cdots\Bigg)\Bigg];\end{aligned} \quad (21)$$

thus,

$$E_{\tau_\phi} \le \sum_{i=1}^{n} \Big[\lambda_i (\vartheta_i - 1) \times \Big(p(\tau_{\phi_i} > 0) + \vartheta_i p(\tau_{\phi_i} > G_{i1}) + \vartheta_i^2 (\vartheta_i + 1) \times \int_{-\infty}^{\delta_{i0}} p(\widetilde{\psi}_i (G_{i2}) < -x_i) \gamma_i (dx_i) + \cdots + \vartheta_i^{\mathcal{F}} (\vartheta_i + 1) \times \int_{-\infty}^{\delta_{i0}} p(\widetilde{\psi}_i (G_{i\mathcal{F}}) < -x_i) \gamma_i (dx_i) + \cdots + \vartheta_i^3 (\vartheta_i + 1) \int_{\delta_{i0}}^{\infty} p(\psi_i (G_{i3}) > -x_i) \gamma_i (dx_i) + \vartheta_i^5 (\vartheta_i + 1) \times \int_{\delta_{i0}}^{\infty} p(\psi_i (G_{i5}) > -x_i) \gamma_i (dx_i) + \cdots + \vartheta_i^{\mathcal{F}+3} (\vartheta_i + 1) \times \int_{\delta_{i0}}^{\infty} p(\psi_i (G_{i(\mathcal{F}+3)}) > -x_i) \gamma_i (dx_i) + \cdots \Big) \Big]. \quad (22)$$

Leads to

$$E_{\tau_\phi} \le \sum_{i=1}^{n} \Big[\lambda_i (\vartheta_i - 1) \times \Big[{}^1\widehat{g}_i + (\vartheta_i + 1) \times \Big(\int_{-\infty}^{\delta_{i0}} {}^1\widehat{w}_i (x_i) \gamma_i (dx_i) + \int_{\delta_{i0}}^{\infty} {}^1\widehat{q}_i (x_i) \gamma_i (dx_i) + \int_{\delta_{i0}}^{\infty} {}^1\widehat{h}_i (x_i) \gamma_i (dx_i) \Big) \Big] \Big]$$
$$= \sum_{i=1}^{n} \lambda_i (\vartheta_i - 1) ({}^1\widehat{g}_i) + \sum_{i=1}^{n} \Big[\lambda_i (\vartheta_i - 1)(\vartheta_i + 1) \times \Big(\int_{-\infty}^{\delta_{i0}} {}^1\widehat{w}_i (x_i) \gamma_i (dx_i) + \int_{\delta_{i0}}^{\infty} {}^1\widehat{q}_i (x_i) \gamma_i (dx_i) + \int_{\delta_{i0}}^{\infty} {}^1\widehat{h}_i (x_i) \gamma_i (dx_i) \Big) \Big], \quad (23)$$

where

$${}^1\widehat{g}_i = p(\tau_{\phi_i} > 0) + \vartheta_i p(\tau_{\phi_i} > G_{i1}) + \vartheta_i^2 \int_{\delta_{i0}}^{\infty} p(\psi_i (G_{i1}) > -x_i) \gamma_i (dx_i),$$
$${}^1\widehat{w}_i (x_i) = \sum_{j=1}^{\infty} \vartheta_i^{2j} p(\widetilde{\psi}_i (G_{i(2j)}) < -x_i), \quad (24)$$
$${}^1\widehat{q}_i (x_i) = \sum_{j=1}^{(\mathcal{F}-1)/2} \vartheta_i^{2j+1} p(\widetilde{\psi}_i (G_{i(2j+1)}) \le -x_i),$$
$${}^1\widehat{h}_i (x_i) = \sum_{j=(\mathcal{F}+1)/2}^{\infty} \vartheta_i^{2j+1} p(\psi_i (G_{i(2j+1)}) > -x_i).$$

Then E_{τ_ϕ} is finite if

$$\int_{\delta_{i0}}^{\infty} \Big[\sum_{j=(\mathcal{F}+1)/2}^{\infty} \vartheta_i^{2j+1} p(\psi_i (G_{i(2j+1)}) > -x_i) + \sum_{j=1}^{(\mathcal{F}-1)/2} \vartheta_i^{2j+1} p(\widetilde{\psi}_i (G_{i(2j+1)}) \le -x_i) \Big] \gamma_i (dx_i), \quad (25)$$

$\int_{-\infty}^{\delta_{i0}} \sum_{j=1}^{\infty} \vartheta_i^{2j} p(\widetilde{\psi}_i(G_{i(2j)}) < -x_i)\gamma_i(dx_i)$, are finite.

By similar way if $\delta_{i0} > 0$, then E_{τ_ϕ} is finite if

$$E_{\tau_\phi} \le \sum_{i=1}^{n} \Big[\lambda_i (\vartheta_i - 1) ({}^2\widehat{g}_i) + \lambda_i (\vartheta_i - 1)(\vartheta_i + 1) \times \Big(\int_{-\infty}^{\delta_{i0}} {}^2\widehat{w}_i (x_i) \gamma_i (dx_i) + \int_{-\infty}^{\delta_{i0}} {}^2\widehat{q}_i (x_i) \gamma_i (dx_i) + \int_{\delta_{i0}}^{\infty} {}^2\widehat{h}_i (x_i) \gamma_i (dx_i) \Big) \Big], \quad (26)$$

where

$$
\begin{aligned}
{}^2\widehat{g}_i &= p\left(\tau_{\phi_1} > 0\right) + \vartheta_i p\left(\tau_{\phi_i} > G_{i1}\right) \\
&\quad + \vartheta_i^2 \int_{\delta_{i0}}^{\infty} p\left(\psi_i\left(G_{i1}\right) > -x_i\right) \gamma_i\left(dx_i\right), \\
{}^2\widehat{\varpi}_i\left(x_i\right) &= \sum_{j=1}^{\mathcal{F}/2} \vartheta_i^{2j} p\left(\psi_i\left(G_{i(2j)}\right) < -x_i\right), \\
{}^2\widehat{q}_i\left(x_i\right) &= \sum_{j=\mathcal{F}/2+1}^{\infty} \vartheta_i^{2j} p\left(\widetilde{\psi}_i\left(G_{i(2j)}\right) \le -x_i\right), \\
{}^2\widehat{h}_i\left(x_i\right) &= \sum_{j=1}^{\infty} \vartheta_i^{2j+1} p\left(\psi_i\left(G_{i(2j+1)}\right) > -x_i\right).
\end{aligned} \tag{27}
$$

And E_{τ_ϕ} is finite if

$$
\begin{aligned}
&\int_{-\infty}^{\delta_{i0}} \left[\sum_{j=1}^{\mathcal{F}/2} \vartheta_i^{2j} p\left(\widetilde{\psi}_i\left(G_{i(2j)}\right) < -x_i\right) \right. \\
&\quad \left. + \sum_{j=\mathcal{F}/2+1}^{\infty} \vartheta_i^{2j} p\left(\widetilde{\psi}_i\left(G_{i(2j)}\right) \le -x_i\right) \right] \gamma_i\left(dx_i\right),
\end{aligned} \tag{28}
$$

$\int_{\delta_{i0}}^{\infty} \sum_{j=1}^{\infty} \vartheta_i^{2j+1} p(\psi_i(G_{i(2j+1)}) > -x_i)$, are finite. □

Lemma 5 (see El-Rayes et al. [9]). *If $a_{\overline{m}} \ge 0$, $a_{\overline{m}+1} \le a_{\overline{m}}$, $\overline{m} = 1, 2, \ldots$, and $\{\overline{d}_{\overline{m}}\}$, $\overline{m} = 1, 2, \ldots$, are a strictly increasing sequence of integer numbers with $\overline{d}_0 = 0$, then, for any $\widetilde{k} = 1, 2, \ldots$,*

$$
\begin{aligned}
\sum_{\overline{m}=\widetilde{k}}^{\infty} \left[\overline{d}_{\overline{m}+1} - \overline{d}_{\overline{m}}\right] a_{\overline{d}_{\overline{m}+1}} &\le \sum_{\overline{m}=\overline{d}_{\widetilde{k}}}^{\infty} a_{\overline{m}} \\
&\le \sum_{\overline{m}=\widetilde{k}}^{\infty} \left[\overline{d}_{\overline{m}+1} - \overline{d}_{\overline{m}}\right] a_{\overline{d}_{\overline{m}}},
\end{aligned} \tag{29}
$$

where $\sum_{\overline{m}=\widetilde{k}}^{\infty}[\overline{d}_{\overline{m}+1} - \overline{d}_{\overline{m}}]a_{\overline{d}_{\overline{m}+1}}$, $\sum_{\overline{m}=\overline{d}_{\widetilde{k}}}^{\infty} a_{\overline{m}}$, and $\sum_{\overline{m}=\widetilde{k}}^{\infty}[\overline{d}_{\overline{m}+1} - \overline{d}_{\overline{m}}]a_{\overline{d}_{\overline{m}}}$ are vectors in formulas and they are taken to be row vectors. These vectors are defined as follows:

$$
\begin{aligned}
&\sum_{\overline{m}=\widetilde{k}}^{\infty} \left[\overline{d}_{\overline{m}+1} - \overline{d}_{\overline{m}}\right] a_{\overline{d}_{\overline{m}+1}} \\
&= \left[\sum_{\overline{m}=\widetilde{k}}^{\infty} \left[\overline{d}_{1(\overline{m}+1)} - \overline{d}_{1\overline{m}}\right] a_{1\overline{d}_{1(\overline{m}+1)}}, \right. \\
&\quad \sum_{\overline{m}=\widetilde{k}}^{\infty} \left[\overline{d}_{2(\overline{m}+1)} - \overline{d}_{2\overline{m}}\right] a_{2\overline{d}_{2(\overline{m}+1)}}, \ldots, \\
&\quad \left. \sum_{\overline{m}=\widetilde{k}}^{\infty} \left[\overline{d}_{n(\overline{m}+1)} - \overline{d}_{n\overline{m}}\right] a_{n\overline{d}_{n(\overline{m}+1)}} \right],
\end{aligned}
$$

$$
\begin{aligned}
&\sum_{\overline{m}=\overline{d}_{\widetilde{k}}}^{\infty} a_{\overline{m}} = \left[\sum_{\overline{m}=\overline{d}_{1\widetilde{k}}}^{\infty} a_{1\overline{m}}, \sum_{\overline{m}=\overline{d}_{2\widetilde{k}}}^{\infty} a_{2\overline{m}}, \ldots, \sum_{\overline{m}=\overline{d}_{n\widetilde{k}}}^{\infty} a_{n\overline{m}} \right], \\
&\sum_{\overline{m}=\widetilde{k}}^{\infty} \left[\overline{d}_{\overline{m}+1} - \overline{d}_{\overline{m}}\right] a_{\overline{d}_{\overline{m}}} \\
&= \left[\sum_{\overline{m}=\widetilde{k}}^{\infty} \left[\overline{d}_{1(\overline{m}+1)} - \overline{d}_{1\overline{m}}\right] a_{1\overline{d}_{1\overline{m}}}, \right. \\
&\quad \sum_{\overline{m}=\widetilde{k}}^{\infty} \left[\overline{d}_{2(\overline{m}+1)} - \overline{d}_{2\overline{m}}\right] a_{2\overline{d}_{2\overline{m}}}, \ldots, \\
&\quad \left. \sum_{\overline{m}=\widetilde{k}}^{\infty} \left[\overline{d}_{n(\overline{m}+1)} - \overline{d}_{n\overline{m}}\right] a_{n\overline{d}_{n\overline{m}}} \right].
\end{aligned} \tag{30}
$$

Now, we will discuss under what conditions of the chosen search plan should be satisfied to make the previous integrals in Theorem 4 are indeed finite. This is a crucial issue related to the existence of a finite search plan.

Theorem 6. *The chosen search plan should satisfy ${}^2\widehat{h}(x) \le L(|x|)$, ${}^2\widehat{q}(x) \le \widetilde{L}(|x|)$, if $\delta_{i0} > 0$ and ${}^1\widehat{\varpi}(x) \le M(|x|)$, ${}^1\widehat{h}(x) \le \widetilde{M}(|x|)$ if $\delta_{i0} < 0$, where $L(|x|)$, $\widetilde{L}(|x|)$, $M(|x|)$, and $\widetilde{M}(|x|)$ are vectors of linear functions given by ${}^1\widehat{\varpi}(x) = [{}^1\widehat{\varpi}_1(x_1), {}^1\widehat{\varpi}_2(x_2), \ldots, {}^1\widehat{\varpi}_n(x_n)]$, ${}^1\widehat{h}(x) = [{}^1\widehat{h}_1(x_1), {}^1\widehat{h}_2(x_2), \ldots, {}^1\widehat{h}_n(x_n)]$, ${}^2\widehat{h}(x) = [{}^2\widehat{h}_1(x_1), {}^2\widehat{h}_2(x_2), \ldots, {}^2\widehat{h}_n(x_n)]$, ${}^2\widehat{q}(x) = [{}^2\widehat{q}_1(x_1), {}^2\widehat{q}_2(x_2), \ldots, {}^2\widehat{q}_n(x_n)]$, $L(|x|) = [L_1(|x_1|), L_2(|x_2|), \ldots, L_n(|x_n|)]$, $\widetilde{L}(|x|) = [\widetilde{L}_1(|x_1|), \widetilde{L}_2(|x_2|), \ldots, \widetilde{L}_n(|x_n|)]$, $M(|x|) = [M_1(|x_1|), M_2(|x_2|), \ldots, M_n(|x_n|)]$ and $\widetilde{M}(|x|) = [\widetilde{M}_1(|x_1|), \widetilde{M}_2(|x_2|), \ldots, \widetilde{M}_n(|x_n|)]$.*

Proof. We will prove this theorem for ${}^2\widehat{h}(x)$ when $\delta_{i0} > 0$, where ${}^2\widehat{h}(x) = [{}^2\widehat{h}_1(x_1), {}^2\widehat{h}_2(x_2), \ldots, {}^2\widehat{h}_n(x_n)] = [\sum_{j=1}^{\infty} \vartheta_1^{2j+1} p(\psi_1(G_{1(2j+1)}) > -x_1), \sum_{j=1}^{\infty} \vartheta_2^{2j+1} p(\psi_2(G_{2(2j+1)}) > -x_2), \ldots, \sum_{j=1}^{\infty} \vartheta_n^{2j+1} p(\psi_n(G_{n(2j+1)}) > -x_n)]$, and we obtain the following cases.

(I) If $x_i > \delta_{i0}$, then we have $[{}^2\widehat{h}_1(x_1), {}^2\widehat{h}_2(x_2), \ldots, {}^2\widehat{h}_n(x_n)] = [{}^2\widehat{h}_1(\delta_{10}), {}^2\widehat{h}_2(\delta_{20}), \ldots, {}^2\widehat{h}_n(\delta_{n0})] + [\sum_{j=1}^{\infty} \vartheta_1^{2j+1} p(-x_1 < \psi_1(G_{1(2j+1)}) \le -\delta_{10}), \sum_{j=1}^{\infty} \vartheta_2^{2j+1} p(-x_2 < \psi_2(G_{2(2j+1)}) \le -\delta_{20}), \ldots, \sum_{j=1}^{\infty} \vartheta_n^{2j+1} p(-x_n < \psi_n(G_{n(2j+1)}) \le -\delta_{n0})]$.

(II) If $0 \le x_i \le \delta_{i0}$, then we have $[{}^2\widehat{h}_1(x_1), {}^2\widehat{h}_2(x_2), \ldots, {}^2\widehat{h}_n(x_n)] = [{}^2\widehat{h}_1(0), {}^2\widehat{h}_2(0), \ldots, {}^2\widehat{h}_n(0)] + [\sum_{j=1}^{\infty} \vartheta_1^{2j+1} p(-x_1 < \psi_1(G_{1(2j+1)}) \le 0), \sum_{j=1}^{\infty} \vartheta_2^{2j+1} p(-x_2 < \psi_2(G_{2(2j+1)}) \le 0), \ldots, \sum_{j=1}^{\infty} \vartheta_n^{2j+1} p(-x_n < \psi_n(G_{n(2j+1)}) \le 0)]$.

(III) If $x_i \leq \delta_{i0}$, then we get $[{}^2\widehat{h}_1(x_1), {}^2\widehat{h}_2(x_2), \ldots, {}^2\widehat{h}_n(x_n)] = [{}^2\widehat{h}_1(0), {}^2\widehat{h}_2(0), \ldots, {}^2\widehat{h}_n(0)] - [\sum_{j=1}^{\infty} \vartheta_1^{2j+1} p(0 < \psi_1(G_{1(2j+1)}) \leq -x_1), \sum_{j=1}^{\infty} \vartheta_2^{2j+1} p(0 < \psi_2(G_{2(2j+1)}) \leq -x_2), \ldots, \sum_{j=1}^{\infty} \vartheta_n^{2j+1} p(0 < \psi_n(G_{n(2j+1)}) \leq -x_n)]$.

From (II) we have $[{}^2\widehat{h}_1(x_1), {}^2\widehat{h}_2(x_2), \ldots, {}^2\widehat{h}_n(x_n)] \leq [{}^2\widehat{h}_1(0), {}^2\widehat{h}_2(0), \ldots, {}^2\widehat{h}_n(0)]$, and for $x_i \geq \delta_{i0}$ $[{}^2\widehat{h}_1(x_1), {}^2\widehat{h}_2(x_2), \ldots, {}^2\widehat{h}_n(x_n)] = [{}^2\widehat{h}_1(\delta_{10}), {}^2\widehat{h}_2(\delta_{20}), \ldots, {}^2\widehat{h}_n(\delta_{n0})] + [\sum_{j=1}^{\infty} \vartheta_1^{2j+1} p(-x_1 < \psi_1(G_{1(2j+1)}) \leq -\delta_{10}), \sum_{j=1}^{\infty} \vartheta_2^{2j+1} p(-x_2 < \psi_2(G_{2(2j+1)}) \leq -\delta_{20}), \ldots, \sum_{j=1}^{\infty} \vartheta_n^{2j+1} p(-x_n < \psi_n(G_{n(2j+1)}) \leq -\delta_{n0})]$, but $[{}^2\widehat{h}_1(x_1), {}^2\widehat{h}_2(x_2), \ldots, {}^2\widehat{h}_n(x_n)] = [{}^2\widehat{h}_1(0), {}^2\widehat{h}_2(0), \ldots, {}^2\widehat{h}_n(0)] + [\sum_{j=1}^{\infty} \vartheta_1^{2j+1} p(-x_1 < \psi_1(G_{1(2j+1)}) \leq 0), \sum_{j=1}^{\infty} \vartheta_2^{2j+1} p(-x_2 < \psi_2(G_{2(2j+1)}) \leq 0), \ldots, \sum_{j=1}^{\infty} \vartheta_n^{2j+1} p(-x_n < \psi_n(G_{n(2j+1)}) \leq 0)]$. Consequently, from Theorem 3, we get $[{}^2\widehat{h}_1(0), {}^2\widehat{h}_2(0), \ldots, {}^2\widehat{h}_n(0)] = [\sum_{j=1}^{\infty} \vartheta_1^{2j+1} p(\psi_1(G_{1(2j+1)}) > 0), \sum_{j=1}^{\infty} \vartheta_2^{2j+1} p(\psi_2(G_{2(2j+1)}) > 0), \ldots, \sum_{j=1}^{\infty} \vartheta_n^{2j+1} p(\psi_n(G_{n(2j+1)}) > 0)] \leq [\sum_{j=1}^{\infty} \vartheta_1^{2j+1} \varepsilon_1^{G_{1(2j+1)}}, \sum_{j=1}^{\infty} \vartheta_2^{2j+1} \varepsilon_2^{G_{2(2j+1)}}, \ldots, \sum_{j=1}^{\infty} \vartheta_n^{2j+1} \varepsilon_n^{G_{n(2j+1)}}] \leq [\vartheta_1^3/(\vartheta_1^2 - 1), \vartheta_2^3/(\vartheta_2^2 - 1), \ldots, \vartheta_n^3/(\vartheta_n^2 - 1)]$, $0 < \varepsilon_i < 1$.

Suppose the following holds:

(i) $\overline{d}_{i\xi_i} = G_{i(2\xi_i+1)}/\sigma_i = \mu_i(\vartheta_i^{2\xi_i+1} - 1)$,

(ii) $\mu_i(\xi_i) = \psi_i(\xi_i\sigma_i)/2 = \sum_{j=1}^{\xi_i} X_{ij}$, where $\{X_{ij}\}$, $j = 1, 2, \ldots$, is a sequence of independent identically distributed random variables,

(iii) $a_i(\xi_i) = p(-x_i/2 < \mu_i(\xi_i) \leq 0) = \sum_{\mathscr{F}=0}^{|x_i|/2} p(-(\mathscr{F}+1) < \mu_i(\xi_i) \leq -\mathscr{F})$,

(iv) $\widetilde{m}$ is an integer such that $\overline{d}_{i\widetilde{m}} = r_{i1}|x_i| + r_{i2}$,

(v) $\widetilde{\alpha}_i = \vartheta_i^2/\mu_i(\vartheta_i - 1)$,

(vi) $U_i(\mathscr{F}, \mathscr{F}+1) = \sum_{\xi_i=0}^{\infty} p(-(\mathscr{F}+1) < \mu_i(\xi_i) \leq -\mathscr{F})$.

Thus, from Theorem 2, we have $a_i(\xi_i)$ is nonincreasing if $\xi_i > \overline{d}_{i\xi_i}$. By applying Lemma 5, we have $[{}^2\widehat{h}_1(x_1), {}^2\widehat{h}_2(x_2), \ldots, {}^2\widehat{h}_n(x_n)] - [{}^2\widehat{h}_1(0), {}^2\widehat{h}_2(0), \ldots, {}^2\widehat{h}_n(0)] = [\sum_{j=1}^{\infty} \vartheta_1^{2j+1} p(-x_1 < \psi_1(G_{1(2j+1)}) \leq 0), \sum_{j=1}^{\infty} \vartheta_2^{2j+1} p(-x_2 < \psi_2(G_{2(2j+1)}) \leq 0), \ldots, \sum_{j=1}^{\infty} \vartheta_n^{2j+1} p(-x_n < \psi_n(G_{n(2j+1)}) \leq 0)] = [\sum_{\xi_1=1}^{\widetilde{m}} \vartheta_1^{2\xi_1+1} a_1(\overline{d}_{1\xi_1}), \sum_{\xi_2=1}^{\widetilde{m}} \vartheta_2^{2\xi_2+1} a_2(\overline{d}_{2\xi_2}), \ldots, \sum_{\xi_n=1}^{\widetilde{m}} \vartheta_n^{2\xi_n+1} a_n(\overline{d}_{n\xi_n})] + [\sum_{\xi_1=\widetilde{m}+1}^{\infty} \vartheta_1^{2\xi_1+1} a_1(\overline{d}_{1\xi_1}), \sum_{\xi_2=\widetilde{m}+1}^{\infty} \vartheta_2^{2\xi_2+1} a_2(\overline{d}_{2\xi_2}), \ldots, \sum_{\xi_n=\widetilde{m}+1}^{\infty} \vartheta_n^{2\xi_n+1} a_n(\overline{d}_{n\xi_n})] \leq [\sum_{\xi_1=1}^{\widetilde{m}} \vartheta_1^{2\xi_1+1}, \sum_{\xi_2=1}^{\widetilde{m}} \vartheta_2^{2\xi_2+1}, \ldots, \sum_{\xi_n=1}^{\widetilde{m}} \vartheta_n^{2\xi_n+1}] + [\widetilde{\alpha}_1 \sum_{\xi_1=\widetilde{m}+1}^{\infty} (\overline{d}_{1\xi_1} - \overline{d}_{1(\xi_1-1)}) a_1(\overline{d}_{1\xi_1}), \widetilde{\alpha}_2 \sum_{\xi_2=\widetilde{m}+1}^{\infty} (\overline{d}_{2\xi_2} - \overline{d}_{2(\xi_2-1)}) a_2(\overline{d}_{2\xi_2}), \ldots, \widetilde{\alpha}_n \sum_{\xi_n=\widetilde{m}+1}^{\infty} (\overline{d}_{n\xi_n} - \overline{d}_{n(\xi_n-1)}) a_n(\overline{d}_{n\xi_n})] \leq [\sum_{\xi_1=1}^{\widetilde{m}} \vartheta_1^{2\xi_1+1}, \sum_{\xi_2=1}^{\widetilde{m}} \vartheta_2^{2\xi_2+1}, \ldots, \sum_{\xi_n=1}^{\widetilde{m}} \vartheta_n^{2\xi_n+1}] + [\widetilde{\alpha}_1 \sum_{\xi_1=\overline{d}_{1\widetilde{m}}}^{\infty} a_1(\xi_1), \widetilde{\alpha}_2 \sum_{\xi_2=\overline{d}_{2\widetilde{m}}}^{\infty} a_2(\xi_2), \ldots, \widetilde{\alpha}_n \sum_{\xi_n=\overline{d}_{n\widetilde{m}}}^{\infty} a_n(\xi_n)] \leq [\sum_{\xi_1=1}^{\widetilde{m}} \vartheta_1^{2\xi_1+1}, \sum_{\xi_2=1}^{\widetilde{m}} \vartheta_2^{2\xi_2+1}, \ldots, \sum_{\xi_n=1}^{\widetilde{m}} \vartheta_n^{2\xi_n+1}] + [\widetilde{\alpha}_1 \sum_{\mathscr{F}=0}^{|x_1|/2} U_1(\mathscr{F}, \mathscr{F}+1), \widetilde{\alpha}_2 \sum_{\mathscr{F}=0}^{|x_2|/2} U_2(\mathscr{F}, \mathscr{F}+1), \ldots, \widetilde{\alpha}_n \sum_{\mathscr{F}=0}^{|x_n|/2} U_n(\mathscr{F}, \mathscr{F}+1)]$.

Since $U_i(\mathscr{F}, \mathscr{F}+1)$ satisfies the conditions of renewal theorem as in Feller [28], then $U_i(\mathscr{F}, \mathscr{F}+1)$ is bounded for all $\mathscr{F}$, $i = 1, 2, \ldots, n$, by a constant so ${}^2\widehat{h}(x) = [q_1(x_1), q_2(x_2), \ldots, q_n(x_n)] \leq [{}^2\widehat{h}_1(\delta_{10}), {}^2\widehat{h}_2(\delta_{20}), \ldots, {}^2\widehat{h}_n(\delta_{n0})] + [{}^2\Lambda_{11}, {}^2\Lambda_{21}, \ldots, {}^2\Lambda_{n1}] + [{}^2\Lambda_{12}|x_1|, {}^2\Lambda_{22}|x_2|, \ldots, {}^2\Lambda_{n2}|x_n|] = [L_1(|x_1|), L_2(|x_2|), \ldots, L_n(|x_n|)] = L(|x|)$, then, ${}^2\widehat{h}(x) \leq L(|x|)$. Similar to the previous analysis way we can prove that ${}^2\widehat{q}(x) \leq \widetilde{L}(|x|)$ and also ${}^1\widehat{w}(x) \leq M(|x|)$, ${}^1\widehat{h}(x) \leq \widetilde{M}(|x|)$ *if* $\delta_{i0} < 0$. Therefore, the proof is completed. □

Theorem 7. *If there exists a finite search plan $\phi(t) = (\phi_1(t), \phi_2(t), \ldots, \phi_n(t)) \in \Phi(t)$, then $E(|X_0|)$ is finite.*

Proof. For $E_{\tau_{\phi_i}} < \infty$, $i = 1, 2, \ldots, n$, we have p (τ_ϕ is finite) $= 1$ and so p (one of τ_{ϕ_i} is finite, $i = 1, 2, \ldots, n$) $= 1$. Therefore,

$$p\left(\bigcup_{i=1}^{n}\left(\tau_{\phi_i} < \infty\right)\right) = \sum_{i=1}^{n} p\left(\tau_{\phi_i} < \infty\right) - \sum_{i,\beta} p\left(\left(\tau_{\phi_i} < \infty\right) \cap \left(\tau_{\phi_\beta} < \infty\right)\right) + \cdots + (-1)^n p\left(\bigcap_{i=1}^{n}\left(\tau_{\phi_i} < \infty\right)\right). \quad (31)$$

Since $\tau_{\phi_i} < \infty$, $i = 1, 2, \ldots, n$, are mutually exclusive events, then

$$\sum_{i=1}^{n} p\left(\tau_{\phi_i} < \infty\right) = 1. \quad (32)$$

Consequently, p (τ_{ϕ_i} is finite) $= 1$ or p (τ_{ϕ_θ} is finite) $= 1$ for all $i \neq \theta$, $i, \theta = 1, 2, \ldots, n$. If p (τ_{ϕ_i} is finite) $= 1$ then $X_0 = \phi_i(\tau_{\phi_i}) - W(\tau_{\phi_i})$ with probability one and $|X_0| \leq |\phi_i(\tau_{\phi_i})| + |W(\tau_{\phi_i})| \leq \tau_{\phi_i} + |W(\tau_{\phi_i})|$ that leads to $E(|X_0|) \leq E_{\tau_{\phi_i}} + E(|W(\tau_{\phi_i})|)$. But $W(\tau_{\phi_i}) < \tau_{\phi_i}$, then $E(|W(\tau_{\phi_i})|) < E_{\tau_{\phi_i}}$, and $E(|X_0|)$ is finite.

On the other hand, if p (τ_{ϕ_θ} is finite) $= 1$ then $X_0 = \phi_\theta(\tau_{\phi_\theta}) - W(\tau_{\phi_\theta})$ with probability one, and by the same manner we can get $E(|X_0|)$ is finite. □

Remark 8. A direct consequence of Theorems 4, 6, and 7 satisfies the existence of a finite search plan if and only if $E(|X_0|)$ is finite.

5. Existence of an Optimal Search Plan

The goal of the searching strategy could minimize the expected value of the first meeting time between one of the searchers and the target. Therefore, the main problem here is to find a search paths $\phi_i(t)$, $i = 1, 2, \ldots, m$. If such a search paths exists, we call it optimal search paths.

Definition 9. Let $\{\phi_{ih}(t)\}_{h\geq 1} \in \Phi_1(t)$, $i = 1, 2, \ldots, m$, be sequences of search plans; we say that $\phi_{ih}(t)$ converges to $\phi_i(t)$ as h tends to ∞ if and only if, for any $t \in I^+$, $\phi_{ih}(t)$ converges to $\phi_i(t)$ uniformly on every compact space (El-Rayes et al. [9]).

Theorem 10. *Let for any $t \in I^+$ and $W(t)$ be a one-dimensional random walk. Then, the mapping,*

$$(\phi_1(t), \phi_2(t), \ldots, \phi_n(t)) \longrightarrow E_{\tau_\phi} \in I^+, \quad (33)$$

is lower semicontinuous on $\Phi_1(t)$.

Proof. Let $B(\phi_i, t)$ be the indicator function of the set $\{\tau_{\phi_i} > t, i = 1, 2, \ldots, n\}$; by the Fatou-Lesbesque theorem, we get

$$\begin{aligned} E_{\tau_\phi} &= E\left[\sum_{t=0}^{\infty} B(\phi_i, t)\right] \\ &= E\left[\sum_{t=0}^{\infty} \lim_{h\to\infty} \inf B(\phi_{ih}(t), t)\right] \\ &\le \sum_{t=0}^{\infty} \lim_{h\to\infty} \inf E\left(1_{\{\tau_{\phi_h} > t\}}\right), \end{aligned} \quad (34)$$

for any sequences $\phi_{ih} \rightarrow \phi_i$ in $\Phi_1(t)$, where $\Phi_1(t)$ is sequentially compact; see El-Rayes et al. [9]. Thus the mapping $(\phi_1(t), \phi_2(t), \ldots, \phi_n(t)) \rightarrow E_{\tau_\phi}$ is lower semicontinuous mapping on $\Phi_1(t)$ and this completed the proof. □

6. Optimal Search Plan

In this section, we will find the optimal search plan that minimize the expected value of the first meeting time.

Since τ_ϕ depends on infimum either one of the functions $\phi_i(t)$, $i = 1, 2, \ldots, n$, then we need to minimize E_{τ_ϕ}. That happens when we obtain the optimal search plan $\phi^*(t) = \{\phi_1^*(t), \phi_2^*(t), \ldots, \phi_n^*(t)\}$, where $\phi_i(t)$ is a function on a distance H_{ij}, $i = 1, 2, \ldots, n$, $1 \le j \le \mathcal{F}/2$. Therefore, it can be seen that the vector $(\phi_1(t), \phi_2(t), \ldots, \phi_n(t))$ are n-distinct objective functions. Since the speed of the n-searchers (search team) is $v_i = 1$, $i = 1, 2, \ldots, n$, then the objective functions $\phi_i(t)$, $i = 1, 2, \ldots, n$, are distinct set of constrains given from Cases 1 and 2. In this problem, we consider the constraint $G_{i(2j-1)} \le t \le G_{i(2j)}$, $i = 1, 2, \ldots, n$, $1 \le j \le \mathcal{F}/2$. Then, the problem takes the form

$$\begin{aligned} \min \quad & (\phi_1(t), \phi_2(t), \ldots, \phi_n(t)) \\ \text{subject to} \quad & t \in T = \Big\{t \in I^+ \mid G_{i(2j-1)} \le t \le G_{i(2j)}, \\ & i = 1, 2, \ldots, n,\ 1 \le j \le \frac{\mathcal{F}}{2}\Big\}, \end{aligned} \quad (35)$$

where t is an n-dimensional vector of the decision variables, $\phi_1(t), \phi_2(t), \ldots, \phi_n(t)$ are n-distinct objective functions of the decision vector of inequality constraints, and T is the feasible set of constrained decisions.

But $v_i = 1$, $i = 1, 2, \ldots, n$, that leads to $t = H_{ij}$, $i = 1, 2, \ldots, n$, $1 \le j \le \mathcal{F}/2$; then the above problem is represented as the following multiobjective nonlinear programming problem (MONLP):

MONLP:

$$\begin{aligned} \min \quad & (\Xi_1(H_{1j}), \Xi_2(H_{2j}), \ldots, \Xi_n(H_{nj})), \\ \text{subject to} \quad & G_{i(2j-1)} \le H_{ij} \le G_{i(2j)}, \\ & H_{ij} \in I^+, \quad i = 1, 2, \ldots, n,\ 1 \le j \le \frac{\mathcal{F}}{2}, \\ \text{where} \quad & \Xi_i(H_{ij}) = H_{i(2j+1)} - \delta_{i0} - [t - G_{i(2j-1)}]. \end{aligned} \quad (36)$$

However, on formulating the MONLP which closely describes and represents the real decision situation, various factors of the real system should be reflected in the description of the objective functions and the constraints. Naturally, these objective functions and the constraints involve many variables and parameters whose possible values may be assigned by the experts. In the conventional approach, such parameters are changed according to the variables in an experimental and/or subjective manner through the experts understanding the nature of the variables.

Substituting $G_{ij} = b\varpi_i(\vartheta_i^j - 1)$ in the above MONLP we have

MONLP(I):

$$\begin{aligned} \min \quad & (\Xi_1(\vartheta_1; \varpi_1), \Xi_2(\vartheta_2; \varpi_2), \ldots, \Xi_n(\vartheta_n; \varpi_n)), \\ \text{subject to} \quad & b\varpi_i(\vartheta_i^{2j-1} - 1) \le H_{ij}, \quad H_{ij} \le b\varpi_i(\vartheta_i^{2j} - 1), \\ & H_{ij} \in I^+, \quad i = 1, 2, \ldots, n,\ 1 \le j \le \frac{\mathcal{F}}{2}, \\ \text{where} \quad & \Xi_i(H_{ij}) = H_{i(2j+1)} - \delta_{i0} \\ & - [t - b\varpi_i(\vartheta_i^{2j-1} - 1)], \end{aligned} \quad (37)$$

where the variable ϑ_i takes +ve numbers greater than 1 and the parameter ϖ_i is the least positive integer such that $\varpi_i(1 - (\vartheta_i - 1)/(\vartheta_i + 1))/2$ is positive number; $H_{ij} \ge 0$ for all $i = 1, 2, \ldots, n$, $1 \le j \le \mathcal{F}/2$. Furthermore, H_{ij} is a function of the variable ϑ_i and the parameter ϖ_i, where

$$\begin{aligned} H_{ij} &= (-1)^{j+1} C_i [G_{ij} + 1 + (-1)^{j+1}] + \delta_{i0} \\ &= (-1)^{j+1} \left(\frac{\vartheta_i - 1}{\vartheta_i + 1}\right) [b\varpi_i(\theta_i^j - 1) + 1 + (-1)^{j+1}] \\ &+ \delta_{i0}. \end{aligned} \quad (38)$$

Then, the previous MONLP(I) takes the form

MONLP(II):

$$\begin{aligned}
\min\quad & \left(\Xi_1(\vartheta_1;\varpi_1),\Xi_2(\vartheta_2;\varpi_2),\ldots,\Xi_n(\vartheta_n;\varpi_n)\right),\\
\text{subject to}\quad & b\varpi_i\left(\vartheta_i^{2j-1}-1\right)\\
&\le (h_1-1)\left(\frac{\vartheta_i-1}{\vartheta_i+1}\right)\\
&\quad\times\left[b\varpi_i\left(\vartheta_i^{j}-1\right)+h_1\right]+\delta_{i0},\\
& b\varpi_i\left(\vartheta_i^{2j}-1\right)\ge (h_1-1)\left(\frac{\vartheta_i-1}{\vartheta_i+1}\right)\\
&\quad\times\left[b\varpi_i\left(\vartheta_i^{j}-1\right)+h_1\right]\\
&\quad+\delta_{i0},
\end{aligned}$$

$$\begin{aligned}
\text{where}\quad \Xi_i(\vartheta_i;\varpi_i) &= (h_3)\left(\frac{\vartheta_i-1}{\vartheta_i+1}\right)\\
&\quad\times\left[b\varpi_i\left(\vartheta_i^{2j-1}-1\right)+h_4\right]\\
&\quad-2\delta_{i0}+(h_2)\left(\frac{\vartheta_i-1}{\vartheta_i+1}\right)\\
&\quad\times\left[b\varpi_i\left(\vartheta_i^{j}-1\right)+h_5\right]\\
&\quad+b\varpi_i\left(\vartheta_i^{2j-1}-1\right),\\
h_1 = 1+(-1)^{j+1},\quad & h_2=(-1)^{j+2},\quad h_3=(-1)^{2j+1},\\
h_4 = 1+(-1)^{2j},\quad & h_5=1+(-1)^{j},\quad \vartheta_i>1,\\
\varpi_i>0\quad \forall i=1,2,\ldots,n,\quad & b=1,2,\ldots,\quad 1\le j\le\frac{\mathcal{F}}{2}.
\end{aligned}\tag{39}$$

Choosing a certain weights $\mho_i$, $i=1,2,\ldots,n$, such that $\sum_{i=1}^{n}\mho_i=1$ and the certain parameters $\varpi_i=1$, $i=1,2,\ldots,n$, then the previous MONLP(II) takes the form

$$\begin{aligned}
\text{MONLP(III)}:\quad \min\quad & \mho_i\left\{(h_3)\left(\frac{\vartheta_i-1}{\vartheta_i+1}\right)\right.\\
&\quad\times\left[b\left(\vartheta_i^{2j-1}-1\right)+h_4\right]\\
&\quad-2\delta_{i0}+(h_2)\left(\frac{\vartheta_i-1}{\vartheta_i+1}\right)\\
&\quad\times\left[b\left(\vartheta_i^{j}-1\right)+h_5\right]\\
&\quad\left.+b\left(\vartheta_i^{2j-1}-1\right)\right\},\\
\text{subject to}\quad & b\left(\vartheta_i^{2j-1}-1\right)\\
&\le (h_1-1)\left(\frac{\vartheta_i-1}{\vartheta_i+1}\right)\\
&\quad\times\left[b\left(\vartheta_i^{j}-1\right)+h_1\right]\\
&\quad+\delta_{i0},\\
& b\left(\vartheta_i^{2j}-1\right)\\
&\ge (h_1-1)\left(\frac{\vartheta_i-1}{\vartheta_i+1}\right)\\
&\quad\times\left[b\left(\vartheta_i^{j}-1\right)+h_1\right]\\
&\quad+\delta_{i0},\\
& 1-\vartheta_i<0,\quad \forall b=1,2,\ldots,\quad 1\le j\le\frac{\mathcal{F}}{2}.
\end{aligned}\tag{40}$$

From the Kuhn-Tucker conditions we have

$$\begin{aligned}
&\frac{1}{2}\left\{\frac{2}{(\vartheta_i+1)^2}\left[(h_3-1)\left[b\left(\vartheta_i^{2j-1}-1\right)+h_4\right]\right.\right.\\
&\qquad\left.+(h_2-1)\left[b\left(\vartheta_i^{j}-1\right)+h_5\right]\right]\\
&\quad+\left(\frac{\vartheta_i-1}{\vartheta_i+1}\right)b\left[(h_3-1)(2j-1)\vartheta_i^{2j-2}+(h_2-1)j\vartheta_i^{j-1}\right]\\
&\quad\left.+b(2j-1)\vartheta_i^{2j-2}\right\}\\
&+u_1\left\{b(2j-1)\vartheta_i^{2j-2}-\frac{(h_1-1)}{\vartheta_i+1}\right.\\
&\qquad\left.\times\left[\frac{2}{\vartheta_i+1}\left[b\left(\vartheta_i^{j}-1\right)+h_1\right]+(\vartheta_i-1)jb\vartheta_i^{j-1}\right]\right\}\\
&+u_2\left\{-b(2j-1)\vartheta_i^{2j-2}+\frac{(h_1-1)}{\vartheta_i+1}\right.\\
&\qquad\left.\times\left[\frac{2}{\vartheta_i+1}\left[b\left(\vartheta_i^{j}-1\right)+h_1\right]+(\vartheta_i-1)jb\vartheta_i^{j-1}\right]\right\}\\
&-u_3=0,
\end{aligned}\tag{41}$$

$$\begin{aligned}
&u_1\left\{b\left(\vartheta_i^{2j-1}-1\right)(\vartheta_i+1)-(h_1-1)(\vartheta_i-1)\right.\\
&\qquad\left.\times\left[b\left(\vartheta_i^{j-1}-1\right)+h_1\right]\right\}=0,
\end{aligned}\tag{42}$$

$$\begin{aligned}
&u_2\left\{(h_1-1)(\vartheta_i-1)\left[b\left(\vartheta_i^{j}-1\right)+h_1\right]\right.\\
&\qquad\left.-b\left(\vartheta_i^{2j-1}-1\right)(\vartheta_i+1)\right\}=0,
\end{aligned}\tag{43}$$

$$u_3\left\{1-\vartheta_i\right\}=0.\tag{44}$$

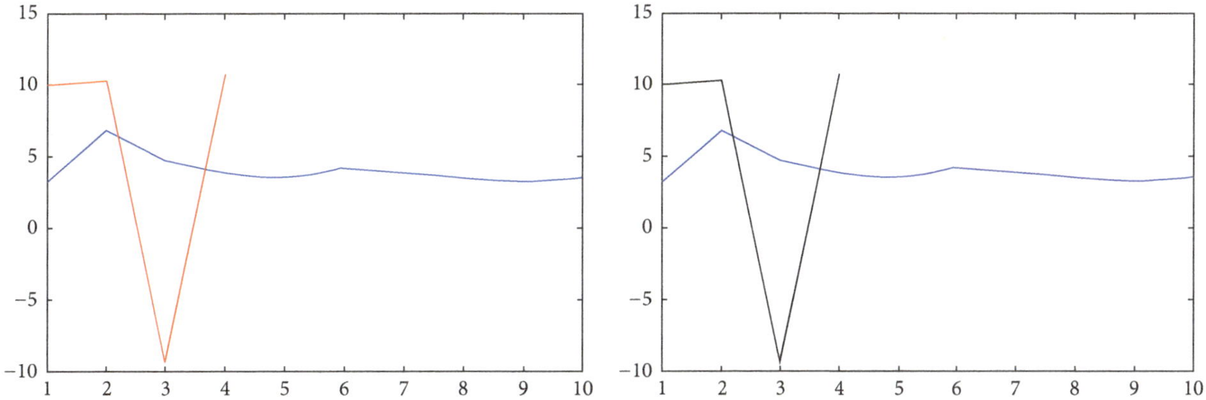

(a) $b = 1$ (b) $b = 2$

Figure 2: (a) and (b) give the optimal expected value of the first meeting time between S_i and the target that moves with discrete random walk in one dimension.

Solving (41)–(44), we found the optimal values of ϑ_i that give the optimal distances H_{ij} for all $b = 1, 2, \ldots$, $1 \le j \le \mathscr{F}/2$, and they are given by

$$\vartheta_i = \sqrt{\frac{\widehat{\Upsilon} + \widehat{\Lambda}}{b(1-2j)\,\vartheta_i^{2j-2}}} - 1, \quad \vartheta_i > 1, \tag{45}$$

where

$$\begin{aligned}\widehat{\Upsilon} &= 2\left[(h_3 - 1)\left[b\left(\vartheta_i^{2j-1} - 1\right) + h_4\right] + (h_2 - 1)\left[b\left(\vartheta_i^{j} - 1\right) + h_5\right]\right],\\ \widehat{\Lambda} &= b(h_3 - 1)(2j-1)\,\vartheta_i^{2j} - b(h_3 - 1)(2j-1)\,\vartheta_i^{2j-2} + b(h_2 - 1)\, j\vartheta_i^{j+1} - b(h_2 - 1)\, j\vartheta_i^{j-1}.\end{aligned} \tag{46}$$

7. An Illustrative Example

In this section, we clarify the effectiveness of this model by considering the following example.

Example 1. Suppose that a criminal drunk (target) leaves its cache and walks up and down through one of n-disjoint streets, totally disoriented. We model the streets as real lines with cache at the point X_0. In addition, assume that the criminal drunk takes unit steps, so we may record his position with an integer. Thus, for example, if he takes 5 steps to the left, he will be at a position $X_0 - 5$. If we consider the previous conditions in Theorems 2 and 3 hold, then the optimal expected value of the first meeting time depends on the optimal values of the distances H_{ij} that the searcher S_i should do it and the optimal values of G_{ij}, $j = 2, 4, 6$. Using the constraint $G_{i(2j-1)} \le t \le G_{i(2j)}$, $i = 1, 2, \ldots, n$, $1 \le j \le \mathscr{F}/2$, from Case 2, such that $j = 2, 4, 6$ and for $b = 1, 2$, in MONLP(III), and also using (38), we obtain the optimal values of ϑ_i that give the optimal values of H_{ij}, G_{ij} in Table 1.

It is clear that the optimal distances H_{ij} that S_i should do it depend on b where $\lambda_i = b\varpi_i$ and $G_{ij} = \lambda_i(\vartheta_i^j - 1)$.

Table 1: The optimal values of ϑ_i, H_{ij} and G_{ij}, $j = 2, 4, 6$, $b = 1, 2$.

b	j	ϑ_i	G_{ij}	H_{ij}
	2	1.8836	0.8836	$0.27075 + \delta_{i0}$
1	4	2.5265	1.5265	$0.6607 - \delta_{i0}$
	6	2.6982	1.6982	$0.7798 + \delta_{i0}$
	2	2.5226	3.0452	$1.3162 + \delta_{i0}$
2	4	2.7897	3.5794	$2.2963 - \delta_{i0}$
	6	2.8652	3.7304	$1.8 + \delta_{i0}$

From the previous numerical calculations and by considering $\delta_{i0} = 10$, we get Figure 2 that shows the projection of the target's motion (one-dimensional random walk) on the S_i path with $b = 1, 2$ where the current position on the line represented in the vertical axis versus the time steps at horizontal axis.

8. Discussion and Conclusions

A multiplicative generalized linear search plan for a one-dimensional random walk target on one of n-disjoint real lines has been presented, where the target initial position is given by a random variable X_0. Therefore, the target will be meet if $E_{\tau_\phi} < \infty$, where $\phi(t) = (\phi_1(t), \phi_2(t), \ldots, \phi_n(t))$ and E_{τ_ϕ} is the expected value of the first meeting time between one of the searchers and the target. We discuss some properties that the search model should satisfy in Theorems 2 and 3. We introduce the proof of conditions that make a search plan finite in Theorem 4, based on the continuity of the search plan and the conditions in Theorems 2 and 3 to show that $E_{\tau_\phi} < \infty$. We provide more analysis by using Lemma 5, Theorems 2 and 3 in Theorem 6 to show that the search plan $\phi(t)$ is finite if the conditions ${}^2\widehat{h}(x) \le L(|x|)$, ${}^2\widehat{q}(x) \le \widetilde{L}(|x|)$, if $\delta_{i0} > 0$ and ${}^1\widehat{w}(x) \le M(|x|)$, ${}^1\widehat{h}(x) \le \widetilde{M}(|x|)$ if $\delta_{i0} < 0$, where $L(|x|)$, $\widetilde{L}(|x|)$, $M(|x|)$, and $\widetilde{M}(|x|)$ are vectors of linear functions. We use Theorem 7 to show that if there exist a finite search plan then the expected value of the target initial position $E(|X_0|)$ is finite. It will also be interesting to see a direct consequence of Theorems 4, 6, and 7 satisfying the existence of a finite search plan if and only if $E(|X_0|)$ is finite. The existence of an optimal search plan has been proved in Theorem 10.

To find the optimal distances H_{ij}, $i = 1,\ldots,n$, $j = 1,2,\ldots$, that give the optimal search plan $\phi^*(t) = (\phi_1^*(t), \phi_2^*(t), \ldots, \phi_n^*(t))$ we solve the multiobjective nonlinear programming problem (MONLP) which contains the variables ϑ_i and the parameters λ_i, $i = 1,\ldots,n$. We use the Kuhn-Tucker conditions to obtain the optimal values of ϑ_i that can give the optimal distances, after substituting (38), and then minimize the expected value of the first meeting time. The effectiveness of this model is illustrated using numerical example.

In future research, it seems that the proposed model will be extendible to the multiple searchers case by considering the combinations of movement of multiple targets.

References

[1] B. O. Koopman, "Search and Screening," OEG Report 56, The Summary Reports Group of the Columbia University Division of War Research (Available from the Center for Naval Analyses), 1946.

[2] S. J. Benkoski, M. G. Monticino, and J. R. Weisinger, "A survey of the search theory literature," *Naval Research Logistics*, vol. 38, pp. 469–494, 1991.

[3] J. R. Frost and L. D. Stone, "Review of Search Theory: Advances and Applications to Search and Rescue Decision Support," Final report CG-D-15-01, US Coast Guard Research and Development Center, 2001.

[4] B. J. MC Cabe, "Searching for one dimensional random walker," *Journal of Applied Probability*, vol. 11, pp. 86–93, 1974.

[5] A. A. Mohamed, "Generalized search for one dimensional random walker," *International Journal of Pure and Applied Mathematics*, vol. 19, pp. 375–387, 2005.

[6] A. B. El-Rayes and A. A. Mohamed, "Searching for a randomly moving target," in *Proceedings of the 3rd ORMA Conference*, vol. 2, pp. 323–329, Egypt, 1989.

[7] B. Fristedt and D. Heath, "Searching for a particl and on the real Line," *Advances in Applied Probability*, vol. 6, pp. 79–102, 1974.

[8] A. Ohsumi, "Optimal search for a markovian target," *Naval Research Logistics*, vol. 38, pp. 531–554, 1991.

[9] A. B. El-Rayes, A. E.-M. A. Mohamed, and H. M. Abou Gabal, "Linear search for a Brownian target motion," *Acta Mathematica Scientia*, vol. 23, no. 3, pp. 321–327, 2003.

[10] A. Mohamed, M. Kassem, and M. El-Hadidy, "Multiplicative linear search for a brownian target motion," *Appli. Mathematical Model*, vol. 35, pp. 4127–4139, 2011.

[11] A. Beck, "On the linear search problem," *Israel Journal of Mathematics*, vol. 2, pp. 221–228, 1964.

[12] A. Beck, "More on the linear search problem," *Israel Journal of Mathematics*, vol. 3, pp. 61–70, 1965.

[13] A. Beck and M. Beck, "Son of the linear search problem," *Israel Journal of Mathematics*, vol. 48, no. 2-3, pp. 109–122, 1984.

[14] A. Beck and M. Beck, "The linear search problem rides again," *Israel Journal of Mathematics*, vol. 53, no. 3, pp. 365–372, 1986.

[15] A. Beck and M. Beck, "Revenge of the linear search problem," *SIAM Journal on Control and Optimization*, vol. 30, no. 1, pp. 112–122, 1992.

[16] A. Beck and D. Newman, "Yet more on the linear search problem," *Israel Journal of Mathematics*, vol. 8, pp. 419–429, 1970.

[17] A. Beck and P. Warren, "The return of the linear search problem," *Israel Journal of Mathematics*, vol. 10, pp. 169–183, 1972.

[18] W. Franck, "On an optimal search problem," *SIAM Review*, vol. 7, pp. 503–512, 1965.

[19] P. Rousseeuw, "Optimal search paths for random variables," *Journal of Computational and Applied Mathematics*, vol. 9, pp. 279–286, 1983.

[20] D. J. Reyniers, "Coordinated search for an object hidden on the line," *European Journal of Operational Research*, vol. 95, no. 3, pp. 663–670, 1996.

[21] D. J. Reyniers, "Co-ordinating two searchers for an object hidden on an interval," *Journal of the Operational Research Society*, vol. 46, no. 11, pp. 1386–1392, 1995.

[22] Z. T. Balkhi, "The generalized linear search problem, existence of optimal search paths," *Journal of the Operations Research Society of Japan*, vol. 30, pp. 399–420, 1987.

[23] Z. T. Balkhi, "Generalized optimal search paths for continuous univariate random variable," *Operations Research*, vol. 23, pp. 67–96, 1987.

[24] A. A. Mohamed, H. M. Abou Gabal, and M. A. El-Hadidy, "Coordinated search for a randomly located target on the plane," *European Journal of Pure and Applied Mathematics*, vol. 2, pp. 97–111, 2009.

[25] A. A. Mohamed, H. A. Fergany, and M. A. El-Hadidy, "On the coordinated search problem on the plane," *Istanbul University Journal of the School of Business Administration*, vol. 41, pp. 80–102, 2012.

[26] A. Mohamed and M. El-Hadidy, "Existence of a periodic search strategy for a parabolic spiral target motion in the plane," *Afrika Matematika*, vol. 24, no. 2, pp. 145–160, 2013.

[27] A. Mohamed and M. El-Hadidy, "Coordinated search for a conditionally deterministic target motion in the plane," *European Journal of Mathematical Sciences*, vol. 2, no. 3, pp. 272–295, 2013.

[28] W. Feller, *An Introduction to Probability Theory and Its Applications*, Wiley, New York, NY, USA, 2nd edition, 1966.

8

Image Watermarking Algorithm Based on Multiobjective Ant Colony Optimization and Singular Value Decomposition in Wavelet Domain

Khaled Loukhaoukha

Department of Electrical and Computer Engineering, Laval University, Québec, QC, Canada 1K 7P4

Correspondence should be addressed to Khaled Loukhaoukha; khaled.loukhaoukha.1@ulaval.ca

Academic Editor: Peiping Shen

We present a new optimal watermarking scheme based on discrete wavelet transform (DWT) and singular value decomposition (SVD) using multiobjective ant colony optimization (MOACO). A binary watermark is decomposed using a singular value decomposition. Then, the singular values are embedded in a detailed subband of host image. The trade-off between watermark transparency and robustness is controlled by multiple scaling factors (MSFs) instead of a single scaling factor (SSF). Determining the optimal values of the multiple scaling factors (MSFs) is a difficult problem. However, a multiobjective ant colony optimization is used to determine these values. Experimental results show much improved performances of the proposed scheme in terms of transparency and robustness compared to other watermarking schemes. Furthermore, it does not suffer from the problem of high probability of false positive detection of the watermarks.

1. Introduction

With the advent of numeric era at the end of 20th century, the exchange of digital documents became a very easy task. This extraordinary technical revolution from analog to numerical technology was not achieved without generating anxiety in terms of the protection of the authors rights since multimedia documents can be quite easily duplicated, modified, and illegally attacked without deterioration. Affected by significant revenue losses multimedia documents author's are motivated more than ever to secure their documents. In this context digital watermarking was introduced: it consists of inscribing invisible (or visible) data into the multimedia documents. This is done in two stages: embedding and extracting process. Digital watermarking schemes for images can be classified into different classes according to embedding domain, embedding rule, imperceptibility, and permanency.

In terms of robustness, the watermarking algorithm can be classified into three categories: fragile, semifragile, and robust. Fragile watermarking is designed to detect any modification in such a way that slight modifications or tampering on the watermarked image will destroy the watermark. This type is employed to ensure the integrity and image authenticity. Conversely, robust watermarking is designed to be resistant against attacks that attempt to remove or destroy the watermark without degrading the visual quality of the watermarked image significantly. Robust watermarking is typically employed for copyright protection and ownership verification. Semifragile watermarking combines the properties of fragile and robust watermarking in order to detect unauthorized manipulations while still being robust against authorized manipulations.

Generally, watermarking algorithms operate either in the spatial domain or in a transform domain such as discrete wavelet transform (DWT) [1–3], discrete Fourier transform (DFT) [4–6], and discrete cosine transform (DCT) [7–9]. Spatial domain watermarking has the advantage of low calculation complexity compared to that in transform domains. However, most watermarking schemes in the scientific literature operate in the transform domain because it provides enhanced imperceptibility and robustness compared to those in spatial domain. However, decomposition of images in a standard basis set using the transform domain does not necessarily leads to the optimal representation of an image.

Therefore, different representations were investigated for watermarking: these include nonnegative matrix factorization (NMF) [10–12] and singular value decomposition (SVD) [13–15]. Watermarking schemes based on SVD are advantageous since slight changes in the singular values do not affect significantly the image quality. Unfortunately, several SVD-based watermarking schemes suffer from typically high probability of false positive watermark detection [16–18].

According to embedding rule, watermarking algorithm can be classified into three categories: multiplicative, additive, or substitution. In multiplicative and additive embedding schemes, the trade-off between imperceptibility and robustness is controlled by a single scaling factor (SSF). Cox et al. [19] suggest to use multiple scaling factors (MSFs) instead of one. They state that single scaling factor may not be applicable for altering all the pixel values of the original image. Determining the optimal values of the multiple scaling factors can be viewed as optimization problem which is unfortunately a difficult problem. In this paper, a multiobjective ant colony optimization (MOACO) is used to solve this difficult problem and to determine these optimal values.

The rest of this paper is organized as follows. Section 2 reviews the basic concepts of the ant colony optimization (ACO). In Section 3, we presented the proposed watermarking algorithm based on discrete wavelet transform (DWT) and singular value decomposition (SVD). The proposed watermarking algorithm using multiobjective ant colony optimization is described in Section 4. The experimental results are provided in Section 5. Finally, the concluding remarks are given in Section 6.

2. Ant Colony Optimization Principle

The ant colony optimization (ACO) was introduced by Dorigo [20] as solution for hard optimization problems. It is inspired by observation of real ant colonies. Ants explore randomly the area surrounding their nest in order to find food. If an ant finds a food source, it evaluates and carries some food to the nest. During the return travels the ant deposits on the ground a chemical substance called pheromone trail. Other ants can smell the pheromone and follow it with some probability. This way, ants can communicate via pheromone and find the optimal path between the food source and the nest. This capability of real ant colony to find optimal paths has led to the definition of artificial ant colonies that can find the optimal solution for hard optimization problems [21] such as the traveling salesman problem. The outline of the generic ACO algorithm is presented in Figure 1.

The first point to take into account in ant colony optimization is how the colony is represented. For continuous variables, a colony of m ants is represented as $m \times n$ matrix C, where $C = [\mathbf{x}_1 \ \mathbf{x}_2 \ \cdots \ \mathbf{x}_m]^T$ such that $\mathbf{x} = [x_1 \ x_2 \ \cdots x_n]$ is a vector of n design variables that corresponds to a single ant. The second point to consider is how to model the pheromone communication scheme. Socha and Dorigo [21] suggest to use a normal distribution for a continuous model implementation:

$$f_{\text{pheromone}}(x) = e^{(-(x-x_{\min})^2/2\sigma^2)}, \tag{1}$$

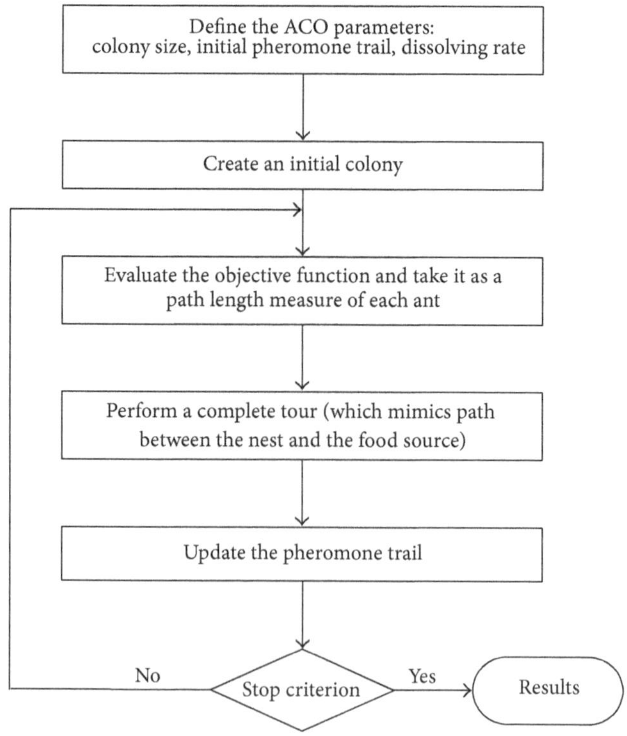

FIGURE 1: Generic ant colony optimization scheme [22].

where $x_{\min}$ is the optimal point found within the design space and the standard deviation σ as an index of the ants aggregation around the current minimum. To initialize the algorithm, $x_{\min}$ is randomly chosen in the design space, using a uniform distribution, and σ is taken at least three times greater than the length of the design space, to uniformly locate the ants within it. As shown in Figure 1, at each iteration, the ACO algorithm updates the values of each design variable, and this is for all the ants of the colony; that is, at each iteration each ant sets the values for the trial solution as per the distribution in (1). At the end, the pheromone distribution over the design space is updated by collecting the information acquired throughout the optimization steps. Since the pheromone is modeled by (1), it is necessary only to update $x_{\min}$ and σ as

$$\sigma = \text{std}(\text{colony}), \tag{2}$$

where std(colony) makes use of the colony of ants (candidate solution) to return a vector containing the standard deviation for each design variable [23]. The accumulation of pheromone increases in the vicinity of the candidate towards the optimum solution. However, to avoid premature convergence, negative update procedures are not discarded: for this, a simple method is used, which consists in dissolving the pheromone. The principle is to spread the amount of pheromone by changing the current standard deviation (for each variable) according to

$$\sigma_{\text{new}} = \gamma \cdot \sigma_{\text{old}}, \tag{3}$$

where $\gamma > 1$ is the dissolving rate.

3. Watermarking Algorithm Based on SVD and DWT

The SVD-DWT watermarking algorithm presented in [14] is described by the following embedding and extraction processes.

3.1. Embedding Process. Consider an original image I of size $N \times N$, and let the watermark W be a binary image of size $M \times M$. The embedding process is as follows.

(1) Decompose the original image I into $3\ell + 1$ sub-bands by applying a ℓ-level discrete wavelet transform (DWT).

(2) Select one sub-band (SB) among the three following sub-bands: HH_ℓ, HL_ℓ, and LH_ℓ.

(3) Compute the inverse DWT of the selected sub-band (SB):

$$X = \mathrm{DWT}^{-1}(SB). \tag{4}$$

(4) Apply a singular value decomposition (SVD) of matrix X:

$$X = U_X \cdot S_X \cdot V_X^T. \tag{5}$$

(5) Encrypt watermark image W using image encryption algorithm to get encrypted watermark denoted by W_C.

(6) Apply a singular value decomposition to watermark matrix W_C.

$$W = U_{W_C} \cdot S_{W_C} \cdot V_{W_C}^T. \tag{6}$$

(7) Compute the one-way hash functions for matrices U_{W_C} and V_{W_C}:

$$\begin{aligned} H_U &= \mathrm{hash}\left(U_{W_C}\right), \\ H_V &= \mathrm{hash}\left(V_{W_C}\right). \end{aligned} \tag{7}$$

(8) Matrices U_{W_C} and V_{W_C} and their hash values H_U and H_V are stored in the private key. (Steps (5) and (7) are necessary to mitigate the problem of false positive detection of watermark. This solution was proposed in [24].)

(9) Compute matrix S_Y according to

$$S_Y = S_X + \alpha \cdot S_{W_C}, \tag{8}$$

where α is the watermark strength factor that controls the trade-off between imperceptibility and robustness of the watermarking scheme. This parameter can be used as single scaling factor (SSF) or multiple scaling factors (MSFs).

(10) Compute matrix Z_W, according to

$$Z_W = U_X \cdot S_Y \cdot V_X^T. \tag{9}$$

(11) Compute the discrete wavelet transform of matrix Z_W,

$$SB_W = \mathrm{DWT}\left(Z_W\right). \tag{10}$$

(12) The watermarked image I_W is computed by applying the inverse ℓ-level discrete wavelet transform to the modified sub-band SB_W and the 3ℓ unmodified sub-bands.

3.2. Extracting Process. The extracting process is summarized by the following steps.

(1) A safety test is first done: hash values of matrices U_{W_C} and V_{W_C}, possibly altered by an attacker as $\widetilde{U}_W$ and $\widetilde{V}_W$, are computed to give $H_{\widetilde{U}}$ and $H_{\widetilde{V}}$. These hash values are compared with these stored during the embedding process:

$$\begin{cases} \text{if } H_U = H_{\widetilde{U}}, \quad \text{and} \quad H_V = H_{\widetilde{V}} \longrightarrow \text{go to 2}, \\ \text{if } H_U \neq H_{\widetilde{U}} \quad \text{or} \quad H_V \neq H_{\widetilde{V}} \longrightarrow \text{stop (probable attack)}. \end{cases} \tag{11}$$

(2) Decompose the original and watermarked images, I and I_W, by applying the ℓ-level discrete wavelet transform.

(3) Select the same sub-band (SB) used in 2 of the watermark embedding process. SB_I and SB_{I_W} are, respectively, the subbands selected for the original and watermarked images.

(4) Compute the inverse discrete wavelet transform of selected subbands SB_I and SB_{I_W}:

$$\begin{aligned} X_I &= \mathrm{DWT}^{-1}\left(SB_I\right), \\ X_{I_W} &= \mathrm{DWT}^{-1}\left(SB_{I_W}\right). \end{aligned} \tag{12}$$

(5) Apply the singular value decomposition on matrices X_I and X_{I_W}:

$$\begin{aligned} X_I &= U_{X_I} \cdot S_{X_I} \cdot V_{X_I}^T, \\ X_{I_W} &= U_{X_{I_W}} \cdot S_{X_{I_W}} \cdot V_{X_{I_W}}^T. \end{aligned} \tag{13}$$

(6) Compute matrix $S_{\widehat{W}_C}$ as follows:

$$S_{\widehat{W}_C} = \frac{S_{X_{I_W}} - S_{X_I}}{\alpha}. \tag{14}$$

(7) Determine the extracted encrypted watermark, $\widehat{W}_C$, by computing

$$\widehat{W}_C = \widetilde{U}_{W_C} \cdot S_{\widehat{W}_C} \cdot \widetilde{V}_{W_C}^T. \tag{15}$$

(8) Decrypt $\widehat{W}_C$ to get the extracted watermark $\widehat{W}$.

3.3. Protection against False Positive Detection. To protect the proposed image watermarking scheme against false positive detection, two countermeasures have been proposed in [24]. The first countermeasure consists of computing a *one-way hash function* for the matrices U_W and V_W. The hash function algorithm such as *message digest 5* (MD5) or *secure hash algorithm 1* (SHA-1) can be used for this purpose. During embedding process, the hash values of matrices U_W and V_W denoted, respectively, by H_U and H_V are stored in a private key. In the extracting process, the hash function is first computed from received (and possibly altered by an attacker) matrices $U_{\widetilde{W}}$ and $V_{\widetilde{W}}$, denoted by $H_{\widetilde{U}}$ and $H_{\widetilde{V}}$. Thus, if $H_U \neq H_{\widetilde{U}}$ or $H_V \neq H_{\widetilde{V}}$, the watermark extracting process is stopped because $U_{\widetilde{W}} \neq U_W$ or $V_{\widetilde{W}} \neq V_W$: otherwise the watermark extracting process is performed. The authors indicate that the hash function test can be also be applied on a combined matrix from U_W and V_W such as $U_W + V_W$ and $U_W \times V_W$.

The second countermeasure consists in encrypting the watermark before the embedding process. The watermark W is encrypted resulting in an encrypted watermark denoted by W_C which will then be embedded in the original image I. Suppose that for watermark extraction process, an attacker uses his own watermark $\widetilde{W}$. Then matrices $U_{\widetilde{W}}$ and $V_{\widetilde{W}}$ will be used instead of proper matrices U_{W_C} and V_{W_C}. The first extracted watermark will be the same as $\widetilde{W}$, but the decryption process must be performed since the embedded watermark is encrypted during the watermark embedding process. Therefore, the extracted watermark $\widehat{W}$ will be a random-like image. Note that, if there is no attack, the first extracted watermark $\widehat{W}_C$ will indeed be the encrypted image, but after decryption, the extracted watermark $\widehat{W}$ will have a high correlation value with the original watermark W. Figures 2 and 3 illustrate an example of one-way hash function and encryption countermeasures, respectively, using Lena as original image I and LOGO as watermark W.

4. Watermarking Algorithm Using Multiobjective Ant Colony Optimization

In general, watermarking schemes are either based on an additive or a multiplicative rule. The embedding rules themselves are usually of the following form, where I and I_W are, respectively, original and watermarked image (or their representation in other domains such as FFT, DCT, and DWT), α is used to control the trade-off between imperceptibility and robustness of an image and generally is used as scaling factor:

$$\begin{aligned} I_W &= I + \alpha \cdot W \longrightarrow \text{additive rule}, \\ I_W &= I \cdot (1 + \alpha \cdot W) \longrightarrow \text{multiplicative rule}. \end{aligned} \tag{16}$$

Cox et al. [19] suggest the use of multiple scaling factors instead of one. They state that a single scaling factor may not be applicable for altering all the values of original image I. Determining the optimal values of these multiple scaling factors is a complex computational problem. To solve it, we propose to use a multi-objective ant colony optimization (MOACO).

TABLE 1: Imperceptibility test results.

Image	Algorithm	PSNR (I, I_W) (dB)	NC $(W, \widehat{W})$
Cameraman	MSF-MOACO	50.118	1.000
	SSF	49.341	1.000
	Xianghong et al. [1]	49.075	1.000
	Liu and Li [3]	45.529	0.998
	Pai and Ruan [25]	55.289	1.000
	Ouhsain and Hamza [12]	53.563	1.000
Goldhill	MSF-MOACO	54.219	1.000
	Using SSF	51.124	1
	Xianghong et al. [1]	49.075	0.999
	Liu and Li [3]	45.320	1.000
	Pai and Ruan [25]	54.844	1.000
	Ouhsain and Hamza [12]	51.892	1.000
Lena	MSF-MOACO	50.942	1.000
	SSF	48.899	1
	Xianghong et al. [1]	49.075	1.000
	Liu and Li [3]	45.472	0.997
	Pai and Ruan [25]	55.186	1.000
	Ouhsain and Hamza [12]	53.192	1.000
Man	MSF-MOACO	51.026	1.000
	SSF	50.735	1.000
	Xianghong et al. [1]	49.075	1.000
	Liu and Li [3]	45.323	1.000
	Pai and Ruan [25]	55.186	0.998
	Ouhsain and Hamza [12]	52.569	0.999

Figure 4 illustrates block diagram of multiobjective optimization which is closed-loop control system. System input is multiple scaling factors and objective measure as system output; this measure is calculated from original image I, watermarked image I_W, watermark W, and the $(T + 1)$ extracted watermarks ($\widehat{W}$ and $\widehat{W}_j$, where $j = \{1, 2, \ldots, T\}$) under attacks.

The steps for applying MOACO into SVD-DWT watermarking scheme are enumerated below.

(1) Define the colony size, the initial pheromone trail, the dissolving rate (σ), the objective function, and a generation number as the algorithm stopping criterion.

(2) Using (17), generate randomly an initial population of ants, which constitute a set of potential solutions. Each ant is denoted by $X = \{x_1, x_2, \ldots, x_n\}$,

$$f_{\text{pheromone}}(x_i) = e^{((x_i - x_i^*)^2 / 2\sigma_i^2)}, \tag{17}$$

where x_i^* is the ith coordinate of the best point found by optimization within the design space at the current iteration and σ_i is the ants aggregation index for the ith coordinate of the design space. At the first iteration, x_i^* is chosen according a uniform

TABLE 2: Correlation coefficient comparison results between MSF-MOACO and SSF algorithms and other algorithms.

Image	Algorithm	SP	GF	CR	CM	SH	SC	HE	QN	RW	CA
Cameraman	MSF-MOACO	0.956	0.972	0.947	0.902	0.989	1	0.985	0.990	1.000	1.000
	SSF	0.726	0.907	0.894	0.729	0.976	1.000	0.959	0.958	0.998	0.997
	Xianghong et al. [1]	0.495	0.847	0.983	0.621	0.580	0.993	0.621	0.750	0.850	0.914
	Liu and Li [3]	0.936	0.872	0.983	0.470	0.940	0.846	0.954	0.743	0.865	0.907
	Pai and Ruan [25]	0.407	0.898	0.975	0.537	0.697	0.989	0.474	0.504	0.867	0.924
	Ouhsain and Hamza [12]	0.710	0.840	0.881	0.775	0.653	0.982	0.920	0.532	0.860	0.902
Goldhill	MSF-MOACO	0.935	0.957	0.935	0.959	0.984	1	0.975	0.972	0.998	1.000
	SSF	0.885	0.936	0.998	0.930	0.950	1.000	0.977	0.965	0.995	0.999
	Xianghong et al. [1]	0.694	0.858	0.983	0.633	0.713	0.986	0.440	0.570	0.850	0.914
	Liu and Li [3]	0.947	0.906	0.980	0.545	0.951	0.870	0.974	0.714	0.867	0.918
	Pai and Ruan [25]	0.374	0.827	0.976	0.293	0.370	0.926	0.313	0.220	0.867	0.912
	Ouhsain and Hamza [12]	0.789	0.835	0.852	0.796	0.675	0.978	0.950	0.523	0.850	0.925
Lena	MSF-MOACO	0.933	0.974	0.918	0.948	0.976	0.999	0.974	0.967	1.000	1.000
	SSF	0.774	0.943	0.867	0.944	0.982	1.000	0.988	0.974	0.998	0.995
	Xianghong et al. [1]	0.616	0.866	0.983	0.640	0.667	0.994	0.587	0.625	0.850	0.914
	Liu and Li [3]	0.926	0.845	0.953	0.541	0.952	0.826	0.987	0.716	0.863	0.912
	Pai and Ruan [25]	0.383	0.920	0.980	0.442	0.681	0.994	0.514	0.393	0.867	0.918
	Ouhsain and Hamza [12]	0.769	0.882	0.853	0.775	0.685	0.983	0.962	0.546	0.849	0.936
Man	MSF-MOACO	0.932	0.952	0.964	0.978	0.966	1.000	0.972	0.970	1	1.000
	SSF	0.588	0.845	0.930	0.905	0.949	0.999	0.977	0.963	0.990	0.998
	Xianghong et al. [1]	0.738	0.876	0.983	0.641	0.692	0.993	0.828	0.456	0.850	0.914
	Liu and Li [3]	0.944	0.876	0.994	0.532	0.966	0.845	0.983	0.803	0.865	0.915
	Pai and Ruan [25]	0.412	0.866	0.981	0.324	0.432	0.983	0.468	0.282	0.867	0.920
	Ouhsain and Hamza [12]	0.745	0.812	0.890	0.756	0.632	0.985	0.945	0.556	0.875	0.952

TABLE 3: Correlation coefficient comparison results between the MSF-MOACO algorithms and other algorithms.

Image	Algorithm	GC	DI	RT	MB	TR
Cameraman	MSF-MOACO	0.970	0.994	0.927	0.989	0.755
	Xianghong et al. [1]	0.847	0.001	0.368	0.428	0.411
	Liu and Li [3]	0.988	0.688	0.665	0.652	0.659
	Pai and Ruan [25]	0.616	0.532	0.421	0.546	0.462
	Ouhsain and Hamza [12]	0.940	0.463	0.201	0.980	0.524
Goldhill	MSF-MOACO	0.979	0.979	0.904	0.974	0.951
	Xianghong et al. [1]	0.915	0.001	0.493	0.516	0.444
	Liu and Li [3]	0.997	0.676	0.702	0.643	0.752
	Pai and Ruan [25]	0.313	0.267	0.271	0.304	0.278
	Ouhsain and Hamza [12]	0.955	0.423	0.322	0.975	0.563
Lena	MSF-MOACO	0.970	0.970	0.916	0.966	0.877
	Xianghong et al. [1]	0.875	0.001	0.446	0.584	0.419
	Liu and Li [3]	0.986	0.686	0.641	0.666	0.683
	Pai and Ruan [25]	0.535	0.435	0.332	0.527	0.374
	Ouhsain and Hamza [12]	0.953	0.452	0.356	0.970	0.550
Man	MSF-MOACO	0.924	0.968	0.847	0.962	0.875
	Xianghong et al. [1]	0.912	0.001	0.518	0.518	0.492
	Liu and Li [3]	0.990	0.683	0.689	0.636	0.766
	Pai and Ruan [25]	0.374	0.316	0.287	0.334	0.292
	Ouhsain and Hamza [12]	0.902	0.432	0.324	0.965	0.543

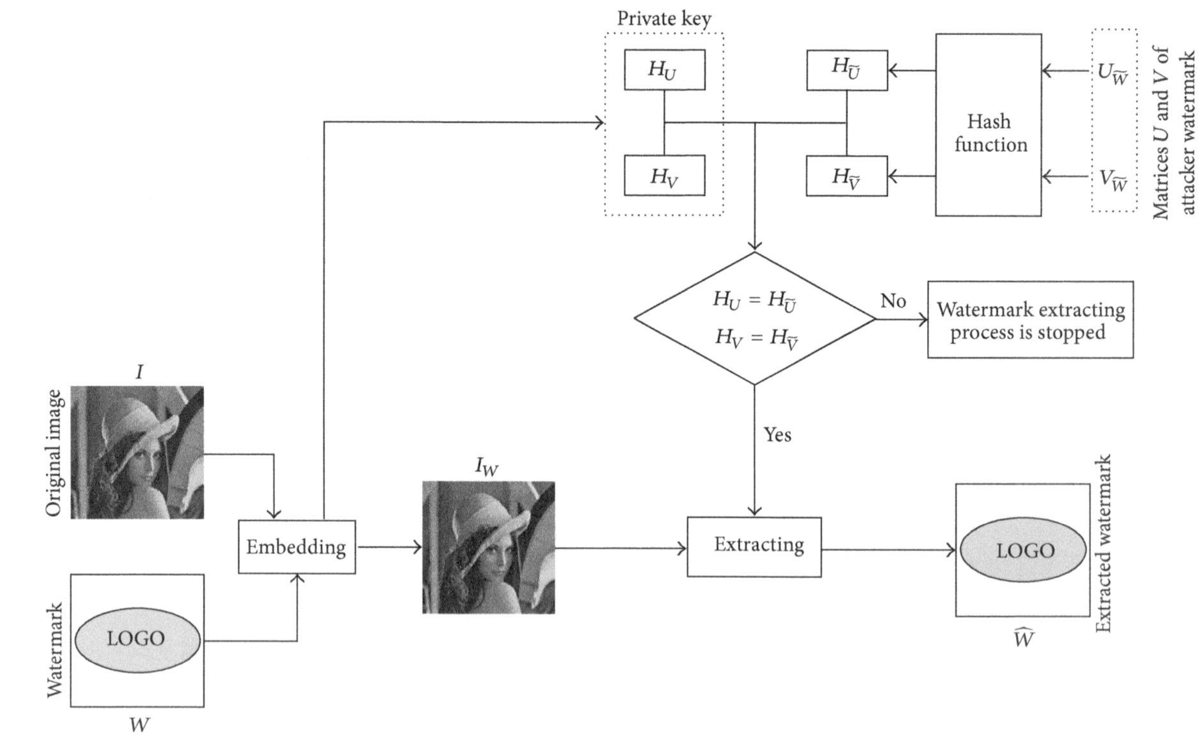

FIGURE 2: Example of the first countermeasure based on one-way hash function.

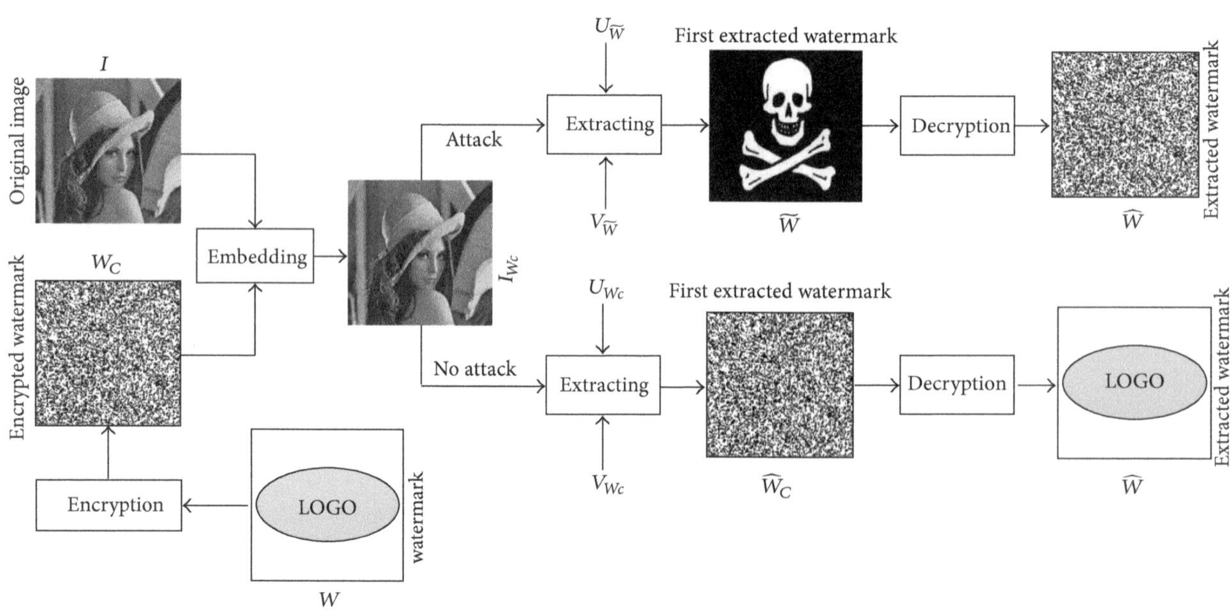

FIGURE 3: Example of the second countermeasure based on encryption.

distribution with σ_i taken as at least 3 times larger than the length of the search interval.

(3) For each ant X of the population, consider the following.

(i) Produce the watermarking image I_W using embedding process (Section 3.1) using the ant X as the watermark strength factor. Note that (8) in watermark embedding process is transformed into. (Note that, α is changed into multiple scaling factors instead of single scaling factor and diag(α) is diagonal matrix created from the vector X, that is, the ant X)

$$S_Y = S_X + \operatorname{diag}(\alpha) \cdot S_{W_C}. \tag{18}$$

(ii) Compute the normalized correlation $NC(I, I_W)$ between original and watermarked images.

(iii) Apply a watermark attack out of a set of T selected attacks upon the watermarked image

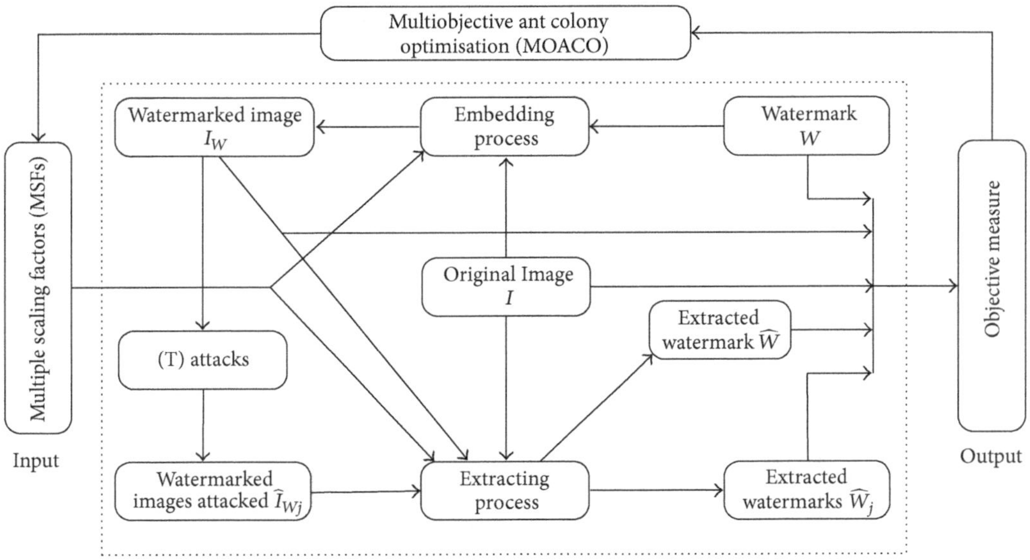

FIGURE 4: : Block diagram of multiobjective optimization.

I_W. This leads to T different attacked watermarked images $\{\widehat{I}_W\}$ for each original watermarked image I_W.

(iv) Extract the watermarks $\widehat{W}_i$ from the attacked watermarked images $\widehat{I}_W$ using the extraction process, as described in Section 3.2, where $i = \{1, 2, \ldots, T\}$.

(v) Compute the normalized correlation between the original watermark W and the set of extracted watermarks $\{\widehat{W}_i\}$, that is, $\{NC(W, \widehat{W}_i)\}$.

(vi) Construct the vector of objective values, $F(X)$, defined as

$$F(p) = \left(\frac{1}{NC\left(I, I_{W_p}\right)} \frac{1}{NC\left(W, \widehat{W}\right)} \frac{1}{NC\left(W, \widehat{W}_{p,1}\right)} \right. \\ \left. \times \frac{1}{NC\left(W, \widehat{W}_{p,2}\right)} \cdots \frac{1}{NC\left(W, \widehat{W}_{p,T}\right)} \right)^T. \quad (19)$$

(vii) Evaluate the vector of objective values according to the *exponential weighted method* for multiobjective optimization [26]:

$$F_{\text{obj}}(X) = \sum_{i=1}^{T+2} \left(e^{p \cdot w} - 1\right) \cdot e^{p \cdot (F(X) - F_0)}, \quad (20)$$

where p, w, and F_0 are positive constants. In experiments, we take $p = 2$, $w = 5$ et $F_0 = 10$.

(4) Find the best ant $X_{\text{best}} = \{x_1^*, x_2^*, \ldots, x_n^*\}$ as the one having the smallest objective value F_{obj}.

(5) Update the pheromone trail distribution using the formula (17); in this step the aggregation index for the ith dimension σ_i is given by

$$\sigma_i = \sqrt{\frac{1}{m} \sum_{i=1}^{m} y_i - \overline{y}}, \quad (21)$$

where $\mathbf{y}$ is the ith column of the colony matrix C, $\overline{y}$ is the mean value of the vector $\mathbf{y}$, and m is the number of colony size.

(6) Save the best ant X_{best} between this generation and the old one generation.

(7) If the generation number is reached, the optimization process of the multiple scaling factors (MSFs) is terminated, else it goes to the next step.

(8) Using the distribution of (17), generate a new population of ants and then go back to 3.

5. Experimental Results

To show the effectiveness of the proposed watermarking algorithm using multiple scaling factors optimized by multiobjective ant colony optimization instead of single scaling factor in terms of imperceptibility and robustness, a number of tests have been carried out using four 256×256 grayscale test images and a 32×32 binary watermark, which are depicted in Figure 5. Note that ℓ is set to be equal to 3.

We conducted several experiments to compare the imperceptibility and robustness of the proposed watermarking algorithm using multiple scaling factors optimized by multiobjective ant colony optimization (i.e., MSF-MOACO) and single scaling factor (SSF) with Xianghong et al. [1], Liu [3], Pai and Ruan [25], and Ouhsain and Hamza [12] algorithms. Table 1 lists the PSNR values of proposed and other algorithms for the same test images. It is clearly shown

(a) Cameraman (b) Goldhill (c) Lena (d) Man (e) Letter A

Figure 5: Original images and binary watermark.

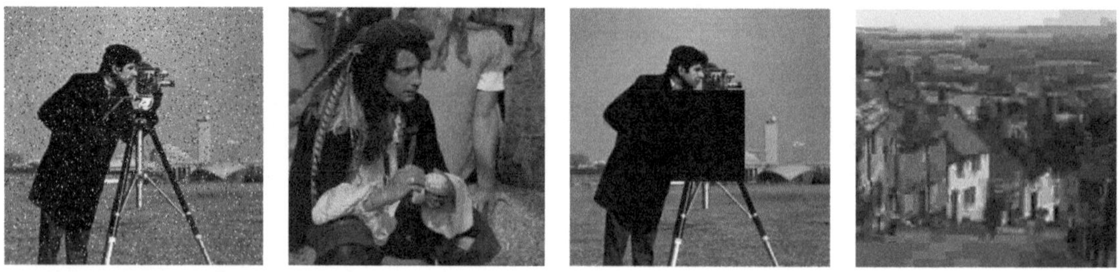

(a) Salt and peppers noise (b) Gaussian filter (c) Cropping (d) JPEG compression

(e) Sharpening (f) Scaling (g) Histogram equalization (h) Gray-scale quantization

(i) Rewatermarked (j) Gamma correction (k) Dithering (l) Motion blur

Figure 6: Watermarked images under different attacks.

that our algorithm using multiple scaling factors (MSF-MOACO) outperforms all other algorithms in terms of the visual quality of the reconstructed watermarked image except Pai and Ruan algorithm [25]. Furthermore, the normalized correlation between watermark W and extracted watermark $\widehat{W}$ of the proposed algorithm (MSF-MOACO) is very close to unity.

For the robustness tests, ten different attacks were selected in conjunction to multiobjective optimization (i.e., $T = 10$). These attacks are salt and peppers noise (with a density of 0.05), Gaussian filtering (3×3), cropping (1/8 of the image center), JPEG compression ($Q = 5$), sharpening, scaling ($256 \rightarrow 512 \rightarrow 256$), histogram equalization, gray-scale quantization (1 bit), re-watermarking using other watermarks that differ from the watermark (Letter A) and collusion attack using five watermarks, which are denoted by SP, GF, CR, CM, SH, SC, HE, QN, RW, and CA, respectively. The normalized correlation values between embedded and extracted watermarks under these attacks are depicted in Table 2, where the MSF-MOACO algorithm outperforms other algorithms [1, 3, 12, 25].

The effectiveness of the proposed algorithm (MSF-MOACO) is also tested against other attacks such as gamma correction ($\gamma = 0.2$), dithering, rotation (25°), motion blur (45°), and translation (25 pixels × 25 pixels), which were denoted by GC, DI, RT, MB, and TR, respectively. Table 3

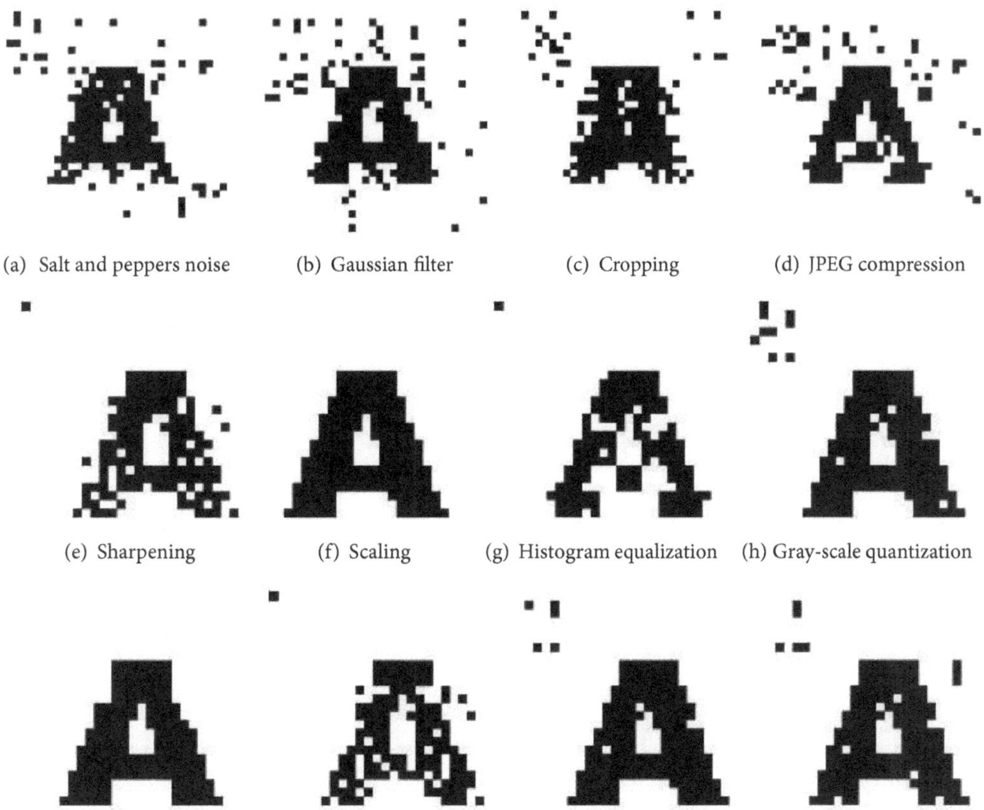

(a) Salt and peppers noise (b) Gaussian filter (c) Cropping (d) JPEG compression

(e) Sharpening (f) Scaling (g) Histogram equalization (h) Gray-scale quantization

(i) Rewatermarked (j) Gamma correction (k) Dithering (l) Motion blur

FIGURE 7: Extracted watermarks under different attacks.

depicts the normalized correlation values under these attacks, where one can observe that MSF-MOACO algorithm is more robust against these attacks compared to other algorithms. Figure 6 shows an example of watermarked images under these attacks, while Figure 7 depicts their corresponding extracted watermarks.

6. Conclusion

In this paper, we present a new watermarking algorithm based on lifting wavelet transform (LWT) and singular value decomposition (SVD) using multiple scaling factors (MSF) optimized by multiobjective ant colony optimization (MOACO). The MSFs are used instead of single scaling factor (SSF) to achieve a highest possible robustness without losing watermark transparency. However, determining the optimal set of multiple scaling factors is a prohibitively complex problem; to solve this problem a multiobjective ant colony optimization is used. Experimental results demonstrate that the proposed watermarking algorithm (MSF-MOACO) outperforms other watermarking algorithms in terms of imperceptibility and robustness. Moreover, MSF-MOACO algorithm showed better imperceptibility and excellent resiliency against a wide range of watermarking attacks such as additive noise, compression, filtering, and geometrical attacks. Furthermore, the problem of false positive detection which affects most SVD watermarking algorithms is resolved using countermeasures based on one-way hash function and watermark encryption proposed in [24].

References

[1] T. Xianghong, L. Lu, Y. Lianjie, and N. Yamei, "A digital watermarking scheme based on DWT and vector transform," in *Proceedings of the International Symposium on Intelligent Multimedia, Video and Speech Processing (ISIMP '04)*, pp. 635–638, October 2004.

[2] L. Li, H. Xu, C. Chang, and Y. Ma, "A novel image watermarking in redistributed invariant wavelet domain," *Journal of Systems and Software*, vol. 84, no. 6, pp. 923–929, 2011.

[3] X.-F. Liu and S.-M. Li, "Image digital watermarking algorithm based on Arnold scrambling and wavelet transform," in *Proceedings of the International Conference on Mechanical and Electronics Engineering*, vol. 130–134, pp. 2928–2931, September 2011.

[4] V. Solachidis and I. Pitas, "Circularly symmetric watermark embedding in 2-D DFT domain," *IEEE Transactions on Image Processing*, vol. 10, no. 11, pp. 1741–1753, 2001.

[5] W. Wang, A. Men, and X. Chen, "Robust image watermarking scheme based on phase features in DFT domain and generalized radon transformations," in *Proceedings of the International Workshop on Education Technology and Computer Science*, pp. 736–739, March 2009.

[6] E. Ganic, S. D. Dexter, and A. M. Eskicioglu, "Embedding multiple watermarks in the DFT domain using low and high frequency bands," in *Security, Steganography, and Watermarking of Multimedia Contents VII*, vol. 5681 of *Proceedings of the SPIE*, pp. 175–184, March 2005, January 2005.

[7] S. D. Lin, S. Shie, and J. Y. Guo, "Improving the robustness of DCT-based image watermarking against JPEG compression," *Computer Standards and Interfaces*, vol. 32, no. 1-2, pp. 51–58, 2010.

[8] A. H. Taherinia and M. Jamzad, "A robust spread spectrum watermarking method using two levels DCT," *International Journal of Electronic Security and Digital Forensics*, vol. 2, no. 3, pp. 280–305, 2009.

[9] J. R. Hernández, M. Amado, and F. Pérez-González, "DCT-domain watermarking techniques for still images: detector performance analysis and a new structure," *IEEE Transactions on Image Processing*, vol. 9, no. 1, pp. 55–68, 2000.

[10] W. Lu, W. Sun, and H. Lu, "Robust watermarking based on DWT and nonnegative matrix factorization," *Computers and Electrical Engineering*, vol. 35, no. 1, pp. 183–188, 2009.

[11] W. Lu and H. Lu, "Robust watermarking based on subsampling and nonnegative matrix factorization," *Informatica*, vol. 19, no. 4, pp. 555–566, 2008.

[12] M. Ouhsain and A. B. Hamza, "Image watermarking scheme using nonnegative matrix factorization and wavelet transform," *Expert Systems with Applications*, vol. 36, no. 2, pp. 2123–2129, 2009.

[13] A. A. Mohammad, A. Alhaj, and S. Shaltaf, "An improved SVD-based watermarking scheme for protecting rightful ownership," *Signal Processing*, vol. 88, no. 9, pp. 2158–2180, 2008.

[14] K. Loukhaoukha and J.-Y. Chouinard, "Hybrid watermarking algorithm based on SVD and lifting wavelet transform for ownership verification," in *Proceedings of the 11th Canadian Workshop on Information Theory (CWIT '09)*, pp. 177–182, May 2009.

[15] R. Liu and T. Tan, "An SVD-based watermarking scheme for protecting rightful ownership," *IEEE Transactions on Multimedia*, vol. 4, no. 1, pp. 121–128, 2002.

[16] R. Rykaczewski, "Comments on 'An SVD-based watermarking scheme for protecting rightful ownership'," *IEEE Transactions on Multimedia*, vol. 9, no. 2, pp. 421–423, 2007.

[17] X. Zhang and K. Li, "Comments on 'an SVD-based watermarking scheme for protecting rightful ownership'," *IEEE Transactions on Multimedia*, vol. 7, no. 3, pp. 593–594, 2005.

[18] L. Xiao, Z. Wei, and J. Ye, "Comments on "Robust embedding of visual watermarks using discrete wavelet transform and singular value decomposition" and theoretical analysis," *Journal of Electronic Imaging*, vol. 17, no. 4, Article ID 040501, 2008.

[19] I. J. Cox, J. Kilian, F. T. Leighton, and T. Shamoon, "Secure spread spectrum watermarking for multimedia," *IEEE Transactions on Image Processing*, vol. 6, no. 12, pp. 1673–1687, 1997.

[20] M. Dorigo, *Optimization, learning and natural algorithms [Ph.D. thesis]*, Dipartimento di Elettronica e Informazione, Politecnico di Milano, Milan, Italy, 1992.

[21] K. Socha and M. Dorigo, "Ant colony optimization for continuous domains," *European Journal of Operational Research*, vol. 185, no. 3, pp. 1155–1173, 2008.

[22] F. A. C. Viana, G. I. Kotinda, D. A. Rade, and V. Steffen Jr., "Tuning dynamic vibration absorbers by using ant colony optimization," *Computers and Structures*, vol. 86, no. 13-14, pp. 1539–1549, 2008.

[23] S. Pourtakdoust and H. Nobahari, "An extension of ant colony system to continuous optimization problems," in *Proceedings of the International Workshop of Ant Colony Optimization and Swarm Intelligence*, pp. 294–301, September 2004.

[24] K. Loukhaoukha and J.-Y. Chouinard, "Security of ownership watermarking of digital images based on singular value decomposition," *Journal of Electronic Imaging*, vol. 19, no. 1, Article ID 013007, 2010.

[25] Y. Pai and S. Ruan, "A high quality robust digital watermarking by smart distribution technique and effective embedded scheme," *IEICE Transactions on Fundamentals of Electronics, Communications and Computer Sciences*, vol. E90, no. 3, pp. 597–605, 2007.

[26] R. T. Marler and J. S. Arora, "Survey of multi-objective optimization methods for engineering," *Structural and Multidisciplinary Optimization*, vol. 26, no. 6, pp. 369–395, 2004.

A Hybrid PSO-Fuzzy Model for Determining the Category of 85th Speed

Abbas Mahmoudabadi[1] and Ali Ghazizadeh[2]

[1] *Department of Industrial Engineering, Payame Noor University (PNU), Shahnaz Alley, Nourian Street, North Dibagi Avenue, Tehran, Iran*
[2] *Technical and Engineering Department, Payame Noor University (PNU), Firouzbakhsh Street, Movahed Danesh Avenue, Aqdasieh, Tehran, Iran*

Correspondence should be addressed to Abbas Mahmoudabadi; mahmoudabadi@phd.pnu.ac.ir

Academic Editor: Ling Wang

The 85th speed of vehicles is one of the traffic engineering parameters used by road safety equipment designers. It is usually used for maintenance activities and designing of warning signs and road equipments. High measuring costs of speed data collection lead decision makers to define a methodology for determining the category of 85th speed using indirect parameters. In this research work, focusing on undivided intercity roads, a hybrid particle-swarm-optimization- (PSO-) fuzzy model has been developed to determine the category of 85th speed. In this model, geometric design parameters including roads' width and length characteristics and roadside land use are considered as input variables whereas the category of the 85th speed is output variable. A set of experimental data is used for evaluating the performance of the proposed model comparing to a well-known model of exponential regression. It is shown that the developed PSO-fuzzy model is capable of determining the category of 85th speed with an accuracy of 96%, while exponential regression can estimate that with up to 84% accuracy. Variable effectiveness procedure shows that the lane width has more direct effect on 85th speed than shoulder width and the number of access points. The percentage of forbidden overtaking is also found to have indirect effect on 85th speed.

1. Introduction

The 85th speed is one of the most important traffic engineering parameters. By definition, the 85th speed is the maximum speed that eighty-five percent (%85) of drivers prefer to drive with a speed less than that. The importance of this parameter is that the majority of drivers drive within their safe considered speed, so the 85th speed is the maximum speed considered as safe speed by most of drivers [1]. The 85th speed is being used in some technical applications such as designing of road safety infrastructure, planning road maintenance programs, and setting warning signs, while construction workers are working on the road in order to provide the traffic with safe passage. Many parameters affect on 85th speed including authorized speed, traffic volume, land topography, and roadside land use condensations [1]. Traditionally, 85th speed is calculated using vehicle's speed on a certain road. Many techniques have been utilized for estimating vehicles' speed, mainly need for electronic devices and infrastructures. Image processing technique [2], combination of fuzzy logic and image processing [3], traffic counting and occupancy data [4], vehicle tracking [5], optical sensors [6], RF-based vehicle detection [7], the field of mechanical engineering by kinematics relationship of measured signals [8], and traffic survey [1] are the well-known techniques on speed data collection. Expensive devices are needed for utilizing either of the previously mentioned techniques. Therefore, developing a low-cost procedure to estimate the category of 85th speed is quite interesting.

Although 85th speed is a measurable parameter, its exact quantity does not play significant role in real applications. So, in the literature, researchers utilized linguistic variables for this kind of parameters. For example, traffic volume is measured precisely; however, it is usually expressed by linguistic words such as heavy, normal, or low traffic [9, 10]. Therefore, it is considerable to utilize fuzzy variables in order to represent traffic parameters corresponding to the 85th speed.

Fuzzy approaches capture the senses of experts about input parameters and convert them to fuzzy outputs using specified fuzzy relationships between input and output variables. The process of developing a fuzzy model has three stages. The first is to define membership functions corresponding to input and output variables, the second is to define rules, and the third is to adjust input-output relationship [11].

For this advantage of fuzzy approaches, they are utilized in many fields of studies such as energy [12, 13], road safety evaluation programs [14], and road control system [15]. In the field of traffic and transportation engineering, many research works have been found in the literature [16]. For example fuzzy approaches have been utilized for developing signal detection theory in order to perceive workers' hazards on subway operations [17], developing fuzzy model to set traffic light timing [18], proposing an expert fuzzy system to predict object collision [19], and assessment of traffic parameters [20]. Fuzzy approaches have also been used to estimate traffic parameters such as traffic jam [10] and passenger car unit [9].

The combinations of fuzzy and heuristic methods have also been used for fuzzy clustering [21], forecasting [22], job scheduling [23], and time series planning [24]. Another application is a combination of fuzzy logic and particle swarm optimization to achieve the best performance of setting fuzzy parameters and improving definition of fuzzy logic relations. For example, a combination of PSO and GA is used in the process of decision making [25], a balanced fuzzy PSO for distinguishing differences between positive and negative information of membership functions [26], an encoding fuzzy model utilized to achieve the structures and parameters [27]. An improved method of optimization using fuzzy logic to integrate the results of particle swarm optimization and genetic algorithm for parameter tuning [28] is also found in this area.

In this research work, a combination of particle swarm optimization and fuzzy approach is used to develop a PSO-fuzzy model to determine the category of 85th speed in intercity roads. The ability of fuzzy approach on defining linguistic variables and PSO ability for parameter tuning are combined in the proposed model, which is capable of determining the category of the 85th speed by experts' points of view and low cost measures. The main contribution in this paper is developing a model to estimate the category of 85th speed of vehicles using a few measurable factors. This concept can be applied in undivided roads, in which there is no enough electronic devices or equipments to collect reliable data particularly corresponding to speed and traffic volume. After introduction, Section 2 presents a very short description on particle swarm optimization technique. Input and output variables corresponding to the model are defined in Section 3, followed by model development in Section 4. Comparison of the results using the experimental data with a well-known exponential regression method of prediction as well as summary and conclusion will be explained at the next on Sections 5 and 6, respectively.

2. Particle Swarm Optimization

One of the well-known evolutionary algorithms used in optimization problems is particle swarm optimization (PSO). Particle swarm optimization (PSO) is a heuristic technique, developed based on bird flocking or fish schooling by Kennedy and Eberhart in 1995 [29]. It is a population-based technique the same as Genetic Algorithms (GA). Although, PSO's theoretical foundation is rather weak, but its application has been proved to be effective [30]. In PSO, feasible solutions, called particles (individuals), are generated by moving particles towards the current optimum particles. The best solution in the current population is called Pbest and in the whole generation is called Gbest [29]. The main idea in this process is to generate new populations by changing the velocity of each particle towards its Pbest and Gbest locations. Running stages continue until stopping criteria are met. At final stage, Gbest is considered as the best solution. Figure 1 illustrates the overall view of particle swarm optimization.

In PSO, a particle is composed of three vectors of X, P, and V, which represent the current location, location of the best solution, and a gradient for moving particles, respectively. It also includes two fitness values: X-Fitness records the fitness of the X-vector and P-Fitness records the fitness of P-vector. Moving each particle is done by simply adding V-vector to the X-vector, while V-vector must have been adjusted by random generated rates [31]. In each iteration, the positions of particles are updated by using (1), while V-vector (velocity) is adjusted by (2)

$$Xid = Xid + Vid, \tag{1}$$

$$\begin{aligned} Vid = {} & Vid + \varphi 1 \times \text{random} \times (Pid - Xid) \\ & + \varphi 2 \times \text{random} \times (Pgd - Xid), \end{aligned} \tag{2}$$

where, Xid is the position of particle, d is dimension, Vid is moving rate, Pid is location of the best solution in iteration i, and Pgd is the location of global best solution. Different amounts of $\varphi 1$ and $\varphi 2$ identify the types of PSO including full model ($\varphi 1, \varphi 2 > 0$), cognition only ($\varphi 1 > 0, \varphi 2 = 0$), social only ($\varphi 1 = 0, \varphi 2 > 0$), and selfless ($\varphi 1 = 0, \varphi 2 > 0$ and $g\#i$) [31].

There are two well-known techniques of controlling velocities called inertia factor and constriction coefficient. When inertia factor is used, (2) is changed to (3) and when constriction coefficient is used (2) is changed to (4) [32]

$$\begin{aligned} Vid = {} & \omega \times Vid + \varphi 1 \times \text{random} \times (Pid - Xid) \\ & + \varphi 2 \times \text{random} \times (Pgd - Xid), \end{aligned} \tag{3}$$

$$\begin{aligned} Vid = {} & K \times [Vid + \varphi 1 \times \text{random} \times (Pid - Xid) \\ & + \varphi 2 \times \text{random} \times (Pgd - Xid)], \end{aligned} \tag{4}$$

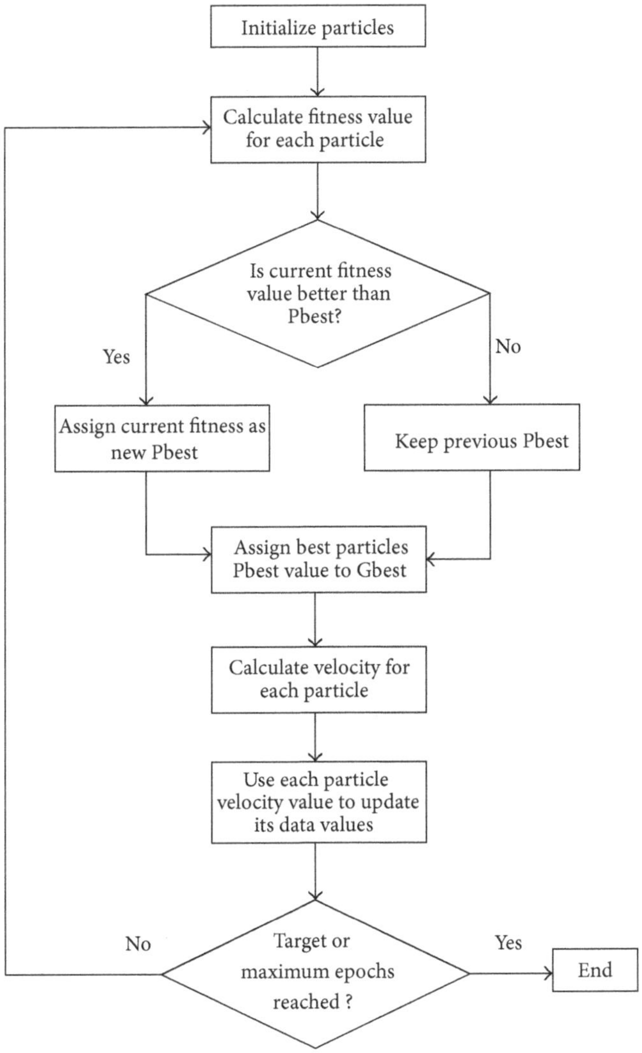

FIGURE 1: Overall view of particle swarm optimization.

where

$$K = \frac{2}{\left|2 - \varphi - \mathrm{sqrt}\left(\varphi^2 - 4 \times \varphi\right)\right|}, \quad \varphi = \varphi 1 + \varphi 2, \ \varphi > 4. \tag{5}$$

Swarm topology is another concern in PSO. In general, a fully connected neighborhood topology named Gbest, in which every particle is a neighbor of every other particle, is used for attracting particles to the best solution. However, each particle has access to the information corresponding to its neighborhood according to a certain swarm topology called local neighbor models (Lbest) [33]. The two most common topologies are ring topology, in which each particle is connected with two neighbors, and wheel topology, in which the individuals are isolated from one to another, and all information is communicated to a focal individual [33].

If there is no centralized concept of a global Gbest, particles select each others as informants and swarm topology is highly dependent on the threshold p, which is constant for all particles in the swarm [33]. Particles are chosen if random value is less than p and Gbest is selected among the chosen particles. The mentioned process of choosing particles is called stochastic star of swarm topology. Finally at the end of PSO, two methods are used for updating particles: synchronous method, in which newly discovered solutions are used more quickly and asynchronous method in which all solutions have equal chance to be selected and updated [30].

3. Input and Output Variables of the Model

In a certain research, done by Iranian transport research center, the 85th speed and its affecting parameters have been collected for 194 locations [1], so available data lead us to develop a model to determine the previously mentioned parameter using indirect measures. In this research work, new definitions have been introduced to represent the proper variables affecting the 85th speed, which is a part of our novelty in this paper. Input parameters which are three variables are defined corresponding to width, length, and road side condition. Each of them is defined as an integration of its corresponding factors. The 85th speed has also been categorized into 5 domains based on experimental applications. The suggested process is explained in four stages as follow.

3.1. Defining Integrated Factors and Parameters. As mentioned before, input variables have been categorized into three main integrated factors including width, length, and roadside land use, while each integrated factor consists of a weighted sum of the corresponding parameters as follows:

(i) width parameters:
  (a) pavement width,
  (b) shoulder width,

(ii) length parameters:
  (a) percentage of forbidden overtaking,
  (b) number of access points,

(iii) roadside land use parameters:
  (a) type of land use (none, residential, industrial, agricultural),
  (b) condensation (poor, medium, condensed).

Some of the previous parameters are measurable or quantitative and some are not. For example, the amount of pavement width is measureable and the amount of road side facilities is not. We need to convert all of the input parameters to fuzzy variables. Some of parameters such as pavement width have a proportional effect on 85th speed that increase maneuverability and speed. On the contrary, other parameters, such as number of access points, have inverse effect on 85th speed which reduce the maneuverability and speed. The previously-mentioned integrated factors, the related parameters, and their effects are shown in Table 1 in which the corresponding symbols and their effects are indicated.

Table 1: The integrated factors and corresponding parameters, types (crisp or fuzzy), symbols, and type of effect on the 85th speed value.

Integrated factor	Parameter	C/F*	Symbol	Effect
Width	Pavement width	C	PW	Direct
	Shoulder width	C	SW	Direct
Length	Overtaking	C	OV	Indirect
	Access point	C	AP	Indirect
Roadside	Type	F	TF	Direct
Land use	Condensation	F	CF	Direct

*C: crisp, F: fuzzy.

Table 2: Rules for defining the fuzzy variable of roadside facilities.

Type	Condensation	LU
Without	—	5
Residential	Poor	4
Residential	Medium	3
Residential	Condensed	2
Industrial	Poor	4
Industrial	Medium	3
Industrial	Condensed	2
Agricultural	Poor	5
Agricultural	Medium	4
Agricultural	Condensed	3

3.2. Defining Fuzzy Variables. For different types and condensations of roadside land use (LU), a number is associated as shown in Table 2. Variable LU represents a combination of both type and condensation of land use in order to be used as input variable. The boundary of each combination has been considered as variable in representation of the fuzzy variable in Table 2, so if a combination of more types of land use or condensation needs to be considered, a simple calculation is proposed to outline the final amount of variable. For example, if roadside land use consists of 20 percent industrial and 80 percent agricultural and condensation is medium, thus variable is calculated by $(0.2 \times 3 + 0.8 \times 4 = 3.8)$.

3.3. Converting Parameters to Fuzzy Input Variables. Integrated variables, which represent land use, width, and length including their parameters, are defined according to (6), (7), and (8), respectively. New variables of $V1$, $V2$, and $V3$ with closed interval of $[0 \;\; 1]$ are considered as inputs to the fuzzy model

$$V1 = \frac{W1 \times \mathrm{LU}}{5} \qquad W1 = [0 \;\; 1], \tag{6}$$

$$V2 = \frac{W2 \times PW + W3 \times SW}{7} \qquad W2, W3 = [0 \;\; 1], \tag{7}$$

$$V3 = \frac{(1/(W4 \times OV + 1)) + (1/(W5 \times AP + 1))}{2} \qquad W4, W5 = [0 \;\; 1]. \tag{8}$$

$W1, \ldots, W5$ weighting factors are considered to be used by the process of particle swarm optimization. New variables

Table 3: Normalized variables of factors by fuzzy sets.

State	Bound	Land use ($V1$)	Width ($V2$)	Length ($V3$)
0	0–0.25	Condensed	Narrow	Low maneuverable
0.5	0.25–0.75	Medium	Medium	Medium maneuverable
1	0.75–1	Poor	Vast	High maneuverable

Table 4: Categories (Domains) of 85th speed.

State	Fuzzy definition	Lower bound (Km/h)	Upper bound (Km/h)
1	Very low	51	60
2	Low	61	75
3	Medium	76	90
4	High	91	100
5	Very high	101	110

$V1, V2$, and $V3$, shown in Table 3, are defined as uniformed fuzzy variables when they have been converted to the fuzzy norms. The formulation of PSO provides an ability to meet the previous constraints, while the feasible solutions are kept while parameters are restricted to closed interval $[0 \;\; 1]$.

3.4. Setting Output Variable. The 85th speed is categorized according to instruction manuals, which are being used for road maintenance and road construction building [34]. In this manual, 85th speed is defined in six domains started from 0 ended to 110 km/h; however, in this research work the range of speed was considered from 50 to 110 km/h, and the scope of 85th speed is limited to five classes as shown in Table 4.

4. Developing PSO-Fuzzy Model

4.1. Main Structure. The proposed PSO-fuzzy model includes two main parts that are PSO and fuzzy parts. The fuzzy model estimates 85th speed based on input variables $V1$, $V2$, and $V3$ while PSO generates input variables based on weighting factors and compares estimated results with corresponding experimental data. The process is an iteration-based one. It means that parameters are updated in each sequence of running the model by checking stopping criteria. The stopping criterion is a constant number of iterations.

A PSO-fuzzy model is required to go through the following steps:

(1) initialization of weighting parameters $(W1, \ldots, W5)$;

(2) converting crisp variables to fuzzy ones using (6), (7), and (8);

(3) running the fuzzy part of the model;

(4) checking stopping criteria;

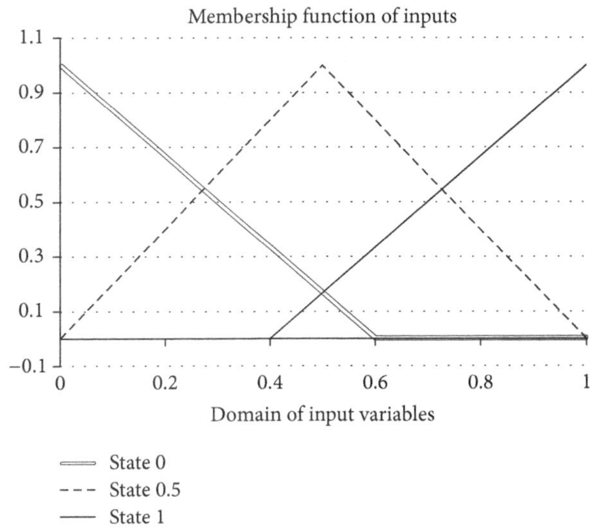

FIGURE 2: Membership functions corresponding to input variables.

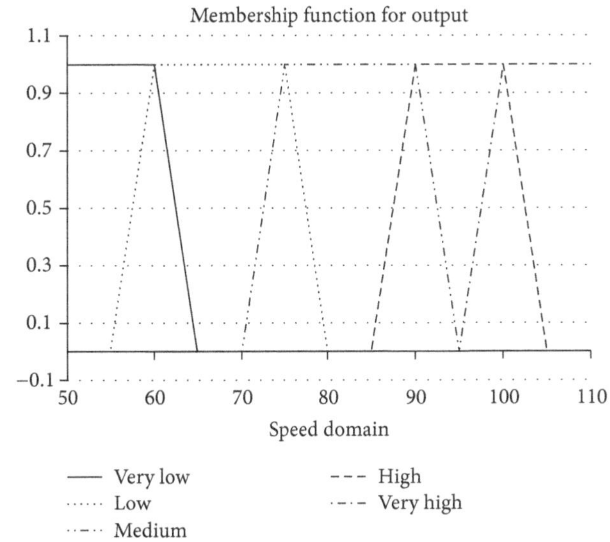

FIGURE 3: Membership function corresponding to 85th speed.

(5) if stopping criteria are met, go to step (7) and show results;

(6) update weighting parameters according to PSO velocity and movement, and then go to step (2);

(7) check and compare results with another method.

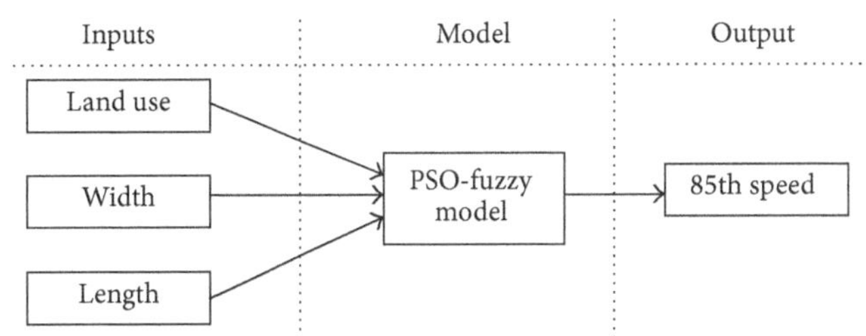

FIGURE 4: Overall view of fuzzy model.

4.2. Defining Membership Functions and Fuzzy Model. The first step of developing the fuzzy model is to define membership functions corresponding to input and output variables. Input variables are defined by one-interval domain, so a triangular membership function is appropriate to define input variables. Output variable is defined into five categories, so a trapezoidal membership function is used to define output variable. For each of the three input variables, a triangular membership functions is defined [11]. Figure 2 shows a sample of membership function utilized for input variables. A trapezoidal membership function is used for output variable as shown in Figure 3. The overall view of fuzzy model is also shown in Figure 4.

4.3. Rule Definition. For this fuzzy model, the corresponding rules are defined according to the traffic experts' points of view which explain the relationships between input and output variables [11]. Since the road width, road length, and land use have three different states, a maximum number of $3 \times 3 \times 3 = 27$ rules can be used to define the relationship between input and output variables. The corresponding states of the rules are shown in Table 5.

4.4. Model Presentation. Minimizing the number of incorrect assignment for defined categories of 85th speed is the main objective function in the proposed PSO-fuzzy model. Assume that N_{mi} is the number of samples assigned to the ith category by model and N_{oi} is the observed one; then objective function would be defined according to the following equation:

$$\text{Min} \quad Z = \sum_{i=1}^{5} \left| N_{mi} - N_{oi} \right| \quad \text{for } i = 1, \ldots, 5 \ \text{(all speed categories)} \tag{9}$$

Subject to: model constraints,

where, new variables defined by (6) to (8) will calculate the normalized variables which are assigned to the 85th speed categories:

$$Nmi = \text{Sum}\left(\text{Rule}\left(V1, V2, V3\right), i\right) \quad \text{for } i = 1, \ldots, 5, \tag{10}$$

$$V1 = \frac{W1 \times \text{LU}}{5}, \tag{11}$$

$$V2 = \frac{W2 \times PW + W3 \times SW}{7}, \tag{12}$$

$$V3 = \frac{(1/(W4 \times OV + 1)) + (1/(W5 \times AP + 1))}{2}. \tag{13}$$

And model constraints are as follows:

$$0 \le V1, V2, V3 \le 1,$$
$$0 \le W1, W2, W3, W4, W5 \le 1. \tag{14}$$

TABLE 5: Rule definition for 85th speed fuzzy model.

Rule	Land use	Width	Length	Speed
1	0	0	0	1
2	0	0	0.5	1
3	0	0	1	2
4	0	0.5	0	1
5	0	0.5	0.5	2
6	0	0.5	1	3
7	0	1	0	2
8	0	1	0.5	2
9	0	1	1	3
10	0.5	0	0	2
11	0.5	0	0.5	2
12	0.5	0	1	3
13	0.5	0.5	0	2
14	0.5	0.5	0.5	2
15	0.5	0.5	1	3
16	0.5	1	0	3
17	0.5	1	0.5	4
18	0.5	1	1	4
19	1	0	0	3
20	1	0	0.5	4
21	1	0	1	4
22	1	0.5	0	4
23	1	0.5	0.5	5
24	1	0.5	1	5
25	1	1	0	4
26	1	1	0.5	5
27	1	1	1	5

Rule function in (10) is the result of fuzzy function derived by Table 5. In the proposed model $W1, \ldots, W5$ are PSO parameters which should be estimated and are restricted in closed interval [0 1]. Closed interval [0 1] should be met in order to keep feasibility solutions. Other parameters are defined in Table 1. Iterative procedure of PSO calculates the amount of each parameter in iterations and converge them to the best solution. A full model of PSO ($\varphi 1, \varphi 2 > 0$) has been developed because Gbest (global optimum) and Pbest (population optimum) are required to be involved in velocity calculation.

Because the convergence speed in the particle swarm optimization algorithm with the constriction coefficient (convergence agent) is much quicker than inertia factor [30], constriction coefficient by a different structure from conventional one is used to control velocities via (15) and (16):

$$\begin{aligned} Vi5 = {} & \text{Max}\left(0, Vi5 + \varphi 1 \times \text{random} \times (Pi5 - Wi5) + \varphi 2\right. \\ & \left. \times \text{random} \times (Pg5 - Xi5)\right) \quad \text{if step} < 0, \end{aligned} \tag{15}$$

$$\begin{aligned} Vi5 = {} & \text{Min}\left(1, Vi5 + \varphi 1 \times \text{random} \times (Pi5 - Wi5) + \varphi 2\right. \\ & \left. \times \text{random} \times (Pg5 - Xi5)\right) \quad \text{if step} > 0, \end{aligned} \tag{16}$$

where,

$$\begin{aligned} \text{Step} = {} & (\varphi 1 \times \text{random} \times (Pid - Xid) \\ & + \varphi 2 \times \text{random} \times (Pgd - Xid)). \end{aligned} \tag{17}$$

A fully connected neighborhood topology has been used as swarm topology in the model; therefore all particles have been connected to each other. All particles are updated in each iteration using synchronous method of updating.

5. Running PSO-Fuzzy Model

5.1. Parameter Calculation and Discussion. A set of 101 experimental data has been used to adjust the PSO-fuzzy model. Data have been practically gathered for speed zone determination in another research work [1], but they are available for checking the proposed algorithm. Although a lot of variables have been gathered in the above research work, the related attributes are used to check the performance of proposed algorithm. The objective function is the accuracy of the model and the percentage of correct assignment of the speed values to the whole observations is defined as accuracy of the model. The error corresponding to the input set of data has been reduced by changing the weighing factors as optimization variables. The number of iterations and population size are two main parameters affecting the improvement of the model. They have significant roles in finding the best values for weighing factors. For each population size and iteration number, the means of the results corresponding to five times of running are shown in Table 6. This table shows that the population size of 100 and the iteration number of 200 to 500 are sufficient for an accuracy of 96%. No more improvement has been observed when the numbers of iteration and population size are increased. Results also revealed that the population size and iteration number dramatically change the parameters, and accuracy. It means that using PSO may be a good technique for parameter tuning in these kinds of objective functions.

5.2. Comparison with Exponential Model. In order to compare the accuracy of results, a conventional method of exponential regression model for parameter estimation has been used. In this model, if the effects of independent variables on dependent variable are power based, the model is called exponential regression. Parameters are calculated based on the minimum of mean or total square errors between observations and model outputs [35]. In this model, independent variables $X_1, X_2, \ldots, X_n$ and dependent variable Y are estimated by the following equation:

$$Y = \beta_0 \times \exp(\beta_1 X_1) \times \exp(\beta_2 X_2) \times \cdots \times \exp(\beta_n X_n). \tag{18}$$

In this equation, coefficients $\beta_0, \beta_1, \beta_2, \beta_3, \ldots,$ and β_n are estimated through minimization of mean or total square

TABLE 6: Results of running PSO-fuzzy model of 85th speed.

Population size	Iteration number	Parameters					Accuracy (percent)
		$W1$	$W2$	$W3$	$W4$	$W5$	
20	100	0.623	0.725	0.302	0.498	0.423	91
	200	0.615	0.749	0.294	0.514	0.427	93
	500	0.684	0.771	0.286	0.519	0.408	94
50	100	0.649	0.789	0.246	0.515	0.424	93
	200	0.632	0.779	0.244	0.512	0.418	94
	500	0.682	0.808	0.254	0.567	0.434	95
100	100	0.598	0.805	0.213	0.532	0.402	95
	200	0.610	0.817	0.204	0.543	0.398	96
	500	0.622	0.844	0.212	0.569	0.389	96

errors between observations and model outputs [36]. Parameters have been estimated for experimental data using solver option in Microsoft Excel and model has been formulated as (19):

$$\text{Speed} = 110.028 \times e^{0.0899PW} \times e^{0.00295SW} \times e^{-0.0991OV} \times e^{-0.0160AP} \times e^{0.0596TR} \times e^{-0.0845CR}. \quad (19)$$

The accuracy of exponential model was 84%, while that of proposed PSO-fuzzy model was 96% which reveals that PSO-fuzzy model is a more accurate model to estimate 85th speed of vehicles in intercity roads. There are two main reasons to achieve better accuracy. The first is using new variables defined as $V1$, $V2$, and $V3$ in Section 3.3 and the second is developing a nonlinear structures which are utilized in other techniques such as neural network approved in other studies [37].

6. Summary and Conclusion

In this paper, a PSO-fuzzy model has been developed and utilized to determine the category of the 85th speed of vehicles in undivided intercity roads. Roadside facilities (land use), width, and length characteristics of roads are considered as input variables and the category of 85th speed as output variable to the fuzzy model, respectively. New definitions of input variables for fuzzy model have been proposed to combine crisp and fuzzy variables in the model. Input variables were converted to normalized format using weighting factors, which are estimated by particle swarm optimization technique. Fully neighborhood particles are considered as topology, while constriction coefficient is used for velocity limitation in synchronous method of updating particles.

A set of 101 experimental data has been used for tunning the weighting factors (PSO parameters). Results show that the algorithm of combining PSO and Fuzzy as well as and new definition of input variables integrating with the crisp and fuzzy variables is capable of estimating the category of 85th speed of the road passing vehicles. The developed model is capable of estimating the 85th speed with a high accuracy of 96%, while conventional method of exponential regression estimates speed categories corresponding to the experimental data with an accuracy of 84%.

Results also reveal that the model is sensitive to population size and iteration number which are basically used by PSO algorithm. As population size and iteration number became larger, better results are achieved. Land use parameter and pavement width represented by weighting factor $W1$ and $W2$, respectively, have significant effect on 85th speed. The percentage of forbidden overtaking represented by $W4$ has indirect effect on 85th speed. However, its effect is not as large as the effects of the land use and pavement width. Shoulder width and number of access points, represented by $W3$ and $W5$, respectively, have less effect on 85th speed. It also shows that pavement width is more important to drivers' speed than shoulder width as well as the number of access points plays a less significant role on 85th speed.

For future researches, it is recommended to develop a more complicated fuzzy model in order to apply more input variables such as awareness campaigns and enforcement activities. In addition, the developed fuzzy model may be utilized for all types of the roads particularly for divided roads and freeways.

References

[1] Transportation Research Institute Andishe Rahian Gostaresh, "Calibration of zoning speed limit in rural roads," Report no. 1, 2007.

[2] G. A. Rezai Rad and J. Mohamadi, "Vehicle speed estimation based on the image," in *Proceedings of the 4th International Conference: Sciences of Electronic, Technologies of Information and Telecommunications*, pp. 1–5, 2007, Hammamet, Tunisia, March 2007.

[3] A. Hussain, K. Shahzad, and C. Tang, "Real time speed estimation of vehicles," *World Academy of Science, Engineering and Technology*, vol. 61, pp. 726–730, 2012.

[4] L. H. Martin, "Estimating vehicle speed from traffic count and occupancy data," *Journal of Data Science*, vol. 2, pp. 231–244, 2004.

[5] S. I. Manjari Gupta and A. Bhattacharyya, "A study on lateral speed estimation methods," *International Journal of Vehicle Autonomous Systems*, vol. 2, no. 1-2, 2004.

[6] D. Bauer, P. Bühler, N. Donath et al., "Embedded vehicle counting system with "silicon retina" optical sensor," in *Workshop on Information Optics*, Toledo, Spain, 2006.

[7] N. Kasem, A. Kosba, and M. Youssef, "An RF-based vehicle detection and speed estimation system," in *Proceedings of the 75th IEEE Vehicular Technology Conference (IEEE VTC '12)*, Yokohama, Japan, 2012.

[8] Y. U. Ali, P. Huei, and H. E. Tseng, "A study on lateral speed estimation methods," *International Journal of Vehicle Autonomous Systems*, vol. 2, no. 1-2, pp. 126–144, 2004.

[9] P. Augural, *Fuzzy Model for Estimating of Passenger Car Unit*, National Institute of Technology, Information Science and Application, Kurukshetra, India, 2008.

[10] P. Panetta, J. Poj, and N. Nanokchai, *Evaluation of Road Traffic Congestion Using Fuzzy Techniques*, King Mongkut South University of Technology, Thonburi, Thailand, 2008.

[11] P. Pongpaibool, P. Jangmchil, and N. Noodwong, *Evaluation of Road Traffic Congestion Using Fuzzy Techniques*, King Mongkut South University of Technology, Thonburi, Thiland, 2008.

[12] L. Ruey-Hsun, T. Sheng-Ren, C. Yie-Tone, and T. Wan-Tsun, "Optimal power flow by a fuzzy based hybrid particle swarm optimization approach," *Electric Power Systems Research*, vol. 81, no. 7, pp. 1466–1474, 2011.

[13] L. Wang and C. Singh, "Balancing risk and cost in fuzzy economic dispatch including wind power penetration based on particle swarm optimization," *Electric Power Systems Research*, vol. 78, no. 8, pp. 1361–1368, 2008.

[14] S. Manna, R. H. Kenneth, M. El-Halwagi, R. Longari, and M. Gentile, *Development of Hierarchical Fuzzy Model for the Evolution of Inherent Safety*, Universidad de las Americas, Pueblea, Mexico, 2004.

[15] C. Yang, S. Fan, Z. Wang, and W. Li, "Application of fuzzy control method in a tunnel lighting system," *Mathematical and Computer Modelling*, vol. 54, no. 3-4, pp. 931–937, 2011.

[16] S. Hoogendoorn, S. L. Hoogendoorn, and H. Schuurman, *Fuzzy Perspective in Traffic Engineering*, Delft university of technology and Ministry of Transport, Delft, The Netherlands, 2000.

[17] Y. Lu, J. Hinze, and Q. Li, "Developing fuzzy signal detection theory for workers' hazard perception measures on subway operations," *Safety Science*, vol. 49, no. 3, pp. 491–497, 2011.

[18] E. Bagheri, M. Feizi, F. Enson, and F. Behnia, *Novel Fuzzy Control Model of Traffic Light Timing at an Urban Intersection*, Department of Computer Science, University of New Branswick, Prediction, Canada, 2004.

[19] J. L. Castro, M. Delgado, J. Medina, and M. D. Ruiz-Lozano, "An expert fuzzy system for predicting object collisions. Its application for avoiding pedestrian accidents," *Expert Systems with Applications*, vol. 38, no. 1, pp. 486–494, 2011.

[20] G. Ficili and D. Panno, "A fuzzy algorithm for combined control of traffic parameters: assessment and key issues," *Computer Communications*, vol. 22, no. 3, pp. 199–210, 1999.

[21] C. F. Juang and C. Y. Wang, "A self-generating fuzzy system with ant and particle swarm cooperative optimization," *Expert Systems with Applications*, vol. 36, no. 3, pp. 5362–5370, 2009.

[22] I. H. Kuo, S. J. Horng, T. W. Kao, T. L. Lin, C. L. Lee, and Y. Pan, "An improved method for forecasting enrollments based on fuzzy time series and particle swarm optimization," *Expert Systems with Applications*, vol. 36, no. 3, pp. 6108–6117, 2009.

[23] H. Liu, A. Abraham, and A. E. Hassanien, "Scheduling jobs on computational grids using a fuzzy particle swarm optimization algorithm," *Future Generation Computer Systems*, vol. 26, no. 8, pp. 1336–1343, 2010.

[24] Y. L. Huang, S. J. Horng, M. He et al., "A hybrid forecasting model for enrollments based on aggregated fuzzy time series and particle swarm optimization," *Expert Systems with Applications*, vol. 38, no. 7, pp. 8014–8023, 2011.

[25] F. Valdez, P. Melin, O. Castillo, and O. Montiel, "A new evolutionary method with a hybrid approach combining particle swarm optimization and genetic algorithms using fuzzy logic for decision making," in *Proceedings of the IEEE Congress on Evolutionary Computation (CEC '08)*, pp. 1333–1339, June 2008.

[26] A. Robati, G. A. Barani, H. Nezam Abadi Pour, M. J. Fadaee, and J. Rahimi Pour Anaraki, "Balanced fuzzy particle swarm optimization," *Applied Mathematical Modelling*, vol. 36, no. 5, pp. 2169–2177, 2012.

[27] L. Zhao, F. Qian, Y. Yang, Y. Zeng, and H. Su, "Automatically extracting T-S fuzzy models using cooperative random learning particle swarm optimization," *Applied Soft Computing Journal*, vol. 10, no. 3, pp. 938–944, 2010.

[28] F. Valdez, P. Melin, and O. Castillo, "An improved evolutionary method with fuzzy logic for combining Particle Swarm Optimization and Genetic Algorithms," *Applied Soft Computing Journal*, vol. 11, no. 2, pp. 2625–2632, 2011.

[29] J. Kennedy and R. Eberhart, "Particle swarm optimization," in *Proceedings of the IEEE International Conference on Neural Networks*, vol. 4, pp. 1942–1948, December 1995.

[30] Q. Bai, "Analysis of particle swarm optimization algorithm," *Computers and Information Science*, vol. 3, no. 1, pp. 180–184, 2010.

[31] J. Kennedy, "Particle swarm: social adaptation of knowledge," in *Proceedings of the IEEE International Conference on Evolutionary Computation (ICEC '97)*, pp. 303–308, IEEE Press, April 1997.

[32] M. Clerc and J. Kennedy, "The particle swarm-explosion, stability, and convergence in a multidimensional complex space," *IEEE Transactions on Evolutionary Computation*, vol. 6, no. 1, pp. 58–73, 2002.

[33] S. Ghosh, D. Kundu, K. Suresh et al., "On some properties of the lbest topology in particle swarm optimization," in *Proceedings of the 9th International Conference on Hybrid Intelligent Systems (HIS '09)*, pp. 370–375, August 2009.

[34] Iranian Management and Planning Organization, *Road Safety Instruction Manual*, vol. 3–7, no. 267, 1st edition, 2005.

[35] P. G. Hoel, *Elementary Statistics*, John Wiley and Sons, New York, NY, USA, 4th edition, 1976.

[36] G. A. F. Seber and C. J. Wild, *Nonlinear Regression*, John Wiley & Sons, New York, NY, USA, 1989.

[37] A. Mahmoudabadi, "Comparison of weighted and simple linear regression and artificial neural network models in freeway accidents prediction," in *Proceeding of the 2nd International Conference on Computer and Network Technology*, pp. 392–396, Bangkok, Thailand, April 2010.

10

Particle Swarm Optimization for Multiband Metamaterial Fractal Antenna

Balamati Choudhury, Sangeetha Manickam, and R. M. Jha

Centre for Electromagnetics, CSIR-National Aerospace Laboratories, Bangalore 560017, India

Correspondence should be addressed to Balamati Choudhury; balamati@nal.res.in

Academic Editor: Ling Wang

The property of self-similarity, recursive irregularity, and space filling capability of fractal antennas makes it useful for various applications in wireless communication, including multiband miniaturized antenna designs. In this paper, an effort has been made to use the metamaterial structures in conjunction with the fractal patch antenna, which resonates at six different frequencies covering both C and X band. Two different types of square SRR are loaded on the fractal antenna for this purpose. Particle swarm optimization (PSO) is used for optimization of these metamaterial structures. The optimized metamaterial structures, after loading upon, show significant increase in performance parameters such as bandwidth, gain, and directivity.

1. Introduction

The non-integral dimensions, recursive irregularity, and space filling capability of fractal antennas make it useful for various applications in wireless communication including miniaturized antenna designs [1]. Their property of being self-similar in the geometry leads to antennas of compact size with simplified circuit designs. Antennas, which have fractal geometry, are self-iterative, exhibiting multiband operation. Fractal antennas are frequency independent and have schemes for realizing low sidelobe designs. An antenna with fractal geometry is preferred to conventional antenna designs due to the iterative behavior of the structure, which is believed to improve the performance factors like gain, bandwidth, return loss and frequency of operation [2].

Metamaterials are artificial structures designed by placing electromagnetic (EM) resonators, such as split ring resonators (SRRs), at regular intervals. The metamaterials have frequency selective response and exhibit unique EM properties such as negative permittivity and permeability, artificial magnetism and negative refractive index, which can be used to improve the performance of antenna [3]. The media composed of metamaterials have tunable effective material parameters, and their electromagnetic response can be adjusted in real time. By using metamaterials as substrates or superstrates for antenna, significant improvements have been observed in the properties of the fractal antenna.

In this paper, an effort is made to use the square *split ring resonator* (SRR) in conjunction with the fractal patch antenna to enhance the directivity, gain, voltage standing wave ratio (VSWR), and bandwidth at multiple resonant frequencies. To serve this purpose, two different types of tunable multiband micro-split metamaterial square SRR are loaded on the fractal antenna [4]. Such an application requires design of the loaded structure at the desired frequency range (equivalent to the resonant frequencies of the fractal antenna, whose performance is to be improved). Towards this, a PSO optimizer is developed, which yields structural parameters at a particular resonant frequency. Particle swarm optimization (PSO) is an evolutionary computational technique based on the movement and intelligence of swarms, which is used for optimizing difficult multidimensional discontinuous problems in a variety of fields including electromagnetics [5, 6]. The optimal structural parameters of square SRR such as length, width, distance between successive rings for a particular resonant frequency are obtained using this efficient optimization technique based on an equivalent circuit analysis.

2. Overview of Metamaterial Fractal Antennas

The term fractal was first coined by Mandelbrot in 1975, in which rough, irregular, or fragmented geometric shape could be subdivided in parts, each of which was reduced-size copy of the whole [7]. A fractal antenna possesses two properties, namely, self-similarity and space filling, and hence the resonance takes place over a wide band as well as at multiple frequencies. These properties of the fractals can be used to develop new configurations for antennas and antenna arrays. Since the fractal antennas are multi-resonant and smaller in size, they are used for wireless applications. Fractal antenna can be designed to receive and transmit over a wide range of frequencies using the self-similar property associated with fractal geometry structures [2]. Space filling fractal geometries such as Hilbert curves are used to design miniature antennas, high-directive radiators, and high-impedance surfaces. This concept was used for design of miniaturized inverted-F antenna for wireless sensor networks [8]. The overall size of the antenna was reduced to 77% by using Hilbert geometry, instead of conventional rectangular patch antenna for the same resonant frequency.

Radar cross section (RCS) reduction in microstrip antenna was achieved by using Koch fractal geometry, without affecting radiation performance of the antenna [9]. An UHF Quasi-Yagi antenna surrounded by fractal metamaterial structures was designed for RFID applications. The operational frequency of this antenna was 2.4 GHz [10]. Similarly, star-shaped fractal patch antenna was used for miniaturization and backscattering radar cross-section (RCS) reduction [11]. Metamaterial unit cells could be miniaturized by incorporating fractal geometries. The effect of various geometrical parameters and the order of fractal curve on the performance of the fractal antenna were also investigated and reported that the fractal geometries of original dimensions give better performance than the miniaturized one [12].

Further, a novel negative-epsilon metamaterial was designed using fractal Hilbert curves with mirror symmetry [13]. The slotted SRRs in the ground plane were abandoned in order to reduce back radiation of the antenna. The numerical analysis and design also shows that a periodic arrangement of Hilbert curve inclusions above a conducting ground plane forms a metamaterial surface with high impedance [14].

Using multi-resonator configuration, a low-loss dispersion-optimized engineered substrate was designed for antenna miniaturization and performance enhancement. The proposed substrate contained four uncoupled inclusions with 3rd order fractal Hilbert structure. A circuit model was developed to analyze and optimize the proposed metamaterial [15]. A metamaterial with zero refractive index was designed over a wide frequency range and used as superstrate for patch antenna [16]. The designed antenna showed an antenna gain improvement by 5 dB.

It is observed from the overview of fractal antenna that Hilbert curves proved to be efficient fractal geometries to generate metamaterial surfaces. Fractal geometries were also used as miniaturized antennas with metamaterial structures as superstrate. In this paper metamaterial structures are loaded on a fractal antenna for increase in the bandwidth at the multiple distinct resonating frequencies.

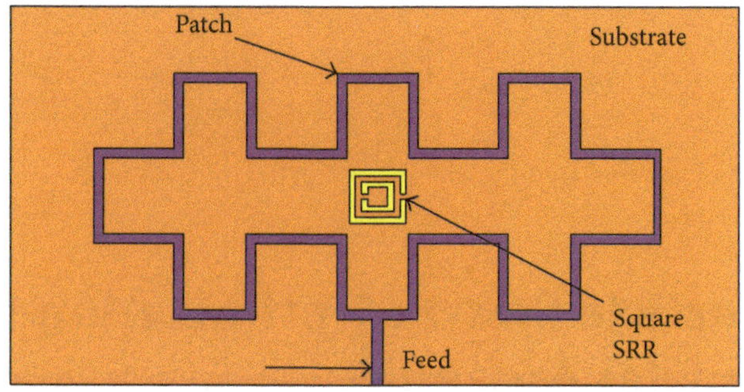

FIGURE 1: Top view of the square SRR loaded fractal antenna.

3. Design of Fractal Antenna

Fractal antennas are widely used for their compact size and light weight in wireless applications. In this work, the fractal antenna is designed with multiple resonant frequencies. Fractal patch antenna is designed on a substrate with dimensions 36 mm × 20 mm × 1.6 mm having a relative permittivity of 4.4 [17]. The patch used in the antenna layout is copper with the area 28 mm × 12 mm. The size of the patch used to feed the antenna is 0.5 mm × 4 mm. Figure 1 shows the schematic diagram of the fractal patch antenna. Further metamaterial structures are loaded on the antenna to improve the performance.

Two different algorithms, namely, PSO and bacteria foraging optimization (BFO), have been implemented towards optimization of square SRR [18]. It was observed that the execution time and accuracy of these algorithms depend on selection of parameters. For this optimization problem, the details of accuracy and execution time for both the algorithms are given in Table 1.

It is readily observed that for the class of problem in hand, PSO is highly convergent and more accurate compared to the BFO. Hence, PSO algorithm is considered in this work for further study of metamaterial antenna design.

4. PSO Optimization for SRR Design

The PSO is a simple, effective, and robust method used for search and optimization in various EM problems. The development of PSO can be illustrated through an analogy similar to a swarm of bees in a field. The goal of a swarm of bees in a field is to find the location with the highest density of flowers [5]. This motivates the engineers to use PSO as an optimization technique [6]. For completeness of the paper, the step-by-step algorithm is described briefly along with the flowchart (Figure 2).

4.1. PSO Algorithm. The PSO algorithm is given here for better understanding of the implementation for optimization of the resonant frequency (and hence extraction of the structural parameters) of a metamaterial SRR described in the following section. The step-by-step procedure is given next.

TABLE 1: Comparison of PSO and BFO for structural optimization of square SRR at a particular desired frequency. The substrate dielectric constant is 3.86, and desired frequency is 2.388 GHz.

Optimization techniques	Accuracy (f_{err})	Gap between rings d (in mm)	Width of rings w (in mm)	Length a (in mm)	Execution time (in seconds)
BFO	0.00161	0.12	0.12	3.96	12.55
PSO	0.000442	0.14	0.10	3.95	1.04

Step 1. Define the solution space. The parameters to be optimized are selected in this space. A minimum ($x_{\min(n)}$) and a maximum ($x_{\max(n)}$) range is defined, where n ranges from 1 to N (dimension of the optimization space).

Step 2. Define a fitness function. The fitness function exhibits a functional dependence that is relative to the importance of each characteristic being optimized.

Step 3. Initialize random swarm location and velocities. Each particle begins at its own random location with a velocity that is random in its direction and magnitude. The personal best (p_{best}) and the global best (g_{best}) are found. The personal best is the position of highest fitness locally, and global best is the position of the highest fitness of the entire swarm.

Step 4. Systematically fly the particles through the solution space. The algorithm acts on each particle one by one, moving it by a small amount and cycling through the entire swarm.

The following steps are encountered on each particle.

(i) Particle's fitness evaluation comparing g_{best} and p_{best}.

(ii) Further depending on the fitness value, the particle's velocity has been updated using the following equation:

$$v_n = w * v_n + c1 * \text{rand}() * (p_{\text{best},n} - x_n) + c2 * \text{rand}() * (g_{\text{best},n} - x_n), \quad (1)$$

where v_n is the velocity of a particle in the nth dimension; w is known as inertial weight (range is between 0.0 and 1.0); x_n is the particle's coordinate in the nth dimension; $c1$ and $c2$ are two scaling factors, which determine the relative pull of p_{best} and g_{best} and rand () is random function in the range [0, 1].

(iii) Once the particle velocity has been updated, the particle has to move to its next location. The velocity is applied for time-step t, and new coordinate x_n is computed for each of the N dimensions according to the following equation:

$$xn = xn + \Delta t * vn. \quad (2)$$

Step 5. For each particle in the swarm, Step 4 is repeated. Every second the snapshot is taken for the entire swarm, so at that time the positions of all particles are evaluated and correction is made to p_{best} and g_{best} values if required.

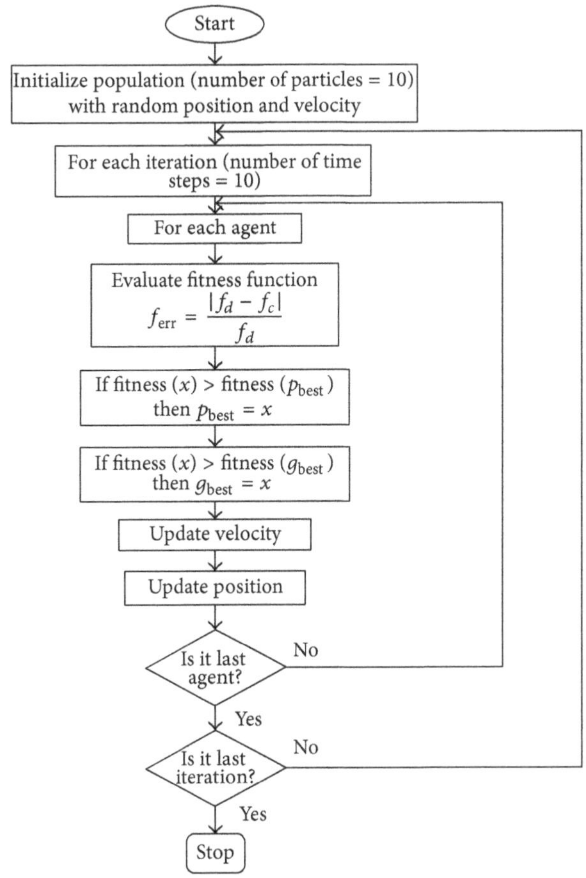

FIGURE 2: Flowchart of the PSO algorithm used for square SRR optimization.

4.2. Implementation of PSO. Square SRR is a metamaterial structure, which consists of double square shaped ring with a gap. This structure is printed on a dielectric substrate of thickness 1.6 mm and permittivity 4.4. The schematic of a square SRR with the dimensions is shown in Figure 3, where a is the side length of the square SRR, w is the width of conductor, d the successive distance between the rings, and g is the gap present in the rings. The equivalent circuit of the square SRR is a parallel LC tank circuit, given in Figure 4. The resonant frequency of the square SRR is obtained by equivalent circuit analysis method. In this method, the distributed network is converted to lumped network (Figure 4) and analysis of the resonant frequency has been carried out [19].

The resonant frequency of the split ring resonator is given by

$$f_r = \frac{1}{2\pi\sqrt{LC}}, \quad (3)$$

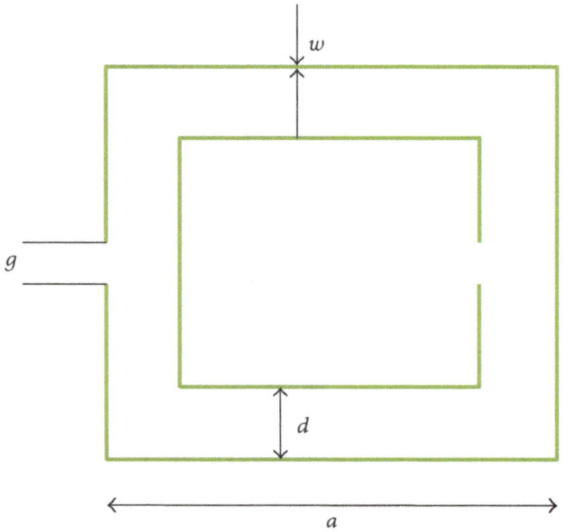

FIGURE 3: Top view of a unit cell of the metamaterial structure.

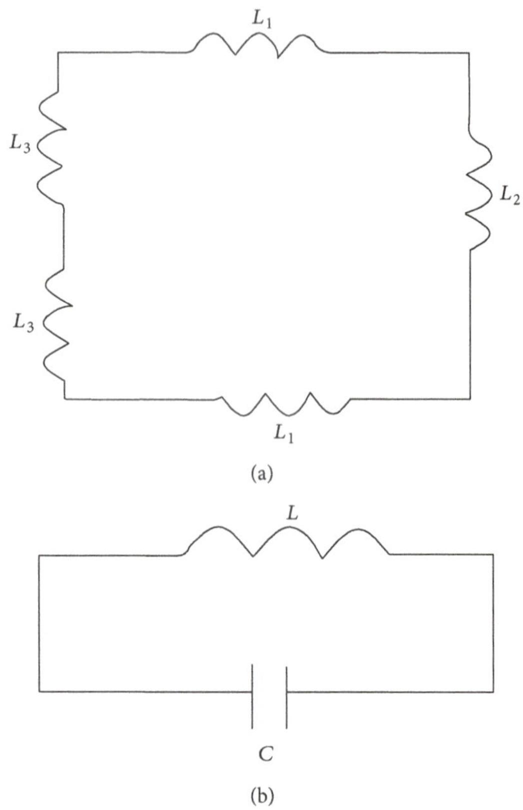

FIGURE 4: (a) Equivalent inductance of the square SRR. (b) Equivalent circuit of the square SRR.

where L is total inductance and C is gap capacitance, which are dependent on structural parameters of square SRR [20]. The expressions for L and C are given below.

The effective inductance of the square SRR is given by [20]

$$L = \frac{4.86\mu_0}{2}(a - w - d)\left[\ln\left(\frac{0.98}{\rho}\right) + 1.84\rho\right], \quad (4)$$

TABLE 2: List of parameters along with their values considered in PSO.

PSO parameters	Value	Use
w	0.25	Inertial weight
$c1$	2.05	Constant 1, to determine p_{best}
$c2$	2.05	Constant 2, to determine g_{best}
N_p	10	Number of particles
N_d	5	Number of dimensions
N_t	20	Number of time steps
$X_{\min}$	0	Scalar, min. for particle position
$X_{\max}$	10	Scalar, max. for particle position
$V_{\min}$	−1.5	Scalar, min. for particle velocity
$V_{\max}$	1.5	Scalar, max. for particle velocity

where ρ is the filling factor which depends on width and thickness of the SRR. The filling factor with respect to Figure 3 is given by

$$\rho = \frac{w + d}{a - w - d}. \quad (5)$$

Similarly, the effective capacitance is given by

$$C_s = \left(a - \frac{3}{2}(w + d)\right) C_{\text{pul}}, \quad (6)$$

where C_{pul} is the per-unit-length capacitance between the rings which is given as

$$C_{\text{pul}} = \varepsilon_0 \varepsilon_{\text{eff}} \frac{K\left(\sqrt{1 - k^2}\right)}{K(k)}. \quad (7)$$

Here ε_{eff} is the effective dielectric constant, which is expressed as

$$\varepsilon_{\text{eff}} = \frac{\varepsilon_r + 1}{2}. \quad (8)$$

$K(k)$ denotes the complete elliptical integral of the first kind with k expressed as

$$k = \frac{d}{d + 2w}. \quad (9)$$

The PSO optimizer acts here as a CAD package which yields the structural parameters such as the length, width, and spacing at a desired resonant frequency. The cost function used for this optimization is

$$f_{\text{err}} = \frac{|f_d - f_c|}{f_d}, \quad (10)$$

where, f_d is the desired frequency and f_c is the frequency arrived at by the equivalent circuit analysis. As per the algorithm mentioned, the different parameters are assigned with respect to the problem. The parameters of the PSO program are given in Table 2. The selection of PSO parameters such as w, $c1$, and $c2$ considered are from the original work by Kennedy and Eberhart [5]. The inertial weight is always

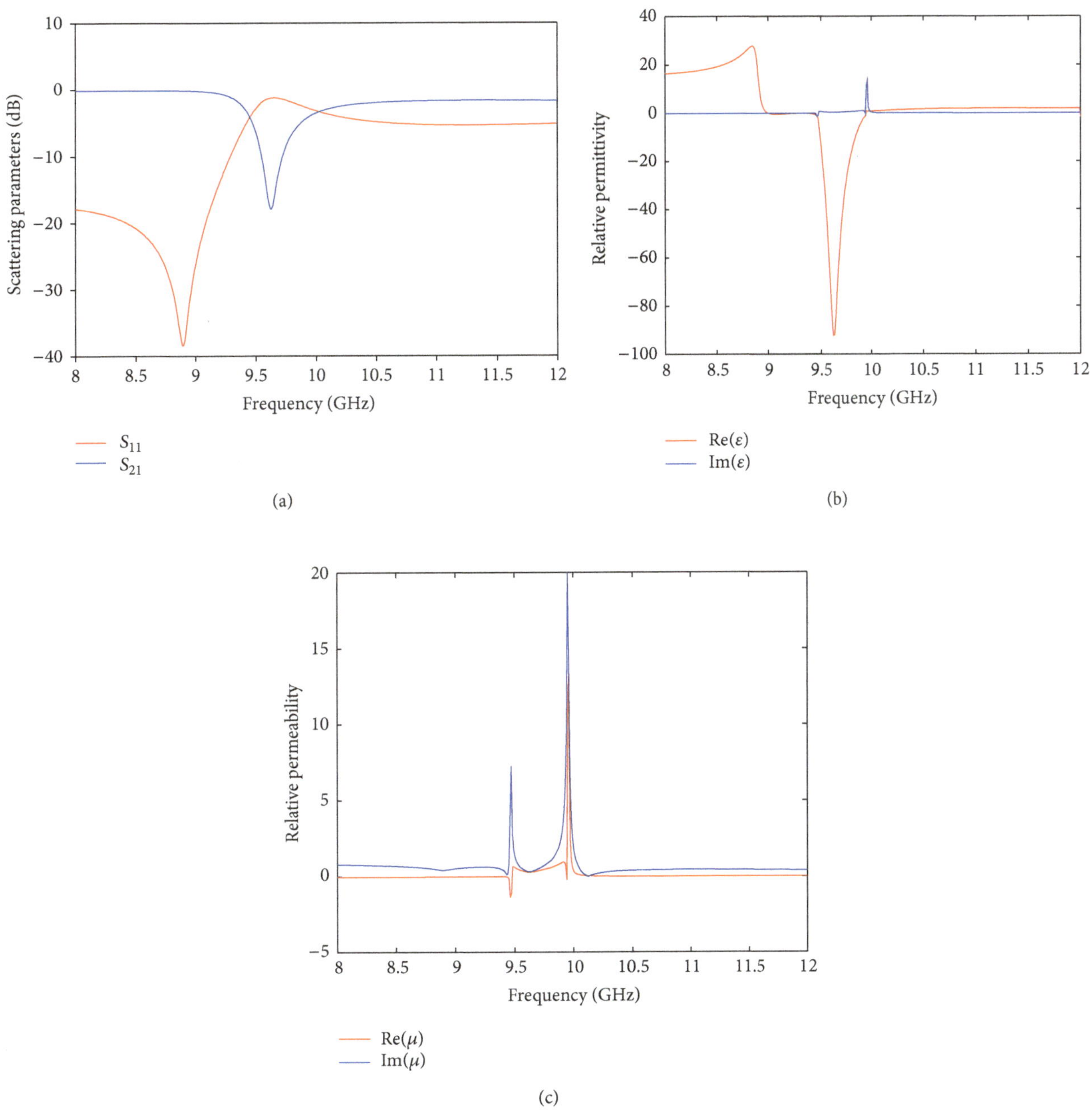

FIGURE 5: (a) Scattering parameters S_{11} and S_{21} parameters of the designed square SRR optimized using PSO algorithm. (b) Relative permittivity of the designed square SRR as optimized by the PSO. (c) Relative permeability of the designed square SRR as optimized by the PSO.

considered as less than one and varies with iteration [6]. The parameter N_d, that is, number of dimensions is taken with respect to square SRR parameters to be optimized such as a, w, d, ε, f_{err}. Similarly, there is no specific criteria for selection of exact values for N_p, N_t and is generally taken by trial and error method with respect to complexity of problem [6]. $X_{\min}$, $X_{\max}$, $V_{\min}$, and $V_{\max}$ are the initial search values, which update with the iteration cycle.

The PSO optimizes the fitness function and extracts the structural parameters. The extracted length a, width w, and the space between the inner and outer ring d are 2.8 mm, 0.3 mm, and 0.3 mm, respectively. These optimized values are used for the design of the square SRR, and the relative permittivity and permeability values are extracted using (5) and (6) [20].

The simulated scattering parameters (S_{11} and S_{21}) of the square SRR placed exactly at the middle of the fractal antenna and the corresponding extracted relative permittivity and permeability are given in Figure 5(a) through Figure 5(c). The metamaterial characteristics of the proposed structure are readily inferred in Figure 5(b), from 9.35 GHz to 9.94 GHz. Similarly, Figure 6(a) shows the top view of a unit cell of the metamaterial structure with micro-splits. The scattering parameters of the second square micro-split SRR placed over

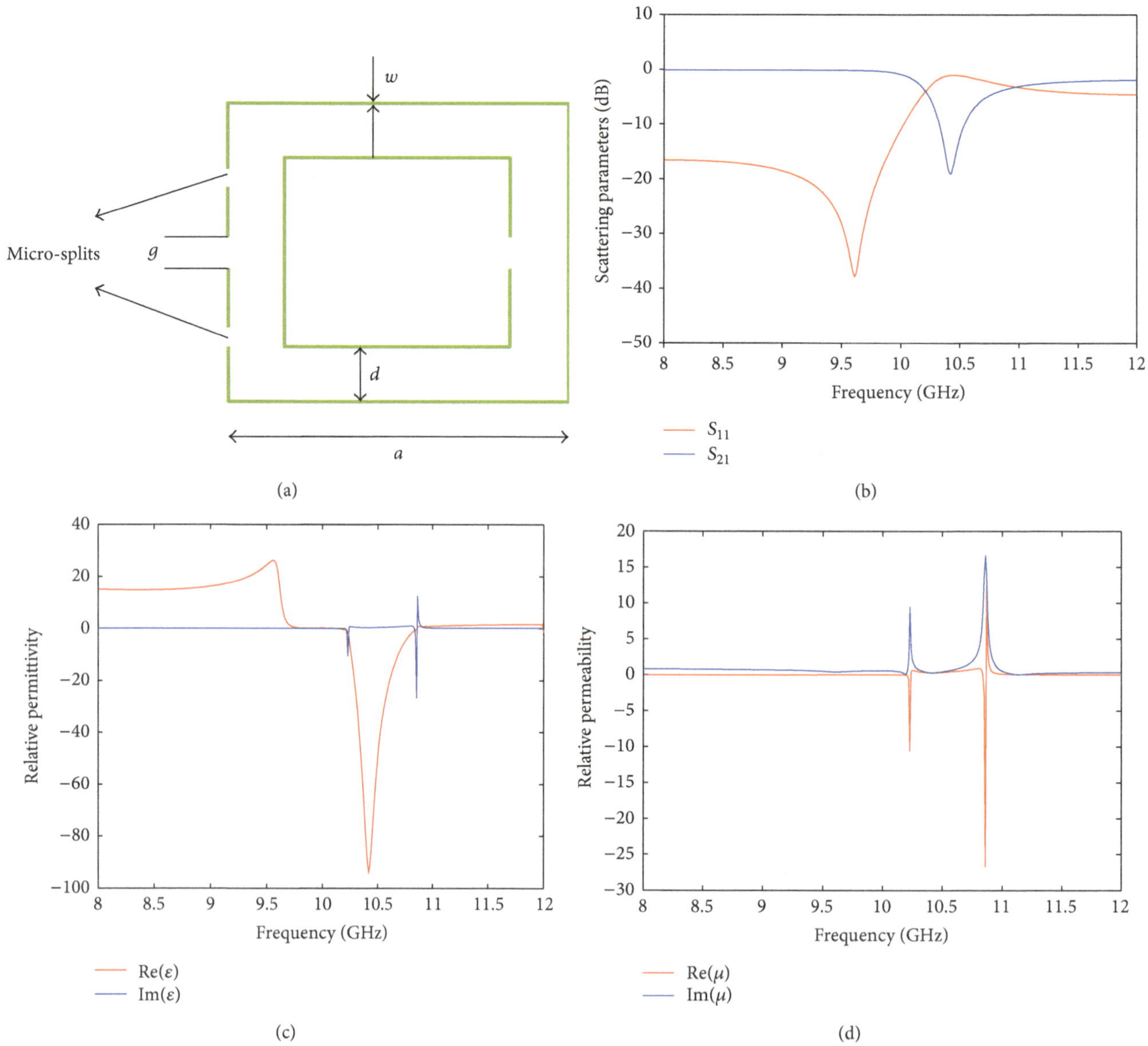

FIGURE 6: (a) Top view of a unit cell of the metamaterial structure with micro-splits (square micro-split SRR), (b) Scattering parameters S_{11} and S_{21} parameters of the designed square micro-split SRR. (c) Extracted permittivity of the designed square micro-split SRR. (d) Extracted permeability of the designed square micro-split SRR.

the fractal antenna and the corresponding extracted relative permittivity and permeability are shown in Figure 6(b) through Figure 6(d):

$$n = \frac{1}{kd}\cos^{-1}\left[\frac{1}{2S_{21}}\left(1 - S_{11}^2 + S_{21}^2\right)\right],$$
$$z = \sqrt{\frac{\left(1 + S_{11}^2\right)^2 - S_{21}^2}{\left(1 - S_{11}^2\right)^2 - S_{21}^2}}, \quad (11)$$
$$\varepsilon = \frac{n}{z},$$
$$\mu = nz,$$

where w = radian frequency, d = thickness of the unit cell.

The PSO algorithm is implemented to optimize the structural square SRR parameters such as width, length, and gap between the splits, for a desired frequency of operation with the overall objective to improve the performance characteristics of the fractal antenna.

5. Performance Enhancement Using Optimized Square SRR

A metamaterial square split ring resonator, whose structural parameters were optimized using particle swarm optimization, was designed and placed at the centre of the fractal patch antenna as shown in Figure 1. After loading a single optimized metamaterial square SRR, it is observed that although the performance of the antenna is increased compared to

TABLE 3: Characteristics of the fractal patch antenna after the addition of metamaterial structure.

Parameters	Resonant frequency of the fractal patch (GHz)					
	3.37	4.96	6.42	7.76	8.93	12.37
Bandwidth (MHz) (without SRR)	40	85	99.5	220	60	420
Bandwidth (MHz) (with SRR)	46	340	120	220	120	460
Return loss (dB) (with SRR)	−12.50	−31.52	−26.8	−20.35	−13.30	−21.26
Directivity (dB) (with SRR)	−3.14	4.86	3.31	7.79	5.69	6.21

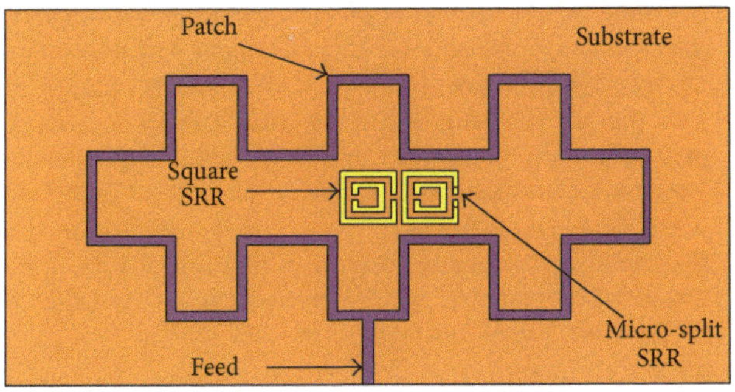

FIGURE 7: Two different square SRR (square SRR and square micro-split SRR) loaded in the fractal antenna elements.

the fractal patch antenna without metamaterial, it is not significant at all the resonant frequencies. Further, in order improve the performance of the fractal patch antenna at all the resonant frequencies, a square SRR with micro-splits in the outer ring is loaded in conjunction with the optimized square SRR (Figure 7). Particle swarm optimization is used for optimization of these metamaterial structures using the algorithm given in Section 4.

The micro-split in the SRR ring helps to resonate at multiple frequencies and improves multiband operation of the antenna. By using two different types of square SRR, the performance parameters of the fractal antenna such as return loss, gain, directivity, bandwidth, and VSWR are improved at multiple resonant frequencies.

The designed metamaterial fractal antenna resonates at six distinct resonant frequencies, which covers both the C and X band (Figure 8). The simulation results show that there is a significant improvement in the antenna characteristics. Table 3 shows the results after placing the metamaterial SRR with micro-splits over the fractal patch antenna.

After placing the metamaterial SRR over the fractal antenna, there is a significant decrease in the return loss and at the same time the bandwidth is widened, with enhancement in the gain and directivity. The radiation efficiency is improved to 40%. It is clearly seen that the gain and directivity of the antenna are improved. There is also an increase in bandwidth at each resonant frequency. Figures 8 and 9 show the simulation results of the fractal patch antenna with optimized metamaterial SRR and micro-split SRR at the middle of the antenna.

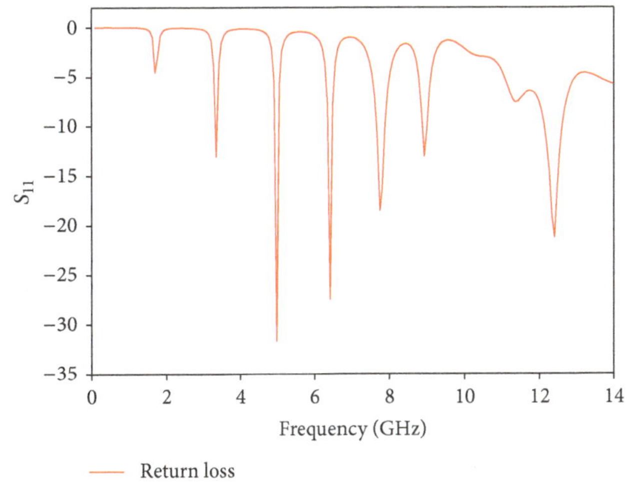

FIGURE 8: Return loss of the metamaterial fractal patch antenna.

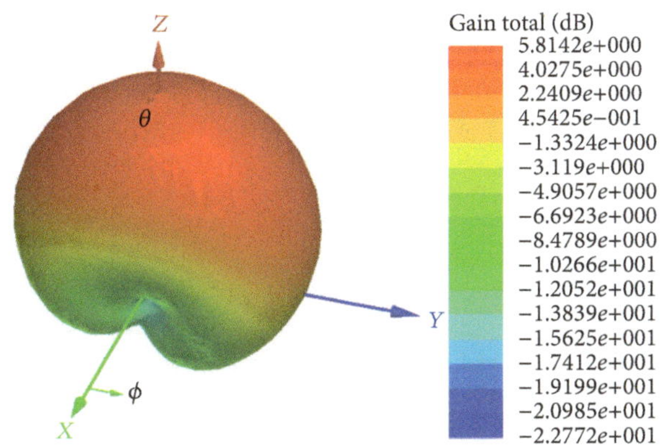

FIGURE 9: Radiation pattern of the metamaterial fractal patch antenna at 7.76 GHz.

6. Conclusion

A multiband metamaterial fractal antenna is designed and simulated, which resonates at six different frequencies covering both C and X band. Two different types of metamaterial structures, namely, a square SRR and a square micro-split SRR, are loaded at the center of the antenna. A PSO based optimizer is used to extract the structural parameters of these metamaterial structures. A comparative analysis of the fractal antenna with and without SRR is reported. It is observed that the radiation efficiency has increased up to 40% after loading the optimized SRRs. The bandwidth at each resonant frequency increases significantly because of the square micro-split SRR. There is also a noticeable improvement in the directivity, as well as the gain at various resonant frequencies.

References

[1] D. H. Werner, R. L. Haupt, and P. L. Werner, "Fractal antenna engineering: the theory and design of fractal antenna arrays," *IEEE Antennas and Propagation Magazine*, vol. 41, no. 5, pp. 37–58, 1999.

[2] T. E. Nur, S. K. Ray, D. Paul, and T. Mollick, "Design of fractal antenna for ultra- wideband applications," *International Journal of Research and Reviews in Wireless Communications*, vol. 1, no. 3, pp. 66–74, 2011.

[3] Y. Rahmat-Samii, "Metamaterials in antenna applications: classifications, designs and applications," in *Proceedings of IEEE International Workshop on Antenna Technology Small Antennas and Novel Metamaterials (IWAT '06)*, pp. 1–4, March 2006.

[4] E. Ekmekci, K. Topalli, T. Akin, and G. Turhan-Sayan, "A tunable multi-band metamaterial design using micro-split SRR structures," *Optics Express*, vol. 17, no. 18, pp. 16046–16058, 2009.

[5] J. Kennedy and R. Eberhart, "Particle swarm optimization," in *Proceedings of the IEEE International Conference on Neural Networks*, pp. 1942–1948, December 1995.

[6] J. Robinson and Y. Rahmat-Samii, "Particle swarm optimization in electromagnetics," *IEEE Transactions on Antennas and Propagation*, vol. 52, no. 2, pp. 397–407, 2004.

[7] B. B. Mandelbrot, *Fractals. Form, Chance and Dimension*, W.H. Freeman, Amsterdam, The Netherlands, 1977.

[8] J. T. Huang, J. H. Shiao, and J. M. Wu, "A miniaturized Hilbert inverted-F antenna for wireless sensor network applications," *IEEE Transactions on Antennas and Propagation*, vol. 58, no. 9, pp. 3100–3103, 2010.

[9] G. Cui, Y. Liu, and S. Gong, "A novel fractal patch antenna with low RCS," *Journal of Electromagnetic Waves and Applications*, vol. 21, no. 15, pp. 2403–2411, 2007.

[10] H. X. D. Araujo, S. E. Barbin, and L. C. Kretly, "Design of UHF Quasi-Yagi antenna with metamaterial structures for RFID applications," in *Proceedings of IEEE Microwave and Optoelectronics Conference*, pp. 8–11, November 2011.

[11] Y. B. Thakare and Rajkumar, "Design of fractal patch antenna for size and radar cross-section reduction," *IET Microwaves, Antennas and Propagation*, vol. 4, no. 2, pp. 175–181, 2010.

[12] V. Crnojević-Bengin, V. Radonić, and B. Jokanović, "Fractal geometries of complementary split-ring resonators," *IEEE Transactions on Microwave Theory and Techniques*, vol. 56, no. 10, pp. 2312–2321, 2008.

[13] M. Palandoken and H. Henke, "Fractal negative-epsilon metamaterial," in *Proceedings of the International Workshop on Antenna Technology: Small Antennas, Innovative Structures and Materials (iWAT '10)*, pp. 1–4, March 2010.

[14] J. McVay, N. Engheta, and A. Hoorfar, "High impedance metamaterial surfaces using Hilbert-curve inclusions," *IEEE Microwave and Wireless Components Letters*, vol. 14, no. 3, pp. 130–132, 2004.

[15] L. Yousefi and O. M. Ramahi, "Miniaturized wideband antenna using engineered magnetic materials with multi-resonator inclusions," in *Proceedings of the IEEE Antennas and Propagation Society International Symposium (AP-S '07)*, pp. 1885–1888, June 2007.

[16] J. Ju, D. Kim, W. J. Lee, and J. I. Choi, "Wideband high-gain antenna using metamaterial superstrate with the zero refractive index," *Microwave and Optical Technology Letters*, vol. 51, no. 8, pp. 1973–1976, 2009.

[17] S. Suganthi, S. Raghavan, and D. Kumar, "Miniature fractal antenna design and simulation for wireless applications," in *Proceedings of the IEEE International Conference on Recent Advances in Intelligent Computational Systems Trivandrum (RAICS '11)*, September 2011.

[18] B. Choudhury, S. Bisoyi, G. Hiremath, and R. M. Jha, "Emerging trends in soft computing for metamaterial design and optimization," in *Proceedings of the Symposium on Metamaterials, International Conference on Computational and Experimental Engineering and Sciences (ICCES '12)*, Crete, Greece, April 2012.

[19] J. W. Fan, C. H. Liang, and X. W. Dai, "Design of cross-coupled dual-band filter with equal-length split-ring resonators," *Progress in Electromagnetics Research*, vol. 75, pp. 285–293, 2007.

[20] F. Bilotti, A. Toscano, and L. Vegni, "Design of spiral and multiple split-ring resonators for the realization of miniaturized metamaterial samples," *IEEE Transactions on Antennas and Propagation*, vol. 55, no. 8, pp. 2258–2267, 2007.

11

Polymorphic Uncertain Linear Programming for Generalized Production Planning Problems

Xinbo Zhang,[1] Feng Zhang,[1] Xiaohong Chen,[2] and Zhong Wan[1]

[1] *School of Mathematics and Statistics, Central South University, Changsha, Hunan 410083, China*
[2] *School of Business, Central South University, Changsha, Hunan 410083, China*

Correspondence should be addressed to Zhong Wan; wanmath@163.com

Academic Editor: Cheng-Hong Yang

A polymorphic uncertain linear programming (PULP) model is constructed to formulate a class of generalized production planning problems. In accordance with the practical environment, some factors such as the consumption of raw material, the limitation of resource and the demand of product are incorporated into the model as parameters of interval and fuzzy subsets, respectively. Based on the theory of fuzzy interval program and the modified possibility degree for the order of interval numbers, a deterministic equivalent formulation for this model is derived such that a robust solution for the uncertain optimization problem is obtained. Case study indicates that the constructed model and the proposed solution are useful to search for an optimal production plan for the polymorphic uncertain generalized production planning problems.

1. Introduction

For a manufacturer, a favorable production planning involves suitably forecasting the demand of products and translating the demand forecast into a plan of production that maximizes the enterprise's profit or minimizes its cost. It has been viewed as a useful approach to formulate the production planning problem into an optimization model by constructing a function of profit or cost as the objective and employing some equalities and inequalities to describe the constraints in the process of production that include the limitations of production resources, the capacity, and the demand of customers.

Actually, in the fields of applied mathematics, operation research, and logistics management sciences, there are many relevant achievements on the problem of production planning. For the recent advancement, for example, see [1–11] and the references therein. It is noted that in the constructed models of production planning, some uncertain factors in the real-world production process, such as the price of product, the customer demand, and the production capacity, are taken into consideration (see [2, 3, 5–11]). For example, in [5], the production capacities and the customer demands are represented by interval numbers, respectively. In [2, 3], the customer demands, the prices of products, and the production capacities are described as fuzzy subsets. In [6], random parameters are used to denote the customer demand and the production capacity.

Very recently, it is noted that, in [12], a polymorphic uncertain model is first employed to formulate a class of optimal design problems of belt drive, and an efficient algorithm, called the two-step based sampling algorithm, is developed to maximize the capacity of V-belt driving. Since polymorphic uncertainty also exists in production planning, different from the existent results, this paper intends to construct a polymorphic uncertain model for the generalized production planning problem, where the amount of production, the amount of inventory, and the unmet demand are the decision variables, the customer demands are represented by interval numbers, and the consumption of raw material and the amounts of all resources available are described as fuzzy subsets. Based on the theory of fuzzy interval program and the modified possibility degree for the order of interval numbers, we shall convert the new model into a deterministic one and find a robust optimal solution for the original uncertain problem by the standard smooth optimization techniques.

The rest of the paper is organized as follows. In the next section, a polymorphic uncertain linear programming (PULP) model for the generalized production planning problem is constructed. In Section 3, an approach is presented to transform the PULP model into a deterministic equivalent formulation and a solution method is presented. Section 4 is devoted to show the validity of the model and the solution method by case study and sensitivity analysis. Final remarks are given in the last section.

2. Model of Generalized Production Planning Problem

In this section, we are going to construct a polymorphic uncertain linear programming (PULP) model for the generalized production planning problem.

We first introduce some denotations.

T: The number of the scheduled production period

I: The number of items (raw materials or finished products)

K: The number of resources

a_{ik}: The amount of resources k required per unit of product i

b_{kt}: The amount of resource k available in period t

d_{it}: The demand for item i in period t

r_{it}: The unit revenue for item i in period t

cp_{it}: The unit variable cost of production for item i in period t

cq_{it}: The unit inventory holding cost for item i in period t

cu_{it}: The unit cost of unmet demand for item i during period t

p_{it}: The production amount of item i during period t

q_{it}: The inventory amount of item i at the end of period t

$\mathfrak{R}$: The set of fuzzy subsets

R: The set of real numbers

u_{it}: The unmet demand amount of item i during period t.

In [5], a deterministic optimization model is presented to formulate the production planning problem:

$$\begin{aligned}
\max \quad & f = \sum_{i=1}^{I}\sum_{t=1}^{T}\left[r_{it}\left(d_{it} - u_{it}\right) - cp_{it}p_{it} - cq_{it}q_{it} - cu_{it}u_{it}\right] \\
\text{s.t.} \quad & \sum_{i=1}^{I} a_{ik}p_{it} \le b_{kt}, \quad \forall k, t, \\
& q_{i,t-1} + p_{it} - q_{it} - u_{it} = d_{it}, \quad \forall i, t, \\
& p_{it}, q_{it}, u_{it} \ge 0, \quad \forall i, t,
\end{aligned} \tag{1}$$

where p_{it}, q_{it}, and u_{it}, for $i = 1, 2, \ldots, I$ and $t = 1, 2, \ldots, T$, are the decision variables. In Model (1), the objective function is to maximize the net revenues of production and minimize the cost of inventory and product shortage. The first type of constraints is on the availability of resource such as the labor, the equipment, and the raw material for the production. The second type of constraints is on the material balance. The last type of constraints in Model (1) is on the nonnegativity of decision variables.

Different from an ordinary model of production planning problem, the inventory amount q_{it} and the unmet demand amount u_{it} are incorporated into Model (1) as decision variables. Thus, it is called a generalized production planning problem.

However, in the real-world process of production, some uncertain factors should be taken into account before the planing of production. It is often that the demand of customer, the consumption of raw material for each product, and the amount of resource available are time-varying. Furthermore, it is unsuitable if all the uncertainties are described by the same mathematical concept such as the random variable (see [6]). An efficient approach is to incorporate the polymorphic uncertainties existing in the problem into the model, and their mathematical descriptions should reflect the features of the time varying parameters as much as possible. In this connection, one can see [12–14].

We now present a polymorphic uncertain linear programming (PULP) model to formulate the problem of production planning as follows:

$$\begin{aligned}
\max \quad & f^{\pm} = \sum_{i=1}^{I}\sum_{t=1}^{T}\left[r_{it}\left(\left[d_{it}^{-}, d_{it}^{+}\right] - u_{it}\right) - cp_{it}p_{it}\right. \\
& \qquad \left. - cq_{it}q_{it} - cu_{it}u_{it}\right] \\
\text{s.t.} \quad & \sum_{i=1}^{I} \tilde{a}_{ik}p_{it} \le \tilde{b}_{kt}, \quad k = 1, 2, \ldots, K, \\
& \qquad t = 1, 2, \ldots, T, \\
& q_{i,t-1} + p_{it} + u_{it} = \left[d_{it}^{-}, d_{it}^{+}\right], \quad i = 1, 2, \ldots, I, \\
& \qquad t = 1, 2, \ldots, T, \\
& p_{it}, q_{it}, u_{it} \ge 0, \quad i = 1, 2, \ldots I, \; t = 1, 2, \ldots, T.
\end{aligned} \tag{2}$$

Different from (1), in Model (2), the amount of resources k required per unit of production of item i, denoted by $\tilde{a}_{ik}$, are assumed to be a triangle fuzzy subset. Similarly, the amount of resources k available in period t, being referred to as $\tilde{b}_{kt}$, are also a triangle fuzzy subset. The demand of product i in period t is regarded as an interval number, denoted by $[d_{it}^-, d_{it}^+]$.

Remark 1. When there exist these uncertain parameters in an optimization model, it is impossible to find an optimal solution in the view point of standard optimization theory. However, in virtue of the method proposed in [12] for the optimal design problem of driving system with polymorphic uncertainties, we can determine an approximate solution of Model (2) by transforming Model (2) into a deterministic one for the given degree of satisfaction as shown in the next section.

3. Deterministic Equivalent Formulation

In this section, we study the deterministic equivalent formulation for the PULP Model (2), which is useful to find a robust optimal solution for an uncertain problem.

We first address how to convert a fuzzy linear program into an interval one.

Consider the following linear programming problem:

$$\begin{aligned} \max \quad & f = \sum_{j=1}^{n} c_j x_j \\ \text{s.t.} \quad & \sum_{j=1}^{n} \tilde{a}_{ij} x_j \le \tilde{b}_i, \quad i = 1, 2, \ldots, m, \\ & x_j \ge 0, \end{aligned} \tag{3}$$

where $\tilde{a}_{ij} \in \mathfrak{R}$ and $\tilde{b}_i \in \mathfrak{R}$, for $i = 1, \ldots, m$ and $j = 1, \ldots, n$, are fuzzy subsets with the same types of membership function and $c_j \in R$, for $j = 1, \ldots, n$, are given constants.

For a fuzzy subset, α-cut is fundamental concept to approximate a fuzzy subset for a given satisfactory level α. For convenience, we state it as follows.

Definition 2 (α-cut). Let $\widetilde{A}$ be a fuzzy subset in the domain X. $\mu_{\widetilde{A}}$ is its membership function. For $0 < \alpha \le 1$, the set

$$\widetilde{A}_\alpha = \{x \in X \mid \mu_{\widetilde{A}}(x) \ge \alpha\} \tag{4}$$

is called the α-cut of $\widetilde{A}$.

By definition, it is clear that if $\mu_{\widetilde{A}}$ is a quasiconcave function, then $\widetilde{A}_\alpha$ is an interval. In the case that $\widetilde{A}$ is a trapezoidal fuzzy subset with a membership function given by

$$\mu_{\widetilde{A}}(x; a, b, c, d) = \begin{cases} 0, & x \le a, \ x > d, \\ \dfrac{x-a}{b-a}, & a < x \le b, \\ 1, & b < x \le c, \\ \dfrac{d-x}{d-c}, & c < x \le d, \end{cases} \tag{5}$$

where $a < b < c < d$, then, from (4), it follows that

$$\widetilde{A}_\alpha = [(1-\alpha)a + \alpha b, (1-\alpha)d + \alpha c]. \tag{6}$$

Throughout this paper, all membership functions are supposed to be quasiconcave. In the following, we present a relation between a fuzzy linear program and an interval one.

From the definition of α-cut and the decomposition theorem in the theory of fuzzy mathematics, we know that

$$\begin{aligned} \tilde{b}_i &= \bigcup_{\alpha \in [0,1]} \alpha (\tilde{b}_i)_\alpha, \\ \sum_{j=1}^{n} \tilde{a}_{ij} x_j &= \bigcup_{\alpha \in [0,1]} \alpha \left(\sum_{j=1}^{n} (\tilde{a}_{ij})_\alpha x_j \right). \end{aligned} \tag{7}$$

Thus, the fuzzy inequality

$$\sum_{j=1}^{n} \tilde{a}_{ij} x_j \le \tilde{b}_i \tag{8}$$

is equivalent to

$$\bigcup_{\alpha \in [0,1]} \alpha \left(\sum_{j=1}^{n} (\tilde{a}_{ij})_\alpha x_j \right) \le \bigcup_{\alpha \in [0,1]} \alpha (\tilde{b}_i)_\alpha. \tag{9}$$

Denote that

$$[a_{ij}^-, a_{ij}^+] = \bigcup_{\alpha \in [0,1]} \alpha (\tilde{a}_{ij})_\alpha, \qquad [b_i^-, b_i^+] = \bigcup_{\alpha \in [0,1]} \alpha (\tilde{b}_i)_\alpha. \tag{10}$$

If x^* solves the following interval program:

$$\begin{aligned} \max \quad & f = \sum_{j=1}^{n} c_j x_j \\ \text{s.t.} \quad & \sum_{j=1}^{n} [a_{ij}^-, a_{ij}^+] x_j \le [b_i^-, b_i^+], \quad i = 1, 2, \ldots, m, \\ & x_j \ge 0, \quad j = 1, 2, \ldots, n, \end{aligned} \tag{11}$$

then x^* is called a solution of Model (3).

Given constants $\alpha_s \in (0, 1]$, $s = 1, 2, \ldots, k$, we can obtain approximate intervals of $[a_{ij}^-, a_{ij}^+]$ and $[b_i^-, b_i^+]$ as follows:

$$[a_{ij}^-, a_{ij}^+] \approx \bigcup_{s=1}^{k} \alpha_s (\tilde{a}_{ij})_{\alpha_s}, \qquad [b_i^-, b_i^+] \approx \bigcup_{s=1}^{k} \alpha_s (\tilde{b}_i)_{\alpha_s}. \tag{12}$$

Thus, the fuzzy inequality

$$\sum_{j=1}^{n} \tilde{a}_{ij} x_j \le \tilde{b}_i \tag{13}$$

is approximated by

$$\sum_{j=1}^{n} \left(\bigcup_{s=1}^{k} \alpha_s (\tilde{a}_{ij})_{\alpha_s} \right) x_j \le \bigcup_{s=1}^{k} \alpha_s (\tilde{b}_i)_{\alpha_s}. \tag{14}$$

To give a definition of robust solution for a fuzzy program, we are in a position to transform an interval linear program into a standard smooth optimization problem by a method of modified possibility degree.

We first present an axiomatic definition of a possibility degree operator.

Definition 3. Let $\mathcal{I}$ denote the set of all interval numbers, and let $\mathcal{R}^I$ denote the set of the relationship between the interval numbers. If a mapping $P : \mathcal{R}^I \rightarrow [0,1]$ satisfies the following conditions:

(1) if $A = [a^-, a^+]$, $B = [b^-, b^+]$, $P(A \leq B) = 1$ if and only if $a^+ \leq b^-$;

(2) if $A = [a^-, a^+]$, $B = [b^-, b^+]$, $P(A \leq B) = 0$ if and only if $a^- \geq b^+$,

(3) $P(A \leq B) = 1 - P(B \geq A)$,

(4) $P(A \leq B) = P(A \geq B)$ if and only if $A = B$,

then P is called a possibility degree operator on $\mathcal{R}^I$.

By definition, it is easy to prove the following operator to be a possibility degree operator in $\mathcal{R}^I$:

$$P([a^-, a^+] \leq [b^-, b^+]) = \begin{cases} 0, & a^- \geq b^+, \\ 0.5\dfrac{b^+ - a^-}{a^+ - a^-}\dfrac{b^+ - a^-}{b^+ - b^-}, & b^- \leq a^- < b^+ \leq a^+, \\ \dfrac{b^- - a^-}{a^+ - a^-} + 0.5\dfrac{b^+ - b^-}{a^+ - a^-}, & a^- < b^- < b^+ \leq a^+, \\ \dfrac{b^- - a^-}{a^+ - a^-} + \dfrac{a^+ - b^-}{a^+ - a^-}\dfrac{b^+ - a^+}{b^+ - b^-} & \\ \quad +0.5\dfrac{a^+ - b^-}{a^+ - a^-}\dfrac{a^+ - b^-}{b^+ - b^-}, & a^- \leq b^- < a^+ \leq b^+, \\ \dfrac{b^+ - a^+}{b^+ - b^-} + 0.5\dfrac{a^+ - a^-}{b^+ - b^-}, & b^- \leq a^- < a^+ < b^+, \\ 1, & a^+ < b^-. \end{cases} \tag{15}$$

Based on the definition of improved interval possibility degree, we are going to prove the following result.

Theorem 4. *Let P be a possibility degree operator defined by (15). Let x^* be a solution of the following interval programming model:*

$$\begin{aligned} \max \quad & f^{\pm} = \sum_{j=1}^{n} c_j^{\pm} x_j \\ \text{s.t.} \quad & \sum_{j=1}^{n} a_{ij}^{\pm} x_j \leq b_i^{\pm}, \quad i = 1, 2, \ldots, m, \\ & x_j \geq 0, \quad j = 1, 2, \ldots, n. \end{aligned} \tag{16}$$

Then, for any $\beta \in [0,1]$ and $\lambda \in [0,1]$, x^* is a solution of the following smooth optimization problem:

$$\begin{aligned} \max \quad & f = \lambda f^- + (1-\lambda) f^+ = \lambda \sum_{j=1}^{n} c_j^- x_j + (1-\lambda) \sum_{j=1}^{n} c_j^+ x_j \\ \text{s.t.} \quad & P\left(\sum_{j=1}^{n} a_{ij}^{\pm} x_j \leq b_i^{\pm}\right) \geq \beta, \quad i = 1, 2, \ldots, m, \\ & x_j \geq 0, \quad j = 1, 2, \ldots, n. \end{aligned} \tag{17}$$

Proof. From the definition of possibility degree in (15), the interval constraint conditions in Model (16) can be converted into its deterministic formulation by introducing the possibility degree parameter β: that is,

$$P\left(\sum_{j=1}^{n} a_{ij}^{\pm} x_j \leq b_i^{\pm}\right) \geq \beta, \quad i = 1, 2, \ldots, m. \tag{18}$$

□

Remark 5. For different sizes of β, there exist 6 cases for evaluating the possibility degree in the following inequality:

$$P\left(\sum_{j=1}^{n} a_{ij}^{\pm} x_j \leq b_i^{\pm}\right) \geq \beta. \tag{19}$$

Actually, (i) if $\beta = 0$, then it is required that

$$\sum_{j=1}^{n} a_{ij}^{-} x_j \geq b_i^{+}, \quad i = 1, 2, \ldots, m; \tag{20}$$

(ii) if $\beta = 1$, then it is needed that

$$\sum_{j=1}^{n} a_{ij}^{+} x_j \leq b_i^{-}, \quad i = 1, 2, \ldots, m; \tag{21}$$

(iii) if $\beta \in (0,1)$, then the cases from 2 to 5 in (15) subsequently occur as an increment of β. Since there are m interval constraints in Model (16), the obtained ordinary constraint is one of the four cases for each interval constraint. So, we should treat the original interval programming problem as 4^m submodels. By comparing the behavior of the solution corresponding to each submodel, a suitable β and the corresponding solution are determined. For example, we fix a medium value of β such that the third case is used to evaluate the possibility degree in (19): then the corresponding submodel turns out to be

$$\begin{aligned} \max \quad & f = \lambda f^- + (1-\lambda) f^+ \\ \text{s.t.} \quad & \frac{b_i^- - \sum_{j=1}^{n} a_{ij}^- x_j}{\sum_{j=1}^{n} a_{ij}^+ x_j - \sum_{j=1}^{n} a_{ij}^- x_j} \end{aligned}$$

$$+0.5\frac{b_i^+ - b_i^-}{\sum_{j=1}^n a_{ij}^+ x_j - \sum_{j=1}^n a_{ij}^- x_j} \geq \beta,$$

$$i = 1, 2, \ldots, m,$$

$$\sum_{j=1}^n a_{ij}^- x_j < b_i^- < b_i^+ \leq \sum_{j=1}^n a_{ij}^+ x_j,$$

$$i = 1, 2, \ldots, m,$$

$$x_j \geq 0, \quad j = 1, 2, \ldots, n. \tag{22}$$

If the level of possibility degree is large enough such that only one case (e.g., case 4 or case 5) needs to be treated and then an optimization problem with interval parameters can be converted into an ordinary linear program. Actually, from (15), it is known that the possibility degree is ascending as β increases. Since a decision-maker in real world always hopes a higher possibility degree, we obtain a deterministic equivalent formulation for the fuzzy linear program (3). Then, based on the above discussion, a concept of robust optimal solution for the PULP Model (2) can be defined.

Definition 6. $x^* \in R^n$ is called a robust optimal solution of the following PULP problem:

$$\begin{aligned}
\max \quad & f^\pm = \sum_{j=1}^n c_j^\pm x_j \\
\text{s.t.} \quad & \sum_{j=1}^n a_{ij}^\pm x_j \leq b_i^\pm, \quad i = 1, 2, \ldots, k, \\
& \sum_{j=1}^n \tilde{a}_{ij} x_j \leq \tilde{b}_i, \quad i = k+1,\ k+2, \ldots, m, \\
& x_j \geq 0, \quad j = 1, 2, \ldots, n,
\end{aligned} \tag{23}$$

if for given $\lambda \in [0, 1]$, $\alpha \in (0, 1]$, and $\beta \in (0, 1)$ large enough, it solves the deterministic optimization problem as below:

$$\begin{aligned}
\max \quad & f = \lambda f^- + (1-\lambda) f^+ \\
\text{s.t.} \quad & P\left(\sum_{j=1}^n a_{ij}^\pm x_j \leq b_i^\pm\right) \geq \beta, \\
& \qquad i = 1, 2, \ldots, k, \\
& P\left(\sum_{j=1}^n (\tilde{a}_{ij})_\alpha x_j \leq (\tilde{b}_i)_\alpha\right) \geq \beta, \\
& \qquad i = k+1,\ k+2, \ldots, m, \\
& x_j \geq 0, \quad j = 1, 2, \ldots, n,
\end{aligned} \tag{24}$$

where $(\tilde{a}_{ij})_\alpha$ and $(\tilde{b}_i)_\alpha$ denote the α-cuts of $\tilde{a}_{ij}$ and $\tilde{b}_i$, respectively.

In the meaning of Definition 6, Problem (24) is called a deterministic equivalent formulation of (23). The following result is directly from Theorem 4 and Definition 6.

Theorem 7. *Suppose that* $\tilde{a}_{ik} = (r_{ik1}, r_{ik2}, r_{ik3})$ *and* $\tilde{b}_{kt} = (e_{kt1}, e_{kt2}, e_{kt3})$ *are triangle fuzzy subsets. Then,*

(1) *for any given* $\alpha \in (0, 1)$, *the cut-sets*

$$\begin{aligned}
(\tilde{a}_{ik})_\alpha &= \left[(1-\alpha) r_{ik1} + \alpha r_{ik2}, (1-\alpha) r_{kt3} + \alpha r_{kt2}\right], \\
(\tilde{b}_{kt})_\alpha &= \left[(1-\alpha) e_{kt1} + \alpha e_{kt2}, (1-\alpha) e_{kt3} + \alpha e_{kt2}\right];
\end{aligned} \tag{25}$$

(2) *the following linear program in the decision variables* p_{it}, q_{it}, *and* u_{it}:

$$\begin{aligned}
\max \quad & f = \lambda f^- + (1-\lambda) f^+ \\
\text{s.t.} \quad & 0.5\left(\sum_{i=1}^I \left[(1-\alpha) r_{ik3} + \alpha r_{ik2}\right] p_{it} - \left[(1-\alpha) e_{kt1} + \alpha e_{kt2}\right]\right)^2 \\
& \times \left(\sum_{i=1}^I (1-\alpha)(r_{ik3} - r_{ik1}) p_{it} \times (1-\alpha)(e_{kt3} - e_{kt1})\right)^{-1} \\
& \leq 1 - \beta, \quad \forall k, t, \\
& \sum_{i=1}^I \left[(1-\alpha) r_{ik1} + \alpha r_{ik2}\right] p_{it} \leq \left[(1-\alpha) e_{kt1} + \alpha e_{kt2}\right], \quad \forall k, t, \\
& \sum_{i=1}^I \left[(1-\alpha) r_{ik3} + \alpha r_{ik2}\right] p_{it} \geq \left[(1-\alpha) e_{kt1} + \alpha e_{kt2}\right], \quad \forall k, t, \\
& \sum_{i=1}^I \left[(1-\alpha) r_{ik3} + \alpha r_{ik2}\right] p_{it} \leq \left[(1-\alpha) e_{kt3} + \alpha e_{kt2}\right], \quad \forall k, t, \\
& q_{i,t-1} + p_{it} + u_{it} \geq d_{it}^-, \quad \forall i, t, \\
& q_{i,t-1} + p_{it} + u_{it} \leq d_{it}^+, \quad \forall i, t, \\
& p_{it}, q_{it}, u_{it} \geq 0, \quad \forall i, t,
\end{aligned} \tag{26}$$

TABLE 1: Values of cp_{it}, cq_{it}, cu_{it}, r_{it}, and $d_{it}^{\pm}$.

$cp_{it}/cq_{it}/cu_{it}/r_{it}/d_{it}^{\pm}$	$t=1$	$t=2$
$i=1$	1/1/2/4/[2, 16]	2/2/3/3/[2, 14]
$i=2$	3/3/1/2/[2, 12]	1/1/2/4/[5, 16]
$i=3$	2/1/2/4/[3, 15]	1/1/1/4/[4, 18]
$i=4$	1/1/2/4/[2, 16]	2/2/3/3/[2, 14]
$i=5$	3/3/1/2/[2, 12]	1/1/2/4/[5, 16]

TABLE 2: Cuts of $\tilde{a}_{ik}$.

$\tilde{a}_{ik}$	$k=1$	$k=2$	$k=3$	$k=4$	$k=5$
$i=1$	(0, 3, 6)	(1, 4, 7)	(3, 6, 9)	(0, 3, 6)	(1, 4, 7)
$i=2$	(3, 6, 9)	(2, 5, 8)	(5, 8, 11)	(3, 6, 9)	(2, 5, 8)
$i=3$	(1, 4, 7)	(5, 8, 11)	(7, 10, 13)	(1, 4, 7)	(5, 8, 11)
$i=4$	(0, 3, 6)	(1, 4, 7)	(3, 6, 9)	(0, 3, 6)	(1, 4, 7)
$i=5$	(3, 6, 9)	(2, 5, 8)	(5, 8, 11)	(3, 6, 9)	(2, 5, 8)

TABLE 3: Cuts of $\tilde{b}_{kt}$.

$\tilde{b}_{kt}$	$k=1$	$k=2$	$k=3$	$k=4$	$k=5$
$t=1$	(3, 6, 9)	(9, 12, 15)	(11, 14, 17)	(5, 8, 11)	(7, 10, 13)
$t=2$	(5, 8, 11)	(7, 10, 13)	(13, 16, 19)	(3, 6, 9)	(9, 12, 15)

is a deterministic equivalent formulation of Model (2), where $\lambda \in [0,1]$, $\alpha \in (0,1]$, and $\beta \in (0,1)$ are chosen large enough and

$$f^{-} = \sum_{i=1}^{I}\sum_{t=1}^{T}\left[r_{it}\left(d_{it}^{-} - u_{it}\right) - cp_{it}p_{it} - cq_{it}q_{it} - cu_{it}u_{it}\right],$$
$$f^{+} = \sum_{i=1}^{I}\sum_{t=1}^{T}\left[r_{it}\left(d_{it}^{+} - u_{it}\right) - cp_{it}p_{it} - cq_{it}q_{it} - cu_{it}u_{it}\right]. \quad (27)$$

It is easy to see that Theorem 7 presents a method to find a robust optimal solution for the uncertain optimization Model (2).

4. Case Study and Sensitivity on the Parameters of Flexibility Degree

In this section, we shall apply the model and the method to solve a problem of production planning.

Suppose that there are 5 items, 5 resources, and 2 periods. The values of the parameters in Model (2) are listed in Table 1.

In this setting, by Theorem 7, a deterministic equivalent formulation of Model (2) is obtained. Actually, the first group of constraints in (26) is equivalent to the following ten inequalities for $k = 1, 2, \ldots, 5$ and $t = 1, 2$, respectively:

$$0.5\left[(6-3\alpha)p_{11} + (7-3\alpha)p_{21} + (9-3\alpha)p_{31} + (6-3\alpha)p_{41} + (7-3\alpha)p_{51} - (3+3\alpha)\right]^2 - (1-\beta)(6-6\alpha)\left(p_{11}+p_{21}+p_{31}+p_{41}+p_{51}\right) \times (6-6\alpha) \le 0,$$

$$0.5\left[(6-3\alpha)p_{12} + (7-3\alpha)p_{22} + (9-3\alpha)p_{32} + (6-3\alpha)p_{42} + (7-3\alpha)p_{52} - (5+3\alpha)\right]^2 - (1-\beta)(6-6\alpha)\left(p_{12}+p_{22}+p_{32}+p_{42}+p_{52}\right)(6-6\alpha) \le 0,$$

$$0.5\left[(9-3\alpha)p_{11} + (8-3\alpha)p_{21} + (11-3\alpha)p_{31} + (9-3\alpha)p_{41} + (8-3\alpha)p_{51} - (9+3\alpha)\right]^2 - (1-\beta)(6-6\alpha)\left(p_{11}+p_{21}+p_{31}+p_{41}+p_{51}\right)(6-6\alpha) \le 0,$$

$$0.5\left[(9-3\alpha)p_{12} + (8-3\alpha)p_{22} + (11-3\alpha)p_{32} + (9-3\alpha)p_{42} + (8-3\alpha)p_{52} - (7+3\alpha)\right]^2 - (1-\beta)(6-6\alpha)\left(p_{12}+p_{22}+p_{32}+p_{42}+p_{52}\right)(6-6\alpha) \le 0,$$

$$0.5\left[(7-3\alpha)p_{11} + (11-3\alpha)p_{21} + (13-3\alpha)p_{31} + (7-3\alpha)p_{41} + (11-3\alpha)p_{51} - (11+3\alpha)\right]^2 - (1-\beta)(6-6\alpha)\left(p_{11}+p_{21}+p_{31}+p_{41}+p_{51}\right)(6-6\alpha) \le 0,$$

$$0.5\left[(7-3\alpha)p_{12} + (11-\alpha)p_{22} + (13-3\alpha)p_{32} + (7-3\alpha)p_{42} + (11-3\alpha)p_{52} - (13+3\alpha)\right]^2 - (1-\beta)(6-6\alpha)\left(p_{12}+p_{22}+p_{32}+p_{42}+p_{52}\right)(6-6\alpha) \le 0,$$

$$0.5\left[(6-3\alpha)p_{11} + (7-3\alpha)p_{21} + (9-3\alpha)p_{31} + (6-3\alpha)p_{41} + (7-3\alpha)p_{51} - (5+3\alpha)\right]^2 (1-\beta)(6-6\alpha)\left(p_{11}+p_{21}+p_{31}+p_{41}+p_{51}\right)(6-6\alpha) \le 0,$$

$$0.5\left[(6-3\alpha)p_{12} + (7-3\alpha)p_{22} + (9-3\alpha)p_{32} + (6-3\alpha)p_{42} + (7-3\alpha)p_{52} - (3+3\alpha)\right]^2 - (1-\beta)(6-6\alpha)\left(p_{12}+p_{22}+p_{32}+p_{42}+p_{52}\right)(6-6\alpha) \le 0,$$

$$0.5\left[(9-3\alpha)p_{11} + (8-3\alpha)p_{21} + (11-3\alpha)p_{31} + (9-3\alpha)p_{41} + (8-3\alpha)p_{51} - (7+3\alpha)\right]^2$$

TABLE 4: Optimal solution for $\lambda = 0.75$, $\alpha = 0.8$, and $\beta =$ **0.9/0.85/0.8**.

Optimal values of the objective function $f =$ **170.0322/170.0285/170.0246**		
p_{it}	q_{it}	u_{it}
$p_{11} =$ **0.1711870/0.2368123/0.2340556**	$q_{11} = 2.0/2.0/2.0$	$u_{11} = 0.0/0.0/0.0$
$p_{12} = 0.0/0.0/0.0$	$q_{12} = 0.0/0.0/0.0$	$u_{12} = 0.0/0.0/0.0$
$p_{21} = 0.0/0.0/0.0$	$q_{21} = 4.980137/4.980137/4.980137$	$u_{21} = 0.0/0.0/0.0$
$p_{22} = 0.1986301/0.1986301/0.1986301$	$q_{22} = 0.0/0.0/0.0$	$u_{22} = 0.0/0.0/0.0$
$p_{31} = 0.0/0.0/0.0$	$q_{31} = 4.0/4.0$	$u_{31} = 0.0/0.0/0.0$
$p_{32} = 0.4138537/0.0/0.0$	$q_{32} = 0.0/0.0/0.0$	$u_{32} = 0.0/0.0/0.0$
$p_{41} =$ **0.7663911/0.1101382/0.1377054**	$q_{41} = 2.0/2.0/2.0$	$u_{41} = 0.0/0.0/0.0$
$p_{42} = 0.0/0.0/0.0$	$q_{42} = 0.0/0.0/0.0$	$u_{42} = 0.0/0.0/0.0$
$p_{51} = 0.0/0.0/0.0$	$q_{51} =$ **4.879848/4.766946/4.682640**	$u_{51} = 0.0/0.0/0.0$
$p_{52} =$ **0.1201523/0.2330544/0.3173601**	$q_{52} = 0.0/0.0/0.0$	$u_{52} = 0.0/0.0/0.0$

Bold values show the notable features of the parameter choices or the numerical results.

TABLE 5: Optimal solution for $\lambda =$ **0.5/0.25**, $\alpha = 0.8$, and $\beta = 0.9$.

Optimal values of the objective function $f =$ **274.0322/378.0322**		
p_{it}	q_{it}	u_{it}
$p_{11} = 0.1711870/0.1711870$	$q_{11} = 2.0/2.0$	$u_{11} = 0.0/0.0$
$p_{12} = 0.0/0.0$	$q_{12} = 0.0/0.0$	$u_{12} = 0.0/0.0$
$p_{21} = 0.0/0.0$	$q_{21} = 4.980137/4.980137$	$u_{21} = 0.0/0.0$
$p_{22} = 0.1986301/0.1986301$	$q_{22} = 0.0/0.0$	$u_{22} = 0.0/0.0$
$p_{31} = 0.0/0.0$	$q_{31} = 3.958614/3.958614$	$u_{31} = 0.0/0.0$
$p_{32} = 0.4138555/0.4138555$	$q_{32} = 0.0/0.0$	$u_{32} = 0.0/0.0$
$p_{41} = 0.7663911/0.7663911$	$q_{41} = 2.0/2.0$	$u_{41} = 0.0/0.0$
$p_{42} = 0.0/0.0$	$q_{42} = 0.0/0.0$	$u_{42} = 0.0/0.0$
$p_{51} = 0.0/0.0$	$q_{51} = 4.879848/4.879848$	$u_{51} = 0.0/0.0$
$p_{52} = 0.1201520/0.1201520$	$q_{52} = 0.0/0.0$	$u_{52} = 0.0/0.0$

Bold values show the notable features of the parameter choices or the numerical results.

$$
\begin{aligned}
&(1-\beta)(6-6\alpha)(p_{11}+p_{21}+p_{31}+p_{41}+p_{51})(6-6\alpha) \\
&\quad \le 0, \\
&0.5\big[(9-3\alpha)\,p_{12}+(8-\alpha)\,p_{22}+(11-3\alpha)\,p_{32} \\
&\quad +(9-3\alpha)\,p_{42}+(8-3\alpha)\,p_{52}-(9+3\alpha)\big]^2 \\
&\quad -(1-\beta)(6-6\alpha)(p_{12}+p_{22}+p_{32}+p_{42}+p_{52})(6-6\alpha) \\
&\quad \le 0.
\end{aligned} \tag{28}
$$

The second group of constraints is corresponding to the second part of inequalities in (26) with $k = 1, 2, \ldots, 5$ and $t = 1, 2$. It reads

$$
\begin{aligned}
&3\alpha p_{11}+(1+3\alpha)\,p_{21}+(3+3\alpha)\,p_{31}+3\alpha p_{41} \\
&\quad +(1+3\alpha)\,p_{51} \le 3+3\alpha, \\
&3\alpha p_{12}+(1+3\alpha)\,p_{22}+(3+3\alpha)\,p_{32}+3\alpha p_{42} \\
&\quad +(1+3\alpha)\,p_{52} \le 5+3\alpha, \\
&(3+3\alpha)\,p_{11}+(2+3\alpha)\,p_{21}+(5+3\alpha)\,p_{31}+(3+3\alpha)\,p_{41} \\
&\quad +(2+3\alpha)\,p_{51} \le 9+3\alpha, \\
&(3+3\alpha)\,p_{12}+(2+3\alpha)\,p_{22}+(5+3\alpha)\,p_{32}+(3+3\alpha)\,p_{42} \\
&\quad +(2+3\alpha)\,p_{52} \le 7+3\alpha, \\
&(1+3\alpha)\,p_{11}+(5+3\alpha)\,p_{21}+(7+3\alpha)\,p_{31}+(1+3\alpha)\,p_{41} \\
&\quad +(5+3\alpha)\,p_{51} \le 11+3\alpha, \\
&(1+3\alpha)\,p_{12}+(5+3\alpha)\,p_{22}+(7+3\alpha)\,p_{32}+(1+3\alpha)\,p_{42} \\
&\quad +(5+3\alpha)\,p_{52} \le 13+3\alpha, \\
&3\alpha p_{11}+(1+3\alpha)\,p_{21}+(3+3\alpha)\,p_{31}+3\alpha p_{41} \\
&\quad +(1+3\alpha)\,p_{51} \le 5+3\alpha, \\
&3\alpha p_{12}+(1+3\alpha)\,p_{22}+(3+3\alpha)\,p_{32}+3\alpha p_{42} \\
&\quad +(1+3\alpha)\,p_{52} \le 3+3\alpha, \\
&(3+3\alpha)\,p_{11}+(2+3\alpha)\,p_{21}+(5+3\alpha)\,p_{31} \\
&\quad +(3+3\alpha)\,p_{41}+(2+3\alpha)\,p_{51} \le 7+3\alpha, \\
&(3+3\alpha)\,p_{12}+(2+3\alpha)\,p_{22}+(5+3\alpha)\,p_{32} \\
&\quad +(3+3\alpha)\,p_{42}+(2+3\alpha)\,p_{52} \le 9+3\alpha.
\end{aligned} \tag{29}
$$

TABLE 6: Optimal solution for $\lambda = 0.75$, $\alpha = \mathbf{0.85/0.9}$, and $\beta = 0.9$.

Optimal values of the objective function $f = \mathbf{170.0035/169.9741}$		
p_{it}	q_{it}	u_{it}
$p_{11} = \mathbf{0.2624145/0.1062255}$	$q_{11} = 2.0/2.0$	$u_{11} = 0.0/0.0$
$p_{12} = 0.0/0.0$	$q_{12} = 0.0/0.0$	$u_{12} = 0.0/0.0$
$p_{21} = 0.0/0.0$	$q_{21} = \mathbf{4.868465/4.886838}$	$u_{21} = 0.0/0.0$
$p_{22} = \mathbf{0.1315347/0.1131623}$	$q_{22} = 0.0/0.0$	$u_{22} = 0.0/0.0$
$p_{31} = 0.0/0.0$	$q_{31} = \mathbf{3.948150/3.937633}$	$u_{31} = 0.0/0.0$
$p_{32} = \mathbf{0.5185039/0.0}$	$q_{32} = 0.0/0.0$	$u_{32} = 0.0/0.0$
$p_{41} = \mathbf{0.2333091/0.1658675}$	$q_{41} = 2.0/2.0$	$u_{41} = 0.0/0.0$
$p_{42} = 0.0/0.0$	$q_{42} = 0.0/0.0$	$u_{42} = 0.0/0.0$
$p_{51} = 0.0/0.0$	$q_{51} = \mathbf{5.000000/4.990076}$	$u_{51} = 0.0/0.0$
$p_{52} = \mathbf{0.0/0.9923663}$	$q_{52} = 0.0/0.0$	$u_{52} = 0.0/0.0$

Bold values show the notable features of the parameter choices or the numerical results.

Similarly, the third group of constraints in (26) is expressed as

$$\begin{aligned}
&(6-3\alpha)\,p_{11} + (7-3\alpha)\,p_{21} + (9-3\alpha)\,p_{31} + (6-3\alpha)\,p_{41} \\
&\quad + (7-3\alpha)\,p_{51} \geq 3 + 3\alpha, \\
&(6-3\alpha)\,p_{12} + (7-3\alpha)\,p_{22} + (9-3\alpha)\,p_{32} + (6-3\alpha)\,p_{42} \\
&\quad + (7-3\alpha)\,p_{52} \geq 5 + 3\alpha, \\
&(9-3\alpha)\,p_{11} + (8-3\alpha)\,p_{21} + (11-3\alpha)\,p_{31} + (9-3\alpha)\,p_{41} \\
&\quad + (8-3\alpha)\,p_{51} \geq 9 + 3\alpha, \\
&(9-3\alpha)\,p_{12} + (8-3\alpha)\,p_{22} + (11-3\alpha)\,p_{32} + (9-3\alpha)\,p_{42} \\
&\quad + (8-3\alpha)\,p_{52} \geq 7 + 3\alpha, \\
&(7-3\alpha)\,p_{11} + (11-3\alpha)\,p_{21} + (13-3\alpha)\,p_{31} + (7-3\alpha)\,p_{41} \\
&\quad + (11-3\alpha)\,p_{51} \geq 11 + 3\alpha, \\
&(7-3\alpha)\,p_{12} + (11-3\alpha)\,p_{22} + (13-3\alpha)\,p_{32} + (7-3\alpha)\,p_{42} \\
&\quad + (11-3\alpha)\,p_{52} \geq 13 + 3\alpha, \\
&(6-3\alpha)\,p_{11} + (7-3\alpha)\,p_{21} + (9-3\alpha)\,p_{31} + (6-3\alpha)\,p_{41} \\
&\quad + (7-3\alpha)\,p_{51} \geq 5 + 3\alpha, \\
&(6-3\alpha)\,p_{12} + (7-3\alpha)\,p_{22} + (9-3\alpha)\,p_{32} + (6-3\alpha)\,p_{42} \\
&\quad + (7-3\alpha)\,p_{52} \geq 3 + 3\alpha, \\
&(9-3\alpha)\,p_{11} + (8-3\alpha)\,p_{21} + (11-3\alpha)\,p_{31} + (9-3\alpha)\,p_{41} \\
&\quad + (7-3\alpha)\,p_{51} \geq 7 + 3\alpha, \\
&(9-3\alpha)\,p_{12} + (8-3\alpha)\,p_{22} + (11-3\alpha)\,p_{32} + (9-3\alpha)\,p_{42} \\
&\quad + (7-3\alpha)\,p_{52} \geq 9 + 3\alpha.
\end{aligned} \tag{30}$$

The last group, associated with the fourth constraint in (26), is specified by

$$\begin{aligned}
&(6-3\alpha)\,p_{11} + (7-3\alpha)\,p_{21} + (9-3\alpha)\,p_{31} + (6-3\alpha)\,p_{41} \\
&\quad + (7-3\alpha)\,p_{51} \leq 9 - 3\alpha, \\
&(6-3\alpha)\,p_{12} + (7-3\alpha)\,p_{22} + (9-3\alpha)\,p_{32} + (6-3\alpha)\,p_{42} \\
&\quad + (7-3\alpha)\,p_{52} \leq 11 - 3\alpha, \\
&(9-3\alpha)\,p_{11} + (8-3\alpha)\,p_{21} + (11-3\alpha)\,p_{31} + (9-3\alpha)\,p_{41} \\
&\quad + (8-3\alpha)\,p_{51} \leq 15 - 3\alpha, \\
&(9-3\alpha)\,p_{12} + (8-3\alpha)\,p_{22} + (11-3\alpha)\,p_{32} + (9-3\alpha)\,p_{42} \\
&\quad + (8-3\alpha)\,p_{52} \leq 13 - 3\alpha, \\
&(7-3\alpha)\,p_{11} + (11-3\alpha)\,p_{21} + (13-3\alpha)\,p_{31} + (7-3\alpha)\,p_{41} \\
&\quad + (11-3\alpha)\,p_{51} \leq 17 - 3\alpha, \\
&(7-3\alpha)\,p_{12} + (11-3\alpha)\,p_{22} + (13-3\alpha)\,p_{32} + (7-3\alpha)\,p_{42} \\
&\quad + (11-3\alpha)\,p_{52} \leq 19 - 3\alpha, \\
&(6-3\alpha)\,p_{11} + (7-3\alpha)\,p_{21} + (9-3\alpha)\,p_{31} + (6-3\alpha)\,p_{41} \\
&\quad + (7-3\alpha)\,p_{51} \leq 11 - 3\alpha, \\
&(6-3\alpha)\,p_{12} + (7-3\alpha)\,p_{22} + (9-3\alpha)\,p_{32} + (6-3\alpha)\,p_{42} \\
&\quad + (7-3\alpha)\,p_{52} \leq 9 - 3\alpha, \\
&(9-3\alpha)\,p_{11} + (8-3\alpha)\,p_{21} + (11-3\alpha)\,p_{31} + (9-3\alpha)\,p_{41} \\
&\quad + (8-3\alpha)\,p_{51} \leq 13 - 3\alpha, \\
&(9-3\alpha)\,p_{12} + (8-3\alpha)\,p_{22} + (11-3\alpha)\,p_{32} + (9-3\alpha)\,p_{42} \\
&\quad + (8-3\alpha)\,p_{52} \leq 15 - 3\alpha.
\end{aligned} \tag{31}$$

Thus, Model (26) is written as

$$\begin{aligned}
\max \quad f = {} & 520 - 416\lambda - p_{11} - q_{11} \\
& - 6u_{11} - 2p_{12} - 2q_{12} - 6u_{12} \\
& - 3p_{21} - 3q_{21} - p_{22} \\
& - q_{22} - 6u_{22} - 2p_{31} - q_{31} - 6u_{31} - p_{32}
\end{aligned}$$

$$
\begin{aligned}
&- q_{32} - 5u_{32} - p_{41} - q_{41} - 6u_{41} - 2p_{42} \\
&- 2q_{42} - 6u_{42} - 3p_{51} - 3q_{51} - p_{52} - q_{52} - 6u_{52} \\
\text{s.t.} \quad & (28)-(31), \\
& q_{11} + p_{12} + u_{12} \geq 2, \\
& q_{21} + p_{22} + u_{22} \geq 5, \\
& q_{31} + p_{32} + u_{32} \geq 4, \\
& q_{41} + p_{42} + u_{42} \geq 2, \\
& q_{51} + p_{52} + u_{52} \geq 5, \\
& q_{11} + p_{12} + u_{12} \leq 14, \\
& q_{21} + p_{22} + u_{22} \leq 16, \\
& q_{31} + p_{32} + u_{32} \leq 18, \\
& q_{41} + p_{42} + u_{42} \leq 14, \\
& q_{51} + p_{52} + u_{52} \leq 16, \\
& p_{it}, q_{it}, u_{it} \geq 0, \quad i = 1, 2, 3, 4, 5, \; t = 1, 2.
\end{aligned}
\tag{32}
$$

Take $\lambda = 0.75$, $\alpha = 0.8$, and $\beta = 0.9$. Tables 2 and 3 present the cuts of the triangle fuzzy subsets $\tilde{a}_{ik}$ and $\tilde{b}_{kt}$, respectively.

With these coefficients shown in Tables 2 and 3, we can obtain an approximate optimal solution of Model (2) by solving problem (32). The numerical results are reported in Table 4.

Next, we are interested in the sensitivity of the obtained optimal solution on the three parameters of flexibility degree, λ, α, and β. By choosing different suitable values of λ, α, and β, we check the changes of optimal solutions.

We first change the values of β by fixing λ and α. The partial numerical results with $\beta = 0.85$ and $\beta = 0.8$ are reported in Table 4.

From Table 4, it is clear that the optimal value of the objective function increases with the increment of β. A part of the nonzero optimal solutions, such as p_{41}, p_{52}, and q_{51}, is seriously changed and another part of the nonzero optimal solutions appears to be more stable.

In Table 5, we report a part of numerical results with two different values of λ. It indicates that the optimal decision-making is basically stable except that the value of the objective function is different.

Finally, in Table 6, we report a part of numerical results with two different values of α. It indicates that the choice of α greatly affects the optimal value of the objective function and the nonzero optimal solutions.

All of the results in Tables 4–6 demonstrate that the zero optimal solutions seem not to be sensitive on the change of the three parameters of flexibility degree(λ, α, and β). From these numerical results, it is concluded that the proposed solution method in this paper can find the relatively robust optimal solution for the original uncertain production planning problem.

5. Final Remarks

In this paper, a polymorphic uncertain linear programming (PULP) model for the generalized production planning problem was constructed where the uncertainty in the real production was taken into consideration.

For an optimization model involved with polymorphic uncertainty, it is impossible to find an optimal solution in the view point of standard optimization theory in general. In virtue of the method proposed in [12], we have found the approximate optimal solution of the generalized production planning problem by deriving the deterministic equivalent formulation.

Case study and numerical analysis have shown that the different parameters of flexibility degree had distinct sensitivity of optimal solution and the zero optimal solutions seem immune to the changing values of the flexibility degree parameters. From the numerical results, it is concluded that the proposed solution method can find the relatively robust optimal solution for the original uncertain production planning problem. Thus, it is helpful in the real-world decision-making.

Conflict of Interests

The authors declare that there is no conflict of interests regarding the publication of this paper.

Acknowledgments

The authors would like to express their thanks to the anonymous referees for their constructive comments on the paper, which have greatly improved its presentation. This research is supported by the Natural Science Foundation of Hunan Province (13JJ3002) and the National Natural Science Foundation of China (Grant nos. 71221061, 71071162).

References

[1] E.-H. Aghezzaf, C. Sitompul, and F. Van den Broecke, "A robust hierarchical production planning for a capacitated two-stage production system," *Computers & Industrial Engineering*, vol. 60, no. 2, pp. 361–372, 2011.

[2] J. Mula, D. Peidro, and R. Poler, "The effectiveness of a fuzzy mathematical programming approach for supply chain production planning with fuzzy demand," *International Journal of Production Economics*, vol. 128, no. 1, pp. 136–143, 2010.

[3] S. A. Torabi, M. Ebadian, and R. Tanha, "Fuzzy hierarchical production planning (with a case study)," *Fuzzy Sets and Systems*, vol. 161, no. 11, pp. 1511–1529, 2010.

[4] R. Q. Zhang, L. K. Zhang, Y. Y. Xiao, and I. Kaku, "The activity-based aggregate production planning with capacity expansion in manufacturing systems," *Computers & Industrial Engineering*, vol. 62, no. 2, pp. 491–503, 2012.

[5] S. A. Abass, M. A. Gomaa, G. A. Elsharawy, and M. S. Elsaid, "Generalized production planning problem under interval uncertainty," *Egyptian Informatics Journal*, vol. 11, no. 1, pp. 27–31, 2010.

[6] L. X. Tang, P. Che, and J. Y. Liu, "A stochastic production planning problem with nonlinear cost," *Computers & Operations Research*, vol. 39, no. 9, pp. 1977–1987, 2012.

[7] D. Kira, M. Kusy, and I. Rakita, "A stochastic linear programming approach to hierarchical production planning," *Journal of the Operational Research Society*, vol. 48, no. 2, pp. 207–211, 1997.

[8] Y. F. Lan, Y. K. Liu, and G. Sun, "An approximation-based approach for fuzzy multi-period production planning problem with credibility objective," *Applied Mathematical Modelling*, vol. 34, no. 11, pp. 3202–3215, 2010.

[9] H. Jung, "A fuzzy AHP-GP approach for integrated production-planning considering manufacturing partners," *Expert Systems with Applications*, vol. 38, no. 5, pp. 5833–5840, 2011.

[10] J. Mula, R. Poler, J. P. García-Sabater, and F. C. Lario, "Models for production planning under uncertainty: a review," *International Journal of Production Economics*, vol. 103, no. 1, pp. 271–285, 2006.

[11] S. P. Sethi, H. Yan, H. Zhang, and Q. Zhang, "Optimal and hierarchical controls in dynamic stochastic manufacturing systems: a survey," *Manufacturing & Service Operations Management*, vol. 4, no. 2, pp. 133–170, 2002.

[12] Z. Wan, S. J. Zhang, and K. L. Teo, "Two-step based sampling method for maximizing the capacity of V-belt driving in polymorphic uncertain environment," *Proceedings of the Institution of Mechanical Engineers C*, vol. 226, no. 1, pp. 177–191, 2012.

[13] Z. Wan, K. L. Teo, L. S. Kong, and C. H. Yang, "A class of mix design problems: formulation, solution methods and applications," *The ANZIAM Journal*, vol. 50, no. 4, pp. 455–474, 2009.

[14] Z. Wan, F. Z. Meng, A. Y. Hao, and Y. L. Wang, "Fuzzy and stochastic parameters-based prediction method for the components of alkali in the sintering process of aluminium," *Fuzzy Systems and Mathematics*, vol. 25, no. 3, pp. 163–167, 2011.

12

Experimental Study of the Interaction between Recrystallization and Precipitation Processes of an AA8011 Commercial Alloy

Ney José Luiggi,[1] María Valera,[1] Jesús P. Rodriguez,[2] and José Prin[3]

[1] *Grupo Física de Metales, Departamento de Física, Núcleo de Sucre, Universidad de Oriente, Cumaná 6101, Venezuela*
[2] *IDECYT, Universidad Simón Rodríguez, Caracas 1060, Venezuela*
[3] *Instituto de Investigaciones de Biomedicina y Ciencias Aplicadas (IIBCA), Universidad de Oriente, Cumaná 6101, Venezuela*

Correspondence should be addressed to Ney José Luiggi; nluiggi51@gmail.com

Academic Editor: B. S. Murty

Phase changes in a commercial AA8011 alloy from different initial microstructure conditions were studied using thermoelectric power (ΔS), differential scanning calorimetry (DSC), and transmission electron microscopy (TEM) techniques with the purpose of obtaining evidence of the interaction between recovery-precipitation and recrystallization-precipitation processes occurring during nonisothermal heating at different rates. Thermoelectric power and its thermal derivative reflect this evidence by a displacement of the characteristic precipitation peaks, the recovery and recrystallization contributions remaining masked by the strong incidence of the iron precipitation on that property, while DSC measurements detect the emergence of new peaks not observed on thermograms of homogenized samples. An exhaustive study of these peaks permits direct differentiation between precipitation and recovery-recrystallization contributions. TEM confirms the interaction between both processes by means of local observations.

1. Introduction

The study of the interaction that occurs between phase precipitation process and recrystallization in deformed materials is a topic of great interest in materials science because this interaction regulates the kinetics of growth of new grains and the formation of phases in the material. The combination of both effects certainly leads to modifications of the physical and mechanical properties of the material.

A plastically deformed material shows a great amount of internal strains, mainly as dislocations, which store part of the energy transferred to the crystal during deformation, the reordering of which helps release said energy. Many mechanisms contribute to that end, where the recovery and recrystallization processes recover and lead to the appearance of new grains through the release of stored energy, the evolution of the recrystallization occurring in a collective manner by nucleation and growth [1–6].

The diffusive process conducive to the phase transformation also takes place at the same temperature ranges where recrystallization occurs, both processes being regulated by the starting microstructure.

Schematically, both a possible evolution of the atomic supersaturation after quenching of a homogenized sample as a function of aging time at a temperature below the solubility limit and the evolution of the density of internal strain as a function of the degree of deformation by rolling in homogenized and quenched samples are shown in Figure 1. Immediately after quenching at a temperature below the solubility limit, the homogenized sample shows the highest atomic supersaturation (signaled with 1 in the figure). As the sample is aged, the atomic supersaturation decreases down to a fully aged sample with the lowest atomic supersaturation (marked with 3). The other curve in Figure 1 shows how the strain concentration increases when a mechanical work is applied. Number 4 corresponds to a sample without deformation,

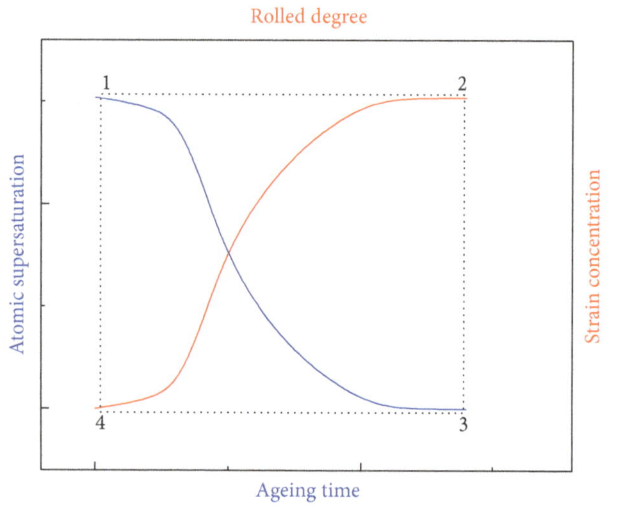

FIGURE 1: Schematic representation of the atomic supersaturation versus aging time and strain concentration versus degree of rolling for homogenized and quenched samples.

while number 2 corresponds to a highly deformed sample, where the density of strains is at a maximum. The four dotted straight lines that intercept both curves represent the extreme conditions of supersaturation and deformation that the initial microstructure of the sample can hypothetically undergo. We may infer from that diagram that, for certain initial microstructural conditions, a simultaneous participation of both processes will ensue. Generally, in the literature, both effects are not separated; in such cases, the results must correspond to one single mechanism and not to the particular mechanisms involved.

The aim of this work is to confirm this interaction by using different techniques and starting microstructures. The commercial alloy AA8011 was chosen for this purpose, where Fe and Si, as main alloying elements, regulate the microstructure and properties of the Al-Fe-Si system. Several works dealing with this alloy have been reported in the literature [7–11]. Garcia-Cordovilla and Louis [12] differentiated exothermal reactions associated with precipitation and recrystallization of this alloy via DSC measurements.

This paper complements the communication reported in work [13] on Al-Fe-Si alloy rolled samples, now considering different starting microstructures achieved by rolling and aging at specific temperatures, whereby the samples studied are characterized, aside from DSC and ΔS, by TEM in order to relate the changes detected in the microstructure of homogenized, deformed, and annealed samples to the mechanisms responsible for the recovery-recrystallization and precipitation processes. The derivative of the thermoelectric power is introduced as a complementary way to identify peaks associated with the particular reactions.

The paper is organized as follows. Section 2 presents the methodology used and defines samples and heat treatments. Section 3 outlines details theoretically associated with each technique employed, and Section 4 presents the results and discussion.

2. Methodology

2.1. Samples Analyzed. Our work was done on commercial as-cast samples of the AA8011 alloy supplied by C.V.G ALCASA of Venezuela in 6 mm thick rolls, milled to 0.5 mm, the composition of which is given in Table 1.

2.2. Thermal Treatments and Equipment. In order to have different starting microstructures, the samples were subjected to different thermal treatments.

(1) Samples were subjected to homogenization for six hours at 620°C, then cooled in aqueous solution kept at 20°C. This treatment guarantees that the starting microstructure will have a high supersaturation of atoms in solid solution and a low concentration of internal stresses. The samples are designated as HS.

(2) Samples were subjected to treatment 1, then annealed to different temperatures in order to guarantee a controlled supersaturation of the solute and a low concentration of internal stresses. The samples are designated as AHS.

(3) Samples were subjected to treatment 1, then deformed by quenching via twin-roll casting to different reductions in thickness. The starting microstructure will display high atom supersaturation and a concentration of internal stresses that increases with the deformation rate. The samples are designated as DHS.

(4) Samples under condition 3 are subjected to artificial annealing for specific amounts of time at different temperatures. This treatment tries to regulate supersaturation of the solute and the concentration of internal stresses of treatment 2. The samples are designated as ADHS.

Thermoelectric power variations measurements were estimated by the Seebeck coefficient in equipment with a $\pm 0.002\,\mu$V/K accuracy [14], whereas DSC measurements were made with a NETZSCH JUPITER STA 499 C, the sensitivity of which allows for the detection of very-low-enthalpy transitions. In both cases, high purity aluminum was used as reference and baseline. Electron microscopy studies were made at IIBCA-UDO-VENEZUELA with a 100 kV HITACHI H600 transmission electron microscope.

3. Fundamental Aspects of the Experimental Techniques Used

In order to benefit from the interpretation of our measurements, we need to consider the basic concepts related to each technique.

3.1. Thermoelectric Power. Our measuring equipment is based on the thermoelectric effect known as the Seebeck effect [15, 16] that measures the variation in thermoelectric potential within a temperature gradient. The potential is generated because of both the distinct electronic distribution on each side of the sample as a result of the heat gradient

TABLE 1: Chemical composition of the AA8011 alloy.

Al	Fe	Si	Mn	Zn	Cr	Cu
Rem (weight %)	0.56	0.40	0.01	0.004	0.003	0.01
Rem (At.%)	0.271	0.385	0.00493	0.00166	0.00156	0.00426

and the chemical potential difference introduced by the union of two different materials (samples and conducting wiring). Three basic relationships allow us to coherently formulate a satisfactory assessment of ΔS [17].

(1) The Mott relation for diffusion thermoelectric power is

$$S_d = \frac{\pi^2 k_b^2}{3\left|e\right|}\left[\frac{1}{\rho}\frac{\partial \rho}{\partial E}\right]_{E=E_F}, \quad (1)$$

where ρ is the resistivity of the sample, E is the electronic energy of the sample, and subindex F is used to specify the Fermi energy of the material being studied.

(2) The Matthiessen relation, particularly as applied to diluted alloys, is

$$\rho = \rho_0 + \sum_{i,o}\left(c_i\rho_i + c_o\rho_o\right), \quad (2)$$

where subscripts 0, i, and o represent the contribution from the thermal vibrations or ideal contribution without matrix defect, the contribution from solute atoms in solid solution, and the contribution from atoms outside the solid solution, respectively. c represents the atomic concentration and the values of ρ_i and ρ_0 for different element of alloying in aluminum are given in [18].

(3) The Nordheim-Gother rule applied to the relative variation of diffusion thermoelectric power relative to the thermoelectric power of the ideal material is

$$\Delta S_d = \frac{\sum \Delta\rho_i \Delta S_i}{\rho}, \quad (3)$$

where ΔS_i represents thermoelectric power specific to each element added to the aluminum matrix and $\Delta\rho_i$ is the resistivity variation generated by each alloying element in aluminum. Once the specific values of both the resistivity and the thermoelectric power of each alloying element are known, the measured value of the diffusion thermoelectric power for particular values of the atom concentrations given in (2) by c_i and c_o can be reproduced.

The literature shows that measurements of thermoelectric power [18] are sensitive to the concentration of the solute in the solid solution; hence, by using the theoretical base stated in [18], we can evaluate the contribution of each element of the alloy to thermoelectric power measured. The thermoelectric power sensitivity to the concentration of atoms in solid solution makes this technique a very useful tool to follow the processes of phase transformations under either isothermal or nonisothermal conditions. The thermoelectric power variation from one temperature to another induces changes in atomic concentration that blend with the structural transformation in such a way that the derivative of the thermoelectric power, as in other techniques, will show peaks and troughs characteristic of the different reactions; that is, the derivative of the thermoelectric power allows for direct observation of the kinetics of phase changes occurring during a specific thermal treatment. It is important for our analysis to indicate that the specific ΔS values [17] for Fe and Si are −8.00 and −0.74 μV/K and that the specific $\Delta\rho$ values are 2.56 and 1.02 $\mu\Omega$ cm/wt%, respectively. With these values, theoretical equations (3) and (4), the particular microstructural conditions, and the contributions of Fe and Si to the TEP can be evaluated.

3.2. Differential Scanning Calorimetry (DSC). DSC studies measure the difference in temperature or the thermal potential registered between the sample being studied and the reference sample, where said measurement is proportional to the heat flow Q registered as a function of the time or temperature. The Q variation is expressed as a function of the specific heat of sample C_P, its mass m, and the heating rate $\Phi = dT/dt$

$$\frac{dQ}{dt} = mC_P\frac{dT}{dt}. \quad (4)$$

The area below the curve between two temperatures corresponds to the changes in the amount of heat of the sample:

$$\Delta Q = \int_{T_1}^{T_2}\left(\frac{dQ}{dt}\right)dt. \quad (5)$$

Calorimetric measurements [19] are sensitive to both endothermic and exothermic reactions occurring during changes in microstructure, being therefore very useful in determining the kinetics of phase changes of materials. The heat flow measured and the reaction enthalpy are related to the extension of the conversion or transformed fraction, via numerical methods, allowing us to deduce the kinetic parameters associated with the reaction.

3.3. Transmission Electron Microscope. The evolution of the microstructure was directly observed with a transmission electron microscope. First, thin disks of 3 and 0.12 mm in diameter and thickness, respectively, were electropolished with the JET technique and then impacted in the center of both sides with an electrolyte stream to reduce thickness until an orifice was made. Thickness around the orifice was less than 0.5 μm, permitting the electron beam of the microscope to pass through. The amount of electrolytes used was 20% perchloric acid and 80% methanol. Micrographs were obtained with a HITACHI Transmission Electron Microscope, model H-600 (accelerating voltage of 100 kV).

4. Experimental Results

4.1. Thermoelectric Power and Derivatives

4.1.1. Continuous and Isochronous Heating. Figure 2(a) shows thermoelectric power as a function of temperature for the AA8011 alloy subjected to different reductions in thickness via rolling during continuous heating at a rate of 2°C/min. The ΔS values taken immediately after rolling, reported in our previous work [13], increase as the rolling rate increases, which shows that the rolling process induces precipitation of the solute and impoverishes the solid solution, no doubt with the participation of a greater dislocation density as the deformation by rolling increases. Figure 2(a) shows a strong increase in ΔS up to 425°C for homogenized HS samples and up to 380°C for severely deformed samples (85% reduction in thickness), while, for intermediate deformations, the peak is orderly placed between those two temperatures. Afterwards, a decline slowly ensues in homogenized samples, the decrease being more pronounced on DHS as the rolling rate increases. A second peak is clearly shown in severely deformed samples at 460°C, followed by a pronounced drop indicating that the atomic agglomerates enter into solid solution. Therefore, the increase of ΔS is associated with the precipitation process, whereas its decrease is associated with the dissolution of precipitates, the inflexion points of said curves representing the maximum transformation rate of the precipitation or dissolution processes, respectively. The derivative of ΔS with respect to temperature shows peaks and troughs for the temperature ranges where measurements of ΔS showed phase precipitation and dissolution, respectively; hence, in the $d\Delta S/dT$ versus T curve, positive peaks completely represent the precipitation process, whereas the process f dissolution of precipitates is represented by a trough, just as in DSC measurements of exothermic and endothermic reactions associated with those processes.

Figure 2(b) shows the derivative of ΔS with respect to temperature for the curves of Figure 2(a). An important reaction is seen for MH samples, the maximum of which is found at 375°C, associated mainly with the precipitation of the α-AlFeSi phase, albeit the presence of the Θ-$AlCu_2$ [3] has been detected in a much lesser proportion at that level, followed by both an endothermic reaction associated with a corresponding dilution and a new exothermic transformation of less magnitude around 470°C, which corresponds mainly to the redistribution of Fe atoms and the generation of the equilibrium phase of the Al_3Fe system. No other transformation seems to occur in the HS samples. The effect from rolling begins to be noticed even for a 25% reduction in sample thickness, where the α-AlFeSi appears at 352°C as a consequence of the dislocation density introduced by rolling, the effect of which is to accelerate the precipitation of that phase. Said acceleration behavior of the precipitation is maintained as the rolling is increased close to a 75% reduction in thickness, where it appears that the number of dislocations introduced is such that said dislocations tend to rearrange themselves in a different manner, slowing the process of acceleration of precipitation of that phase to a temperature near 332°C. This behavior is different in the precipitation of the

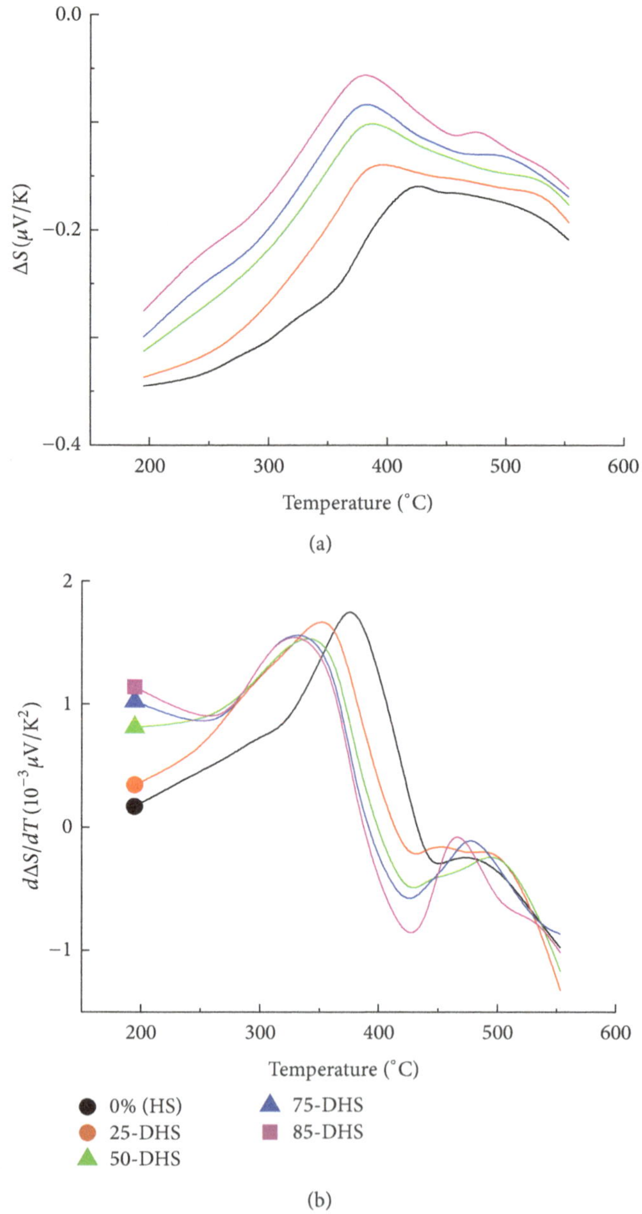

FIGURE 2: (a) Thermoelectric power and (b) derivative of the thermoelectric power versus temperature for the homogenized AA8011 alloy, quenched and deformed at different rolling degrees, then annealed continuously at 2°C/min.

Al_3Fe phase, the precipitation of which seems to slow down with the increase of rolling to a 50% reduction in thickness, to later accelerate for severe deformations. It is also worth noting that, for severe deformations, an initial endothermic transformation occurs towards 250°C, which is associated with the recovery process and the dissolution of the Guinier-Preston zones. This behavior, not observed in HS samples, is typical of the DHS samples and makes us ponder whether the recovery process occurs under 250°C, a process that increases according to the degree of rolling. Two processes coexist in the first exothermic reaction: the precipitation of the α-AlFeSi phase and the recrystallization of grains from the release of energy absorbed during rolling. As for the equilibrium phase, the changes observed indicate an interaction between the precipitation of said phase and the final growth stage of grains. Note that above 500°C the qualitative behavior of all curves is similar, indicating that the recrystallization process has been completed.

Figure 3 shows the results of ΔS for samples subjected to isochronous heating for 10 minutes at each temperature. The intention behind such heating is to monitor the effect of each temperature on microstructure changes that occur in the HS and DHS samples studied. Figure 3(a) shows that for temperatures below the appearance of the first peak, the behavior observed is similar to the one shown previously, even though the position of said peak seems to be slightly changed due to the isochronous treatment of the samples. Like in the previous case for HS and DHS samples, the peak is associated with the precipitation of the α-AlFeSi phase; however, the release of energy stored must occur simultaneously in the DHS samples, although not enough for a complete recrystallization to occur, the reason being that the measurements of thermoelectric power still keep showing a single peak associated with both processes, the precipitation of Fe masking the particular effect that the dislocations may have on that property [9, 13]. However, at higher temperatures, the dissolution of the α phase, as well as the precipitation of the Al_3Fe phase, seems to present a more uniform behavior in agreement with our previous explanation. Towards 486°C, an important and uniform decrease in ΔS begins in the HS and DHS samples, indicating that the recrystallization process has been completed. Figure 3(b) shows the derivative ΔS as a function of temperature. The differences not detected in Figure 3(a) are detailed. The two exothermic transformations inherent to the precipitation of both the α-AlFeSi and the Al_3Fe phases are present, as well as the endothermic reactions that lead to the dissolution of the Guinier-Preston zones in the first inverted peak and to the redistribution of dislocations in severely deformed samples and, in the second peak, to the dissolution of the α-AlFeSi phase. This figure highlights the effect of isochronous heating at high temperatures which tends, in every case, to uniform both the precipitation and the bringing of the Al_3Fe equilibrium phase into solid solution. This happens, among other reasons, because 10 minutes at such temperatures is seemingly sufficient to complete the recrystallization process. It is important to note that while the precipitation peak of the α phase is placed at 383°C, the joint peak of precipitation and recrystallization occurs between 330°C for 25% reduction in thickness and 314°C for 85% reduction in thickness, respectively, in total agreement with the value reported by Lima [10].

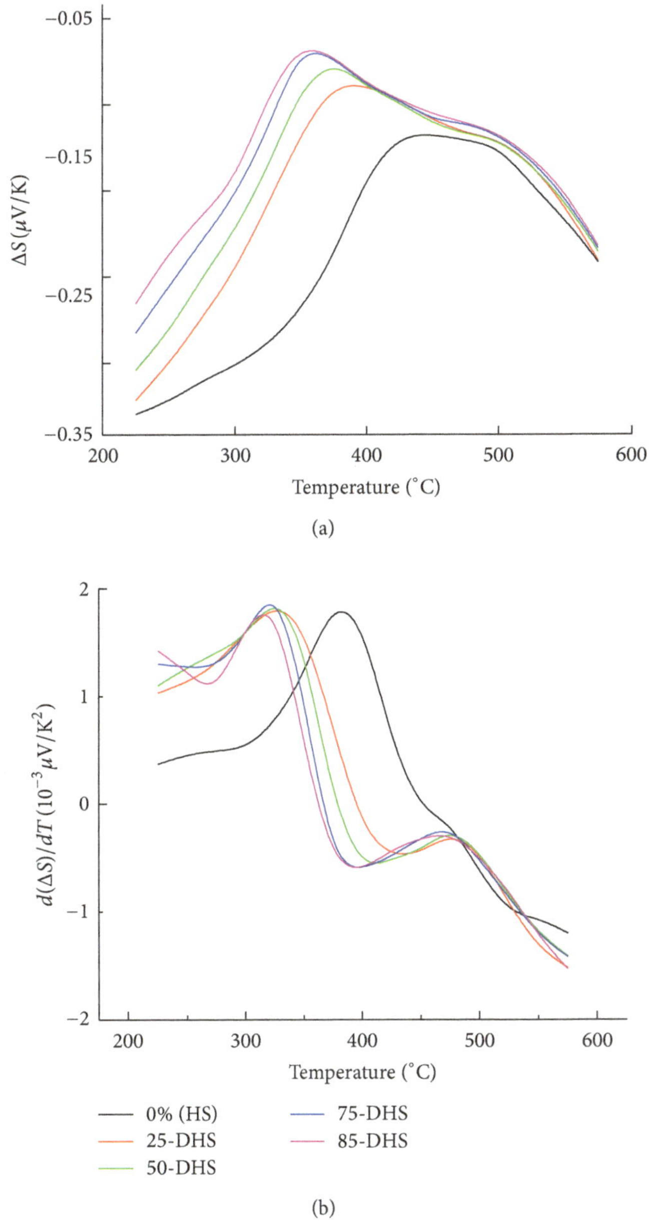

FIGURE 3: (a) Thermoelectric power and (b) derivative of the thermoelectric power versus temperature for the homogenized AA8011 alloy, quenched and deformed at different rolling degrees, then heated isochronously during 10 minutes at each temperature.

4.1.2. Isothermal Preaging at 300°C in Highly Rolled Samples (85%). Preaging for different time periods at 300°C and measuring under isochronous 25-minute heating conditions seek to decrease the amount of solute in the solid solution, so that a precipitation-recrystallization interaction different from that of the previous cases is promoted. The values of ΔS, once the corresponding preaging of the samples has been done, and before the heating begins, are shown in Table 2. We can see how ΔS increases as the rolled sample stays at 300°C for a longer period of time, indicating a solute-impoverished aluminum matrix.

Figure 4(a) shows the evolution of ΔS as a function of temperature under the conditions mentioned above. We can observe how preaging affects the first stages of the curve, given that, for a longer time at 300°C, a greater precipitation of Fe and Si is promoted, impoverishing the solid solution and recovering the microstructure by eliminating dislocations and defects introduced by rolling.

Once the peak associated with both precipitation of the α phase and recrystallization has been reached, the combined effect of preaging and isochronous treatment tends to show a behavior similar to ΔS on each curve, meaning, in light of the treatment applied, that only before and during the precipitation of the α-AlFeSi phase is there a strong interaction between the elements that define the microstructure. The main peaks are placed at a temperature near 335°C for low preaging times, corresponding to the precipitation-recrystallization reaction, and at 322°C for high preaging times, where undoubtedly a great amount of recrystallization and precipitation has occurred (mainly for 500 min at 300°C with 90% of the precipitation—according to Luiggi [9]). The second peak, regardless of preaging and associated with the precipitation of the stable Al_3Fe phase, occurs at 485°C. The transition between both reactions shows a difference in ΔS of $-0.1\,\mu V/K$ only explained by a significant introduction of Si atoms in the solution and a few Fe atoms, which regroup when temperature increases in order to generate said stable phase. The derivative of ΔS versus T (not shown) established, just like in earlier cases, the transformation details. Figure 4(b) shows an amplified version of the behavior of ΔS in the first heating stage for high preaging times. In it the appearance of an intermediate, well-defined peak can be observed for preaging periods greater than or equal to 50 minutes before the precipitation peak of the α-AlFeSi phase. The size of ΔS, in the beginning, indicates that every Si atom and, to a great extent, the Fe atoms have precipitated, whereas the first decrease in ΔS corresponds to the dissolution of the Guinier-Preston zones and possibly to the recovery of the microstructure. The peaks around 275°C reflect great recovery and atom rearrangement activity inherent to recrystallization, while the peaks towards 320°C correspond to the formation of the α-AlFeSi phase, which occurs, depending on the variation magnitude of ΔS, as the result of structural rearrangement rather than as impoverishment of the solid solution. The fact that the peak associated with the precipitation of the α-AlFeSi phase does not appear displaced in this figure but for lower preaging times is a consequence of the previous energy relaxation of the system produced by preaging, which is greater for longer preaging times.

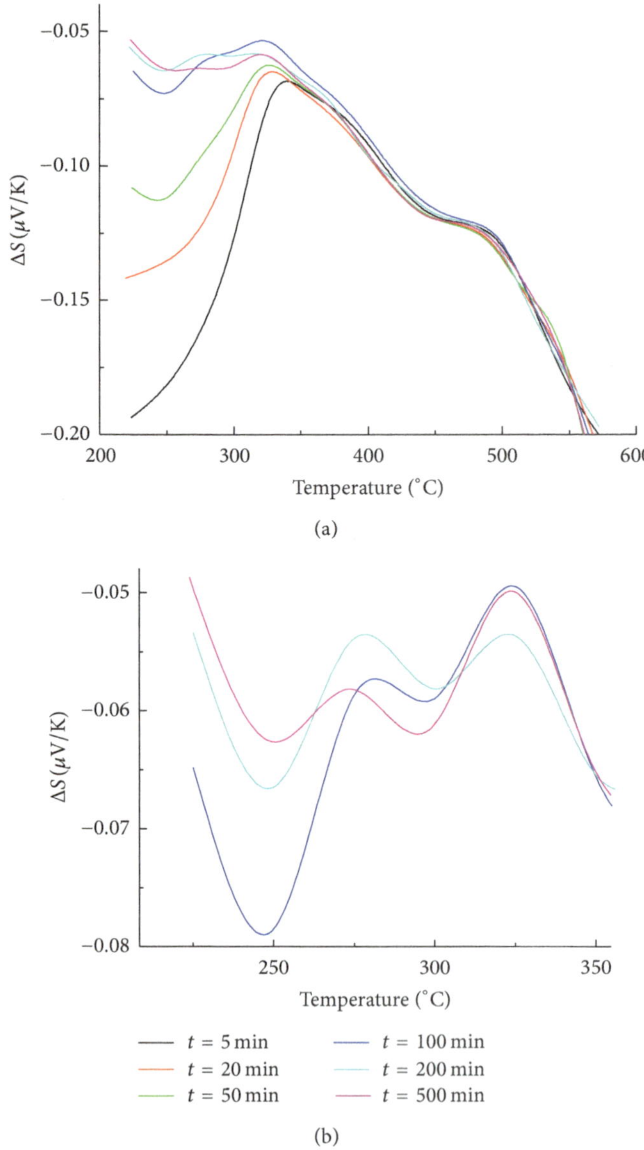

FIGURE 4: (a) Thermoelectric power versus temperature for 85-DHS samples preaged during different times at 300°C, then annealed isochronously for 25 minutes at each temperature. (b) Amplified initial part of curve (a) for high preaging times at 300°C.

4.1.3. Preaging for 10 Minutes at Different Temperatures in Severely Rolled Samples (85%). The objective of this experience is to regulate the extent of atom supersaturation and to reduce the concentration of internal stresses as we increase temperature. The homogenized sample rolled to 85% reduction in thickness presents a combination of concentration of internal stresses and high atom supersaturation; hence, the 10-minute stay at each temperature fosters the precipitation of solute at that temperature, as well as recovery and recrystallization. The initial values for ΔS, after preaging at different temperatures, are shown in the 4th row of Table 2, where we observe that ΔS increases as preaging temperature also increases, just up to $T = 350$°C, to then diminish irregularly around $-0.14\,\mu V/K$ for preaging at greater temperatures, which means that, up to 350°C, a decrease of the solute in the solid solution occurs simultaneously due to the precipitation of α and to a decrease in the concentration of internal stresses by the relaxation of the deformed matrix due to the release of energy by elimination of dislocations during preaging. The decrease in ΔS at higher temperatures implies both a lesser fraction of solute brought into the solution and atom redistribution at those temperatures.

Figure 5(a) shows the variation in ΔS as a function of temperature for preaging at temperatures above 250°C which, aside from confirming the results of Table 2, allows us to visualize the behavior during the isochronous heating already mentioned. Two types of behavior limited by the curve at 350°C are reported. Below such temperature, two exothermic transformations are reported, as discussed in the previous cases where reduction in atom concentration and internal

TABLE 2: Thermoelectric power of MHD samples with thickness reduction to 85%. Preaged at 300°C for different times (2nd row) and for 10 minutes at different T (4th row). Measurements made immediately after treatment.

Preaging time (min)	0	5	20	50	100	200	500
ΔS (V/K)	−0.3090	−0.2058	−0.1455	−0.1070	−0.0675	−0.0625	−0.0615
Preaging T (°C)	As-Cast	250	300	350	400	450	500
ΔS (V/K)	−0.4255	−0.2371	−0.1726	−0.1050	−0.1444	−0.1378	−0.1580

strain are not enough to introduce modifications regarding the HS samples. At 350°C, the evolution of ΔS appears different from that of previous cases, with a maximum located 70°C below the maxima reported in the two preceding curves and a slower transition between the two exothermic reactions, which begs the assumption of a strong interaction between recrystallization and precipitation in the α phase. For 350°C < T < 500°C, three exothermic transformations clearly occur: the first one near 275°C, the second one around 340°C, and the third at 475°C. While the last two reactions have been present in every kinetics study, the first peak seems to be a direct result from the relaxation of the deformed matrix, where 10 minutes at those temperatures, along with the respective isochronous treatment, is not sufficient to produce complete recrystallization of the sample studied.

Also note that preaging at 500°C tends to reproduce the behavior of the homogenized sample. In every curve shown, transformation towards 475°C appears bereft of the effects from recrystallization; hence, we consider that the matrix relaxation process was completed in its entirety. Figure 5(b) shows the corresponding derivatives confirming the above: a sequence of transformations vindicating the precipitation of α phase and β phase, the first one interacting directly with the recrystallization process and the second, independent of it, in some cases generating exothermic transformations within the temperature recrystallization range that can only be associated with that process. In conclusion, the curves shown confirm three aspects: (1) values of ΔS reported can be in agreement with the theory only if the alloying Fe actively participates in the conformation of the existing phases; (2) precipitation at temperatures below 400°C of the α-AlFeSi phase and its interaction with the recrystallization process; (3) precipitation of the Al_3Fe phase at temperatures above 400°C, where the growth process of the recrystallized grains seems not to modify its kinetics of precipitation.

4.2. Differential Scanning Calorimetry. All DSC measures were made during continuous heating under different microstructure conditions.

4.2.1. Deformation Effect from Rolling. Figure 6 shows a graph of heat flow versus temperature for the AA8011 alloy under three initial microstructure conditions: homogenized (HS), homogenized and rolled to 50% in thickness reduction (50-DHS), and homogenized and rolled up to 85% in thickness reduction (85-DHS). Figure 6(a), where the heating rate is $\Phi = dT/dt = 5$°C/min, shows the presence of three exothermic reactions at 90°C, 278°C, and 466°C for the HS sample (empty circles), each corresponding to the formation of Guinier-Preston zones, the α-AlFeSi phase, and the β-Al_3Fe phase [3, 13], respectively. This behavior tends to stay for low deformations, though a clear variation in entropy is manifest, as seen by the widening of the exothermic peaks. For 85-DHS, there is a clear effect on each of the peaks. The characteristic peak of the precipitation of the GP zones, though manifested, is diffuse, implying that an important redistribution of solute atoms is produced during rolling, inhibiting part of the exothermic reaction associated with the GP zones, though its position seems invariable regardless of temperature.

The exothermic reaction that characterizes the precipitation of the α phase in HS gives way to two exothermic reactions, one at 288°C and the other at 328°C, where both peaks are at temperatures above the value reported for the homogenized condition, implying that rolling slows down the precipitation process of the α phase. Even though both peaks appear at temperatures lower than those obtained with thermoelectric power measurements, they are without a doubt a consequence of both the recovery and recrystallization process, as well as of the precipitation process.

Given that, generally, recrystallization precedes precipitation [3, 12], it may be established that the first one is inherent to recrystallization, and the second is essential also to precipitation of the α phase, therefore proving the interaction between precipitation and recrystallization. Similarly, the peak for the β precipitation is also broken down into two exothermic reactions at 454°C and 502°C, respectively, the value for the β peak remaining interlocked in homogenized condition between both reactions. Again, with the data currently available it is difficult to assign any given particular reaction to each transformation; however, given the high temperature at which they occur where, according to our previous ΔS studies, recrystallization has already occurred, the peaks can thus be associated with the β phase precipitation, the Θ precipitation, and the growth of recrystallized grains. Possibly, this very last one is present in both exothermic events. Note that unlike our previous work [13] where the peak was associated with recrystallization, in the present work, which is a product of other techniques, such a contention must be discarded and the mechanisms involved that must be established are, besides the precipitation of the β and Θ phases, the growth of recrystallized grains. The endothermic reactions inherent to the reversion of the processes described above are displaced towards the low temperatures as the degree of rolling is increased, indicating that deformation creates pathways for an easy reversion of the phases formed.

Figure 6(b) depicts the same study above but at a heating rate of 10°C/min, plotted from 200°C. The exothermic

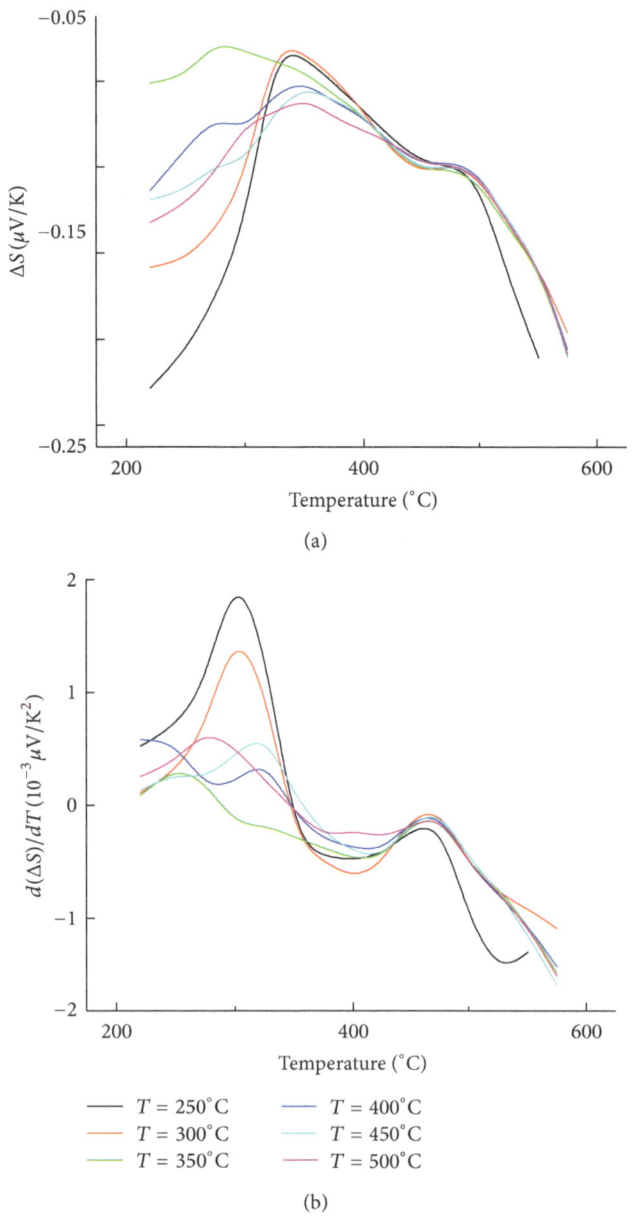

FIGURE 5: (a) Thermoelectric power and (b) derivative of the thermoelectric power versus temperature for the homogenized AA8011 alloy, for 85-DHS samples, subjected to preaging for 10 minutes at different temperatures.

reaction inherent to precipitation of the GP zones is not manifested in the thermogram, although the endothermic reaction inherent to its dissolution is reported, the behavior of which is similar to the one already proposed, indicating an increase in the dissolution rate as deformation increases. For HS samples, temperatures of the characteristic peaks for precipitation of the α and β phases are at 300°C and 471°C, respectively, while the endothermic trough that announces the end of the decomposition of the α phase is at 410°C. For a reduction in thickness of 50%, a tendency towards the appearance of two exothermic reactions ensues around 300°C, one of which is associated with precipitation and the

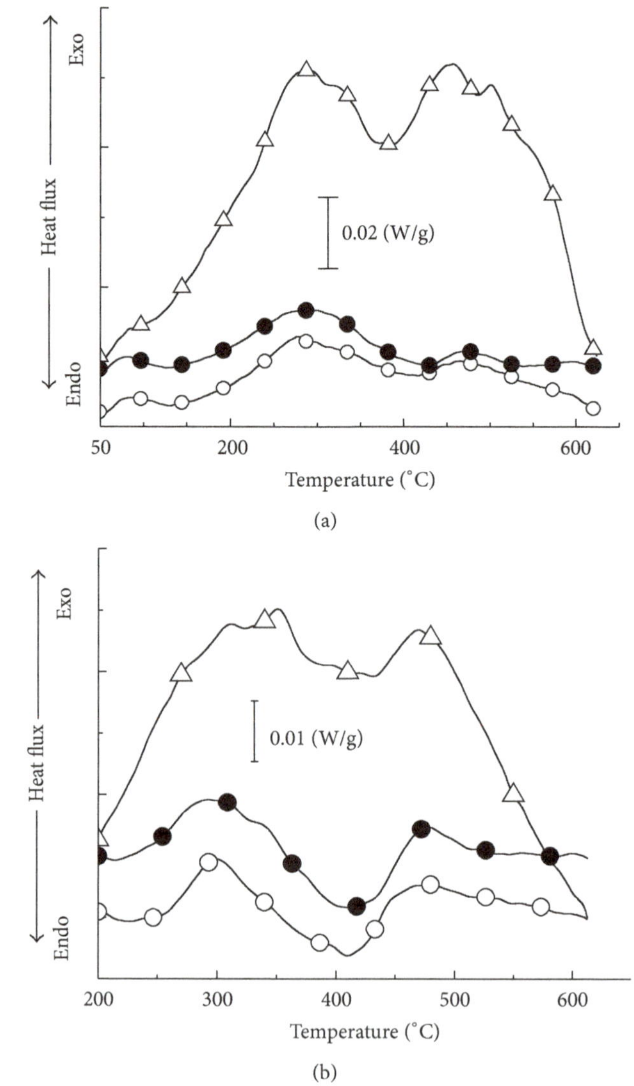

FIGURE 6: Heat flow versus temperature for the AA8011 alloy under three initial microstructures: ○ homogenized (HS), ● homogenized and rolled to a 50% reduction in thickness (50-DHS), and △ homogenized and rolled to an 85% reduction in thickness (85-DHS). (a) Φ = 5°C/min (from [13]) and (b) Φ = 10°C/min.

other with recrystallization, whereas one single peak appears in the exothermic reaction observed at higher temperatures. For an 85% reduction in thickness, these effects are stronger, while the two first exothermic reactions appear at 311°C and 352°C, with a greater apparent entropy for the second peak. These two reactions are associated with the recrystallization and precipitation process of the α phase. The transition towards the β phase is slower than in previous cases, showing one single exothermic peak at 471°C. Undoubtedly, this peak is associated mainly with both β phase precipitation and growth of recrystallized grains.

4.2.2. Effect of the Heating Rate (Φ) *on Homogenized and Deformed Microstructures.* Figure 7(a) shows homogenized samples of thermograms made at different heating rates Φ: 5, 10, 20, and 40°C/min. The inset shows Φ = 5°C/min in

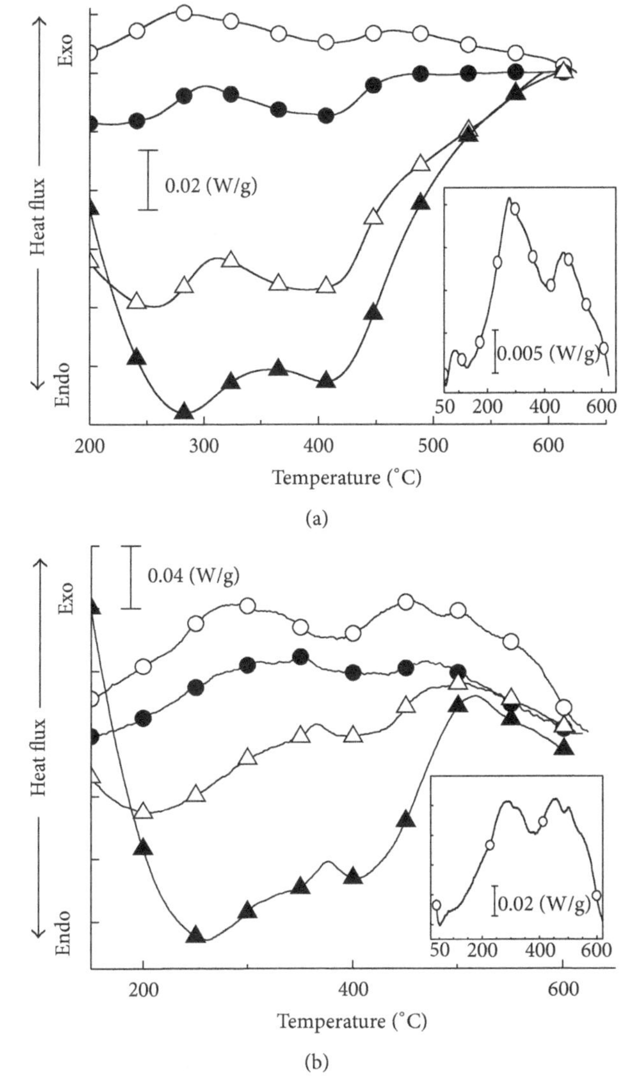

FIGURE 7: Effect of the heating rate on the heat flow. (a) Homogenized samples. (b) Severely deformed samples (from [13]): ○ 5°C/min, ● 10°C/min, △ 20°C/min, and ▲ 40°C/min.

order to show the whole temperature range studied. Here, we again highlight the three exothermic reactions at 90°C, 278°C, and 466°C that correspond to the formation of the Guinier-Preston zones, the α-AlFeSi phase, and the β-Al_3Fe phase, respectively. The effect of incrementing Φ is that of displacing said exothermic reactions towards high temperatures, having the effect that some reactions will not be detected because Φ is higher than the reaction rate, just as it happens with precipitation of the GP zones at high Φ. Note that even if, at high Φ, the evolution of the kinetics of the α phase can be followed, the same does not occur with the β phase because a collapse associated with the casting of the sample seems to occur, not allowing us to ascertain its kinetics at such high temperatures.

Figure 7(b) shows the effect of heating at different Φ rates on homogenized and rolled samples with 85% reduction in thickness (85-DHS). The inset highlights the thermogram at 5°C/min covering the whole temperature range, the effect of deformation in the first heating stage, and the subsequent appearance of four exothermic transformations, the interpretation of which was given in part (b) of the preceding figure. For Φ > 5°C/min, thermograms maintain the double exothermic reaction around the reaction that leads to precipitation of the α phase, obviously displaced because both the nucleation and recrystallization processes are thermally triggered so that the higher the Φ value, the higher the temperatures to which these peaks, not always well defined, are displaced. Note also that the two exothermic reactions that appear at high T for Φ = 5°C/min disappear for higher heating rates, and only one exothermic reaction appears, just like in the thermoelectric power measurements, near 475°C for 10°C/min, at 485°C for 20°C/min and at 516°C for 40°C/min. This behavior may be explained by the difference existing between the reaction rate of the β phase precipitation and the growth rate of the recrystallized grains which, in addition to the increment in heating rate, generate the superimposition of both mechanisms. Again, as supported by previous ΔS studies, the interpretation of the recovery and recrystallization processes in the present work is slightly different from the one given in [13].

4.2.3. Effect of Preaging on Samples Severely Deformed at Different Temperatures. Figure 8 shows thermograms of 85-DHS samples, preaged at fixed temperatures across different times, measured at Φ = 5°C/min. Figure 8(a) shows the results for 20 min preaging at 300°C (solid circles) and at 50 minutes (solid triangles) and without preaging (empty circles). It is expected that, in aluminum alloys, preaging in HS samples will erase the exothermic transformation that characterizes the precipitation of the Guinier-Preston zones [20]. Such an effect would also be expected for deformed samples, however, depending on the alloying element, temperature, preaging time, and deformation degree. We can observe in the figure that, for 85-DHS samples, an exothermic transformation still appears near 90°C, which becomes greater as the preaging time increases. One possible explanation for such behavior may be found in the presence of dislocation entanglements distributed in such a way that they hinder free atom diffusion and hence the less favorable entropic answer when the sample is more deformed. Preaging for 20 minutes at 300°C tends to reduce such ripples and to improve the diffusion pathways. This effect seems to appear in the fusion of the double recrystallization and precipitation peak which, just like in thermoelectric power, generates one single, ample, and extended peak. The $\alpha \rightarrow \beta$ transition zone also becomes slow due to the same treatment, so that, finally, the exothermic reaction from the β phase precipitation is generated, the position of which seemingly stays without change at 460°C after preheating. These effects tend to smoothen, when the sample is preheated for 50 minutes at 300°C, while only a few signs of the recrystallization process are left.

Figure 8(b) shows the effect of preaging at 400°C for 20, 50, and 100 minutes. These thermograms confirm the behavior noted at 300°C regarding the effect of preaging

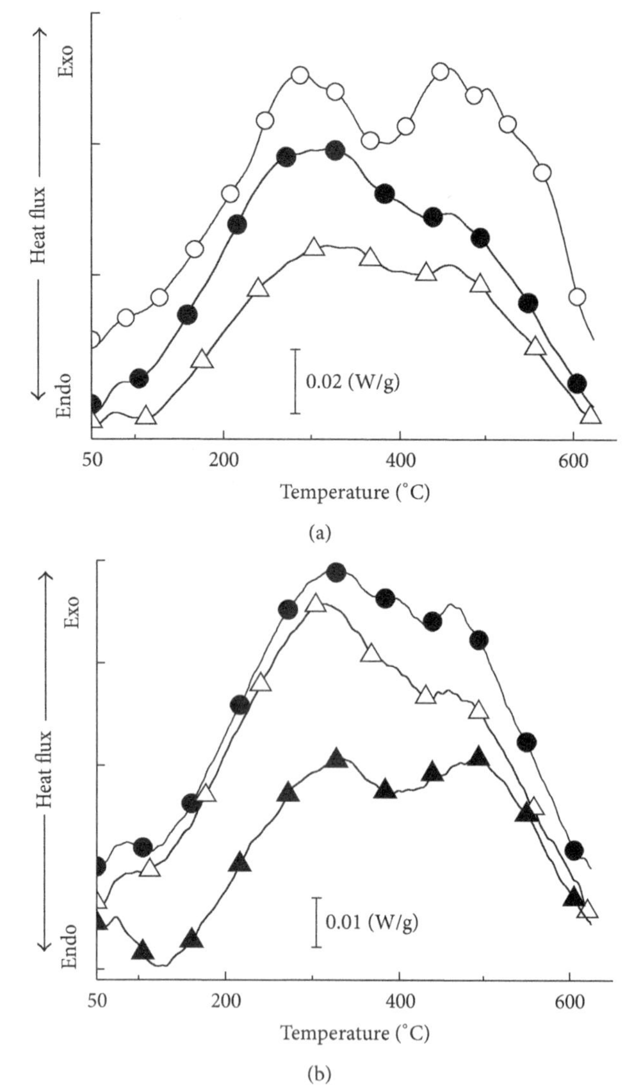

Figure 8: Effect of preaging on samples severely deformed at 5°C/min. (a) T = 300°C and (b) T = 400°C. ○ without preaging, ● 20 min at T, △ 50 min at T, and ▲ 100 min at T.

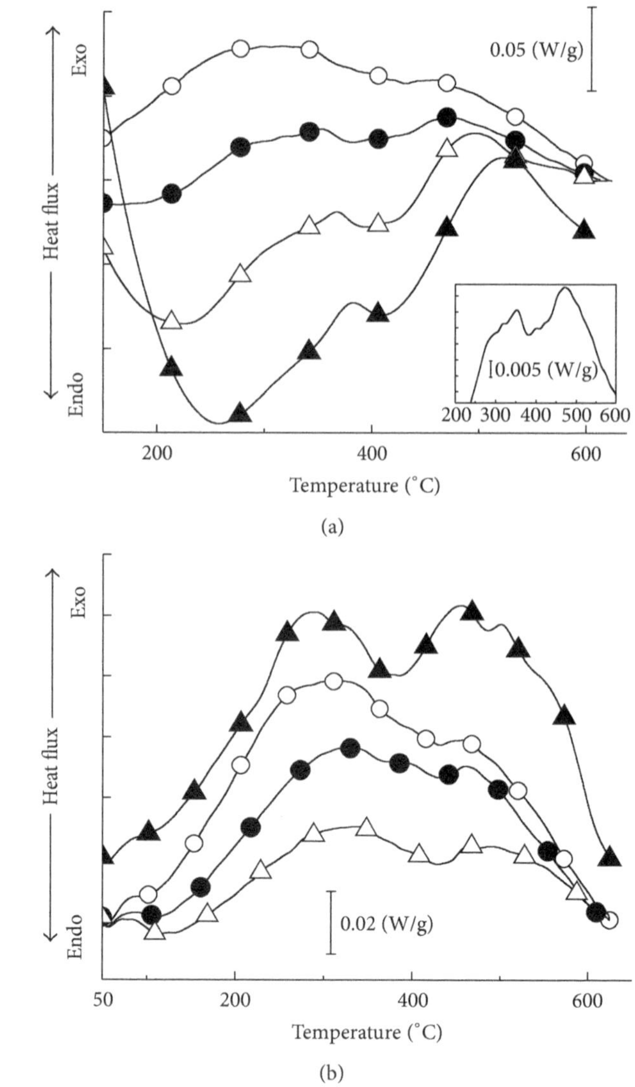

Figure 9: Thermograms of 85-DHS samples. (a) Effect of the heating rate for samples pre-heated for 20 minutes at 300°C (same notation as Figure 7). (b) 20 min preheating at different temperatures measured at 5°C/min. ▲ without preheating. ○ T = 300°C. ● T = 400°C. △ T = 500°C.

around the first exothermic reaction. A notable difference occurs at the reversion level of the first peak for 100 min preaging at 400°C, where most certainly the recrystallization process has been completed during preaging. The behavior of the second and third exothermic reaction tends to be free of recrystallization. These thermograms are evidence of the precipitation-recrystallization interaction sequence starting from the severely deformed sample (Figure 8(a)) through different stages where atom supersaturation and the concentration of internal stresses are reduced, tending, specifically in the case for 100 minutes at 400°C, to have a behavior close to that of the homogenized sample, conveying the idea that full recrystallization has occurred. Even though it is not shown, preaging at 500°C was also checked at different times, where it was evident that the reactions associated with recrystallization were not present, in agreement with our previous analysis.

4.2.4. 20 min Preaging at Different Heating Rate and Temperatures in 85-DHS Samples. The thermogram of Figure 9(a) shows the effect of different Φ on samples severely deformed and preaged for 20 minutes at 300°C. The thermogram at 5°C/min was discussed before. The inset shows the amplified version of the thermogram for Φ = 10°C/min, where the transformation details are highlighted. In it, two peaks of high enthalpy are observed at 352°C and 470°C, respectively; an exothermic reaction of lesser magnitude at 315°C; and another small exothermic reaction at 400°C in the $\alpha \rightarrow \beta$ transition. This behavior confirms the precipitation-recrystallization interaction, where recrystallization corresponds to the peak at 315°C, prior to the precipitation of the α phase, in total agreement with Gokhale et al. [3]. The reactions that occur at higher temperatures correspond to the growth of recrystallized grains and to the β phase, respectively. Note the movement of each of the reactions towards higher temperatures as Φ increases, indicating that they occur by a diffusion mechanism. For high Φ values, the exothermic reactions that take place at high temperatures tend to combine into a single one, due to the difference between the reaction rates of both mechanisms.

Figure 9(b) shows the effect of preaging for 20 minutes at different temperatures (300°C, 400°C, and 500°C), including a reference without preaging, heated at 5°C/min. This figure confirms the effect of preheating on each of the phases: as the pre-heating temperature becomes higher, both the atom supersaturation and the dislocation density tend to abate, which translates into better defined exothermic reactions due to the greater energy accumulated in the absence of or little preheating time. In this figure, near 90°C, the exothermic reaction associated to the recovery-precipitation (GP zones) interaction is maintained, becoming more defined as the preheating temperature increases, due to the progressive completion of the recovery process. The corresponding exothermic reactions, due to the coexistence of the α phase precipitation and to recrystallization, both well defined in the thermogram without preaging (solid triangles), tend to appear as a single reaction for preaging processes at 500°C, indicating that, at that temperature, the recrystallization process has been completed and the peak reported is due to the precipitation. This also applies to the reactions associated with the β phase precipitation and the growth of recrystallized grains, which are transformed into one single peak because, at that temperature, the dislocation density is very low. An important aspect to be noted in this figure is the slight displacement of the precipitation peaks as the preaging temperature increases, confirming the existence of a lesser supersaturation and a smaller dislocation density in these cases.

The results obtained via DSC are coherent with the ones obtained via thermoelectric power, even though the temperatures at which these transformation peaks occur differ slightly, due to the response of the microstructure to the measurement technique used.

4.3. Transmission Electron Microscopy (TEM). The observations of the AA8011 alloy via TEM allow a microanalysis and ultrastructural analysis of the samples under different microstructure conditions generated by the different thermal and mechanical treatments, allowing direct access to the substructures of dislocations inside subgrains or at the grain boundaries, as well as to the phases and precipitates that are present and which have promoted interactions among themselves for the formation of the final structure of the material. A micrographic study of the AA8011 alloy was made by Gokhale et al. [3] under microstructure conditions different from the present work. They use heat-rolled samples, identify the recrystallization peak by DSC, and then age the sample at around the peak temperature. In the present work, the samples, whether deformed or not, are subjected to nonisothermal heating at a rate of 5°C/min up to the temperature of the peak, and once the desired condition is achieved, the sample is quenched in a cold bath. Some

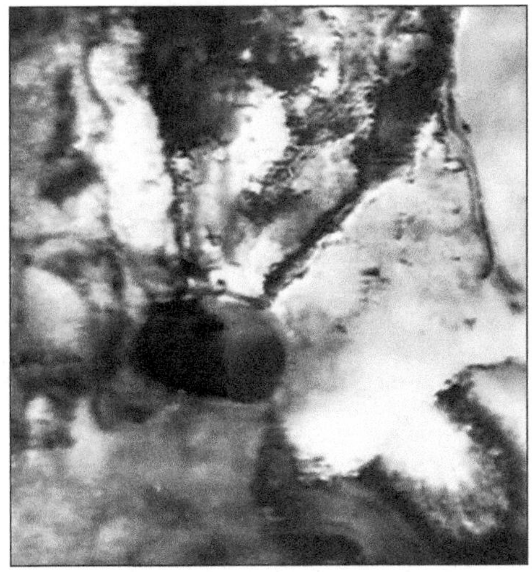

Figure 10: TEM micrograph of a cast AA8011 sample showing presence of particles, dislocations, grains, subgrains, and grain joints.

interesting work on deformed aluminum microstructures and derived alloys is to be found in the literature. The microscopy study exposed below is not intended to identify the phases present; however, it is well established that these phases are composed mainly of Al-Fe and Al-Fe-Si globular morphology (α-AlFeSi) and acicular (β-AlFeSi) [21–24].

4.3.1. Sample Cast, Homogenized, and Rolled to Different Reductions in Thickness

(a) Cast Sample. Earlier research [23, 24] determined that, in the cast sample, there are zones of fragmented segregation and second-phase particles with a redistribution of segregates in the aluminum matrix, where the skeletal phase of the type known as "Chinese structure" can be appreciated. The micrograph of Figure 10 corresponds to a cast sample where a variety of linear defects and dislocation substructures are observed interacting with each other. Furthermore, high deformation zones typical of this type of cast samples are observed, which generate much potential energy (stored) that will eventually trigger precipitation and recrystallization mechanisms in the alloy. Also, particles of likely ovoid shape and size around 6 μm are detailed, anchored to the lattice matrix, and interacting with a distribution of subgrains and dislocations, both inter- and transsubgranularly. These effects are typical of alloys subjected to a hot-rolling process. The anchor of the particle considerably limits later subgrain development, which affects the mechanical properties of the material, depending on the distribution and size of the grains.

(b) Homogenized Samples. We have taken the work of Torres et al. [24] as reference for the analysis of the homogenized sample. They observed intermetallic phases with a distribution in the matrix core; particles with spherical (α-AlFeSi) and needle-like (β-AlFeSi) shape were reported. They also observed the presence of dispersed silicon particles, which embrittles the alloy by directly affecting its finish. Also, the

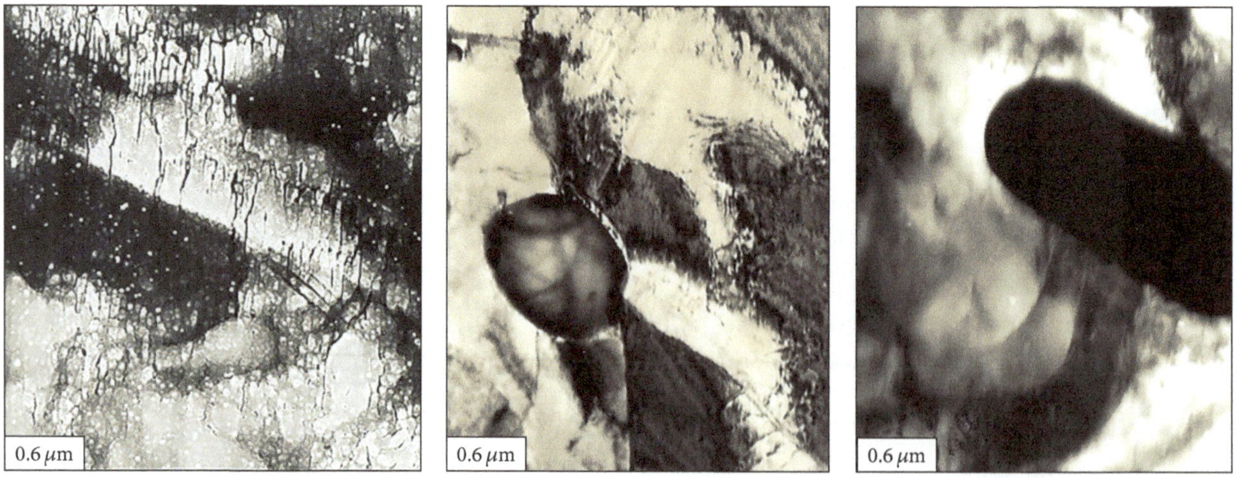

(a) (b) (c)

Figure 11: TEM micrographs of a 50-DHS sample, showing dislocation substructures and elongated particles with dislocations entangled in and around the grains.

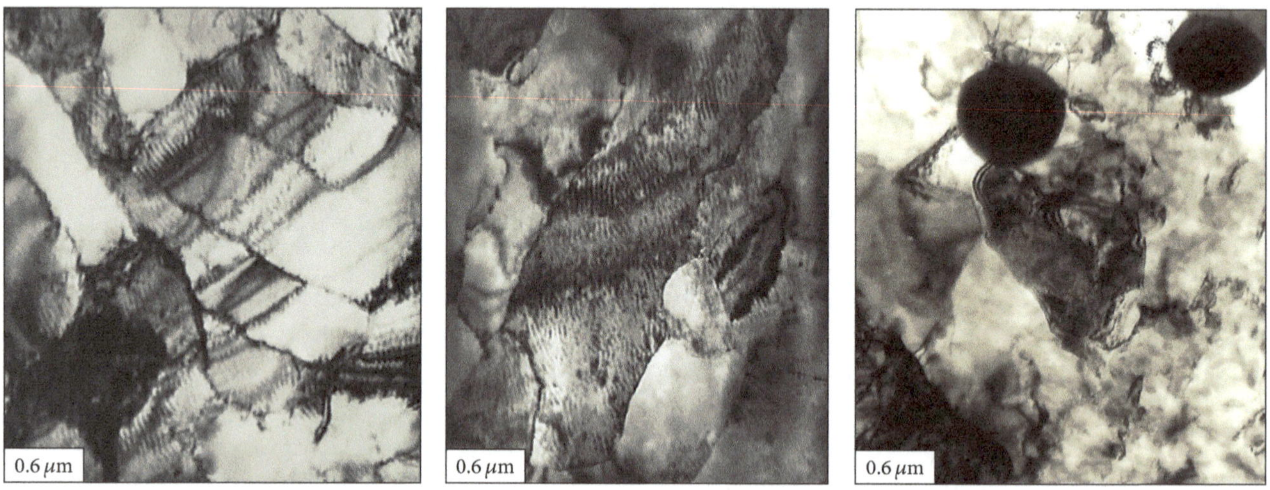

(a) (b) (c)

Figure 12: TEM micrographs of an 85-DHS sample, showing distribution of deformed grains and an elevated dislocations density.

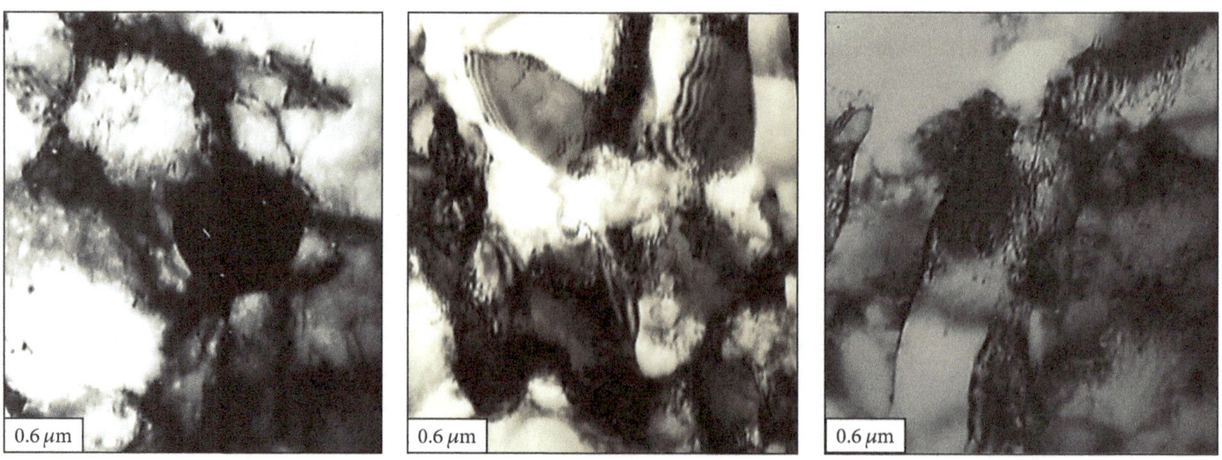

(a) (b) (c)

Figure 13: TEM micrograph of a 50-DHS sample, heated up to 90°C at a rate of 5°C/min.

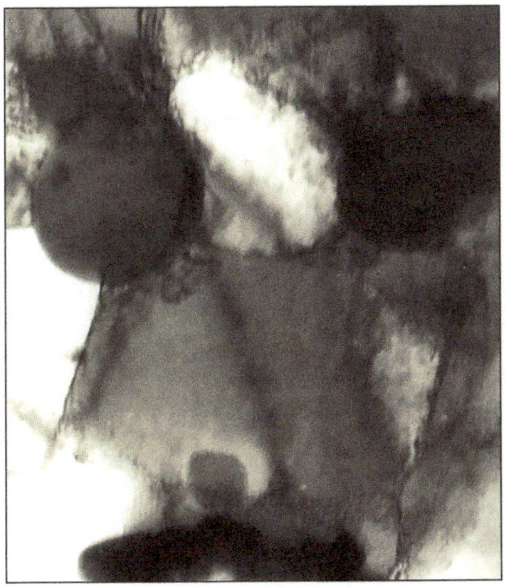

FIGURE 14: TEM micrograph of a 50-DHS sample heated up to 278°C at a rate of 5°C/min.

matrix shows low density dislocation zones and a grain distribution of different sizes.

(c) 50-DHSR Samples. Figure 11 shows the effect of rolling on a homogenized sample, subsequently rolled to a 50% reduction in thickness. Micrographs in Figures 11(a) and 11(b) show subgrains of different sizes lengthened by rolling, with well-defined contours. Also, numerous discontinuities tending to form dislocation substructures with different orientations at grain boundaries are observed in the matrix. This high dislocation density generates a very high accumulated energy potential that might greatly affect the recrystallization and precipitation process. Likewise, Figure 11(c) shows an elongated particle of approximately 12 μm in which grains of irregular shape are anchored, with dislocations entangled in and around the grains, indicating a clear precipitate-grain-subgrain-dislocation interaction. Therefore, the dislocation density increases its distribution in disarray, arranging itself as a cobweb or jungle in the core of the matrix.

(d) 85-DHS Samples. Micrograph in Figure 12(a) shows the distribution of deformed grains and subgrains, oriented by the rolling process, with a high dislocation density that tends to redistribute, both inter- and trans-granularly in the material. Micrograph in Figure 12(b) shows the presence of irregular grains and subgrains of different sizes and high dislocation density on the grain edges or boundaries, a product of the high activity during formation. Likewise, micrograph in Figure 12(c) shows two spherical particles anchored, interfering with the trajectory of the dislocations and hindering the complete formation of grains and subgrains in the aluminum matrix. This anchorage of particles obeys the fractionation of particles of greater size into smaller ones due to the deformation process and to the reorganization suffered by the dislocations. However, when the degree of deformation is high, 85% in our case, the dislocation density increases and its redistribution is not uniform, the array of the grains becoming an entanglement in the core of the matrix; nonetheless, when deformation is too severe, the grains may annihilate each other or order themselves, providing evidence of grain boundaries as a product of polygonization.

4.3.2. 50-DHS and Annealed Samples (ADHS) at the DSC Peak Temperatures. Micrograph in Figure 13 shows samples annealed at a rate of 5°C/min up to a temperature near the temperature of the peak reported in DSC thermograms studied before. The microstructure for samples heated up to 90°C highlights (Figure 13(a)) the presence of a precipitate of spherical shape of approximately 6 μm in diameter, with a relatively uniform subgrain distribution and a high density of dislocation substructures. The presence of few small precipitates is also noted.

Figures 13(b) and 13(c) highlight remnants of the rolling process with elongated grains as a product of the orientation of deformation and a distribution of dislocations around grains parallel to their interior. Besides the aforesaid microstructural elements, the appearance of bands where dislocations of varied thickness, subgrains, and bands interact is also observed. Apparently, the annealing temperature is not sufficiently high to produce a redistribution of the dislocation substructure.

Figure 14, on the other hand, shows micrographs for T = 278°C corresponding to the presence of grains and subgrains of different sizes. Spherical particles interacting with a significant number of dislocations are observed, meaning that the recovery process unfolds slowly. Besides, dislocation structures are observed because the annealing temperature is not sufficiently high to produce a redistribution of dislocation structures.

Even though the precipitates have not been identified at this level, a direct interaction between these and the dislocation structures is seen, implying that the recovery-recrystallization process is happening.

Figure 15 shows a micrograph corresponding to samples heated up to T = 466°C, in which particles on the boundaries or edges of grains, dislocations interacting with the particles, and reorganization of dislocations are observed (Figure 15(a)) in the aluminum matrix, acquiring an orderly configuration. Figure 16(b) shows details of the same zone with dislocation lines superimposed on a recrystallized microstructure.

4.3.3. 85-DHS and Annealed Samples (ADHS) at the DSC Peaks Temperature. Figure 16 shows two aspects of the samples heated up to 90°C. Figure 16(a) highlights the presence of lengthened grains with an important dislocation density, in the shape of bands, in the grain boundaries, and in the grain itself; however, unlike in the previous study, there are no dislocation entanglements. Figure 16 shows particles interfering with varied irregular grains and dislocations that try to surround the particles.

Figure 17 shows micrographs corresponding to a heating of up to T = 455°C. Micrograph in Figure 17(a) shows the coexistence of a spherical particle, around 5 μm in diameter, with a well-defined grain border, with an outstanding variety of inner dislocations and the appearance of recrystallized grains. Micrograph in Figure 17(b) shows the microstructure

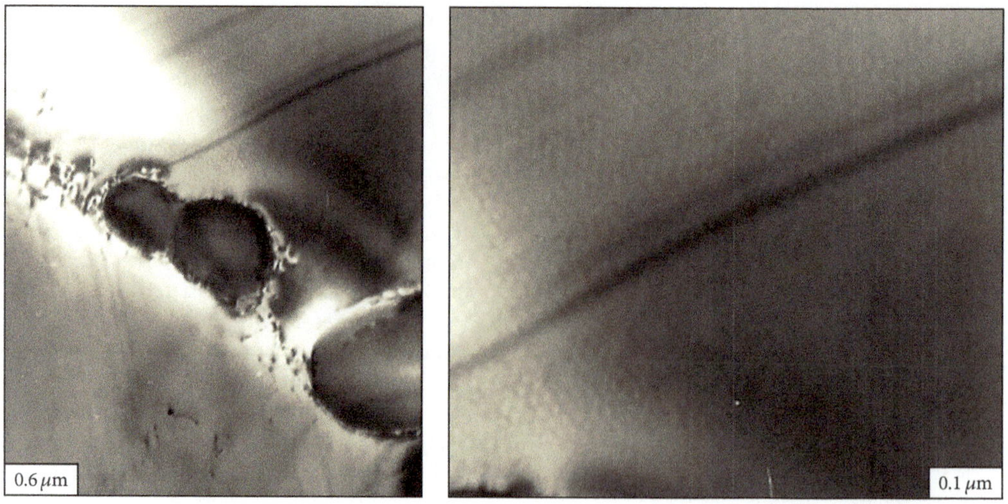

(a) (b)

FIGURE 15: TEM micrographs of a 50-DHS sample heated up to 466°C at a rate of 5°C/min.

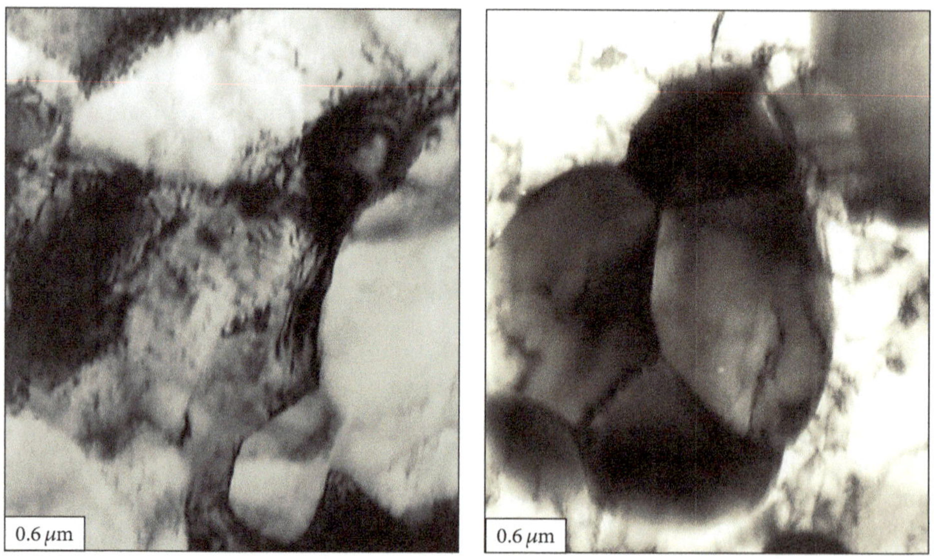

(a) (b)

FIGURE 16: TEM micrographs of an 85-DHS sample, heated up to 90°C at a rate of 5°C/min.

in the region of confluence of several grains or triple point. Very small precipitates interacting with the aluminum matrix, possibly dispersoids, are also observed.

Figure 18 shows a microstructure at $T = 502^{\circ}$C. A big ovoid-shaped particle is shown (Figure 18(a)) which seems not to interact strongly with the matrix and visibly recrystallized grains, while, in Figure 18(b), the presence of a great number of subgrains is highlighted; some are precipitates and some disorderly dislocations with different conformations are seen.

From the present microscopic study, we may infer that, as the degree of rolling increases, the degree of rolling-borne deformation induces, up to a value, a growing and heterogeneously distributed dislocation density, originating a mess or jungle of dislocations at the core of the matrix, to be later reordered and even eliminated in severe deformations.

Continuous heating up to a certain temperature fosters both the release of energy stored in the form of dislocations and their rearrangement, in turn prompting the recrystallization and interaction with the precipitated phase at those temperatures. Definitely, the micrographs provided confirm that deformation creates preferential sites for the onset of the recrystallization and of the phases formed at specific temperatures, so that interaction of both processes is produced from the search of equilibrium. Deformation by rolling is a significant factor in the recrystallization and precipitation process, and the likelihood that the material reaches equilibrium as a consequence of the release of energy

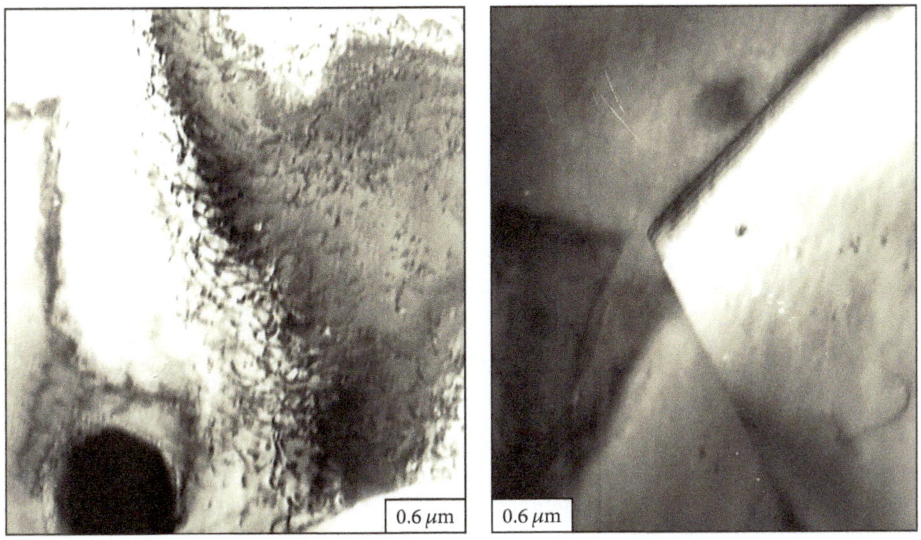

(a) (b)

FIGURE 17: TEM micrographs of an 85-DHS sample, heated up to 475°C at a rate of 5°C/min.

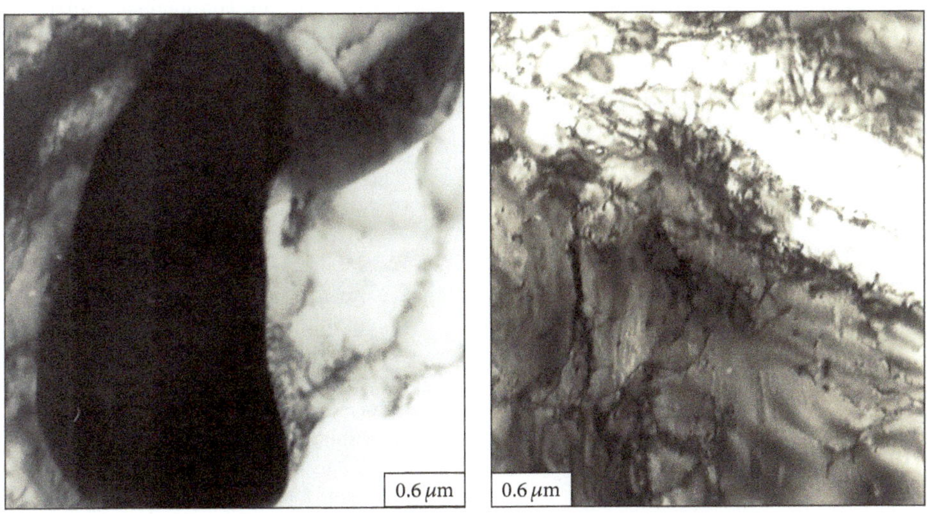

(a) (b)

FIGURE 18: TEM micrographs of an 85-DHS sample heated up to 502°C at a rate of 5°C/min.

absorbed during the work will depend on temperature and annealing time.

5. Conclusions

The AA8011 alloy has been exhaustively studied under different microstructures by thermoelectric power, differential scanning calorimetry, and transmission electron microscopy in order to prove the occurrence of precipitation and recrystallization under different conditions, prompting the following conclusions.

(1) The three techniques used reflect their sensitivity to the microstructural changes studied, the recrystallization and precipitation processes differing on the basis of the conditions of the samples, and the initial microstructure remaining regulated by the preaging treatments.

(2) Thermoelectric power studies establish the existence of two transformations in the HS samples associated with precipitation of the α-AlFeSi below 400°C and β-Al_3Fe near 470°C, their transition occurring basically because of the rearrangement of Fe atoms in the precipitated phase. Rolling accelerates precipitation of α phase (DHS samples) up to a defined degree of deformation, suggestive of the reorganization of the dislocation structure and the interaction between the recrystallization and precipitation processes. The β phase is less affected by the rolling process and the temperature range where precipitation occurs coinciding with that of recrystallized grains growth.

The derivative of the thermoelectric power appears as a powerful tool for identifying microstructural changes, be it by recrystallization or by precipitation. In most cases, by means of this technique, both recrystallization and precipitation of α phase coincide on a single peak, broader than that inherent to precipitation.

(3) From the DSC studies, we conclude that the behavior obtained for ΔS is confirmed, including some additional points. For example, at low heating rates and in highly rolled samples, additional exothermic reactions appear. In this work, the reaction close to the precipitation peak of α phase is associated with recrystallization, while the reaction happening close to the β phase is associated with the growth of recrystallized grains. Details of precipitation in the Guinier-Preston zones are included.

The evolution of the heat flow at different heating rates confirms that both the nucleation process and recrystallization are mechanisms thermally active, occurring by diffusion. Preliminary details of the activation energy inferred from this type of study are presented by Luiggi [13].

(4) TEM micrographs confirm the premises supporting the present work, reflecting that the rolling process introduces in the material a dislocation density that grows as the degree of deformation increases. At a low strain rate, the dislocations are distributed heterogeneously and disorderly, forming a dislocation entanglement, while, for high strain, the dislocation distribution tends to show a certain order. This fact explains the differences of behavior of DSC and ΔS for the different microstructures studied.

The micrographs show an interaction between precipitates and dislocations, reporting a total or partial recrystallization at the temperatures deduced from ΔS and DSC experiments.

A study elaborating on recrystallization interaction and on phase kinetics evolution in severely deformed samples at lower heating will be undertaken, to delve in the deconvolution of the kinetics where both processes occur simultaneously.

Conflict of Interests

The authors declare that there is no conflict of interests regarding the publication of this paper.

Acknowledgments

This work is supported by the Office of Academic Planning at the Universidad de Oriente through POA Project PN 5.5/2010. The authors would also like to acknowledge Carlos Mota and his company Traduce, CA, for the translation of this paper.

References

[1] P. R. Rios, F. Siciliano Jr., H. R. Z. Sandim, R. L. Plaut, and A. F. Padilha, "Nucleation and growth during recrystallization," *Materials Research*, vol. 8, no. 3, pp. 225–238, 2005.

[2] E. P. Rocha Lima, R. A. Sanguinetti Ferreira, N. Freitas de Quadros, and Y. Prasad Yadava, "Estudo dos aspectos cinéticos e morfológicos durante recristalização da liga de alumínio aa 8011," *Revista Iberoamericana de Ingeniería Mec*, vol. 10, no. 1, pp. 131–137, 2006.

[3] A. M. Gokhale, C. V. Iswaran, and R. T. DeHoff, "Use of the stereological counting measurements in testing theories of growth rates," *Metallurgical Transactions A*, vol. 10, no. 9, pp. 1239–1245, 1979.

[4] F. J. Humphreys and M. Hatherly, *Recrystallization and Related Annealing Phenomena*, Pergamon Press, Oxford, UK, 2nd edition, 2004.

[5] R. A. Vandermeer, R. A. Masumura, and B. B. Rath, "Microstructural paths of shape-preserved nucleation and growth transformations," *Acta Metallurgica et Materialia*, vol. 39, no. 3, pp. 383–389, 1991.

[6] R. E. S. Puchi, B. Fajardo, and J. V. Valera, "Recrystallization of commercial twin-Roll Cast Aluminum-iron-silicon Alloy Homogenized at 853 K," in *Proceedings of the 4th International Conference on Aluminum Alloys*, T. H. Sanders Jr. and E. A. Starke Jr., Eds., vol. 1, pp. 18–25, Atlanta, Ga, USA, 1994.

[7] W. H. Hildebrandt, "Differential scanning calorimetry evaluations of recrystallization behavior in aluminum sheet," *Metallurgical Transactions A*, vol. 10, no. 8, pp. 1045–1048, 1979.

[8] N. J. Luiggi, "Thermoelectric power study of non-isothermal precipitation in commercial (8011) aluminium-iron-silicon alloys," *Zeitschrift fuer Metallkunde*, vol. 88, no. 9, pp. 728–732, 1997.

[9] A. Ney José Luiggi, "Characterization by thermoelectric power of a commercial aluminum-lron-silicon alloy (8011) during isothermal precipitation," *Metallurgical and Materials Transactions A*, vol. 29, no. 11, pp. 2669–2677, 1998.

[10] E. P. R. Lima, *Estudo da cinética de recristalização da liga da aluminio AA8011 na condição H18 [M.S. thesis]*, Universidade Federal de Pernambuco-Brasil, 2002.

[11] S. Komatsu, M. Ikeda, T. Muramatsu, and M. Matsuo, "Estimation of solute Fe concentration in commercial pure aluminum," in *International Workshop on the Effect of Iron and Silicon in Aluminium and Its Alloys*, pp. 31–55, May 1989.

[12] C. Garcia-Cordovilla and E. Louis, "A differential scanning calorimetry study of recrystallization and its interaction with precipitation in Al-Fe-Si commercial alloys (AA1145 and AA8011)," *Journal of Materials Science*, vol. 21, no. 3, pp. 971–979, 1986.

[13] N. J. Luiggi A., "A preliminary study of the phase transformations in rolled Al-Fe-Si alloy," *Metallurgical and Materials Transactions A*, vol. 41, no. 13, pp. 3271–3275, 2010.

[14] R. Borrelly and J. L. Bouvier-Volaille, "Appareil de caractérisation des alliages par mesure de leur pouvoir thermoélectrique," *Traitements Thermiques*, vol. 221, pp. 43–45, 1988.

[15] F. J. Blatt, P. A. Schroeder, C. L. Foiles, and D. Greig:, *Thermoelectric Power of Metals*, Plenum press, New York, NY, USA, 1976.

[16] M. Perez, V. Massardier, and X. Kleber, "Thermoelectric power applied to metallurgy: principle and recent applications," *International Journal of Materials Research*, vol. 100, no. 10, pp. 1461–1465, 2009.

[17] N. J. Luiggi, "Isothermal precipitation of commercial 3003 AI alloys studied by thermoelectric power," *Metallurgical and Materials Transactions B*, vol. 28, no. 1, pp. 125–133, 1997.

[18] R. Borrelly, P. Merle, and D. Adenis, "Seebeck effect study of precipitation in 3004," in *Proceedings of the Technical Sessions by the TMS Light Metals Committee at the 118 TMS Annual Meeting*, pp. 703–712, March 1989.

[19] R. J. Willson, "Calorimetry," in *Principles of Thermal Analysis and Calorimetry*, P. J. Haines, Ed., pp. 129–161, Oakland Analytical Services, Farnhurn, Surrey, UK, 2002.

[20] L. He, H. Zhang, and J. Cui, "Effects of pre-ageing treatment on subsequent artificial ageing characteristics of an Al-1.01Mg-0.68Si-1.78Cu alloy," *Journal of Materials Science and Technology*, vol. 26, no. 2, pp. 141–145, 2010.

[21] C. Y. Yu, P. L. Sun, P. W. Kao, and C. P. Chang, "Evolution of microstructure during annealing of a severely deformed aluminum," *Materials Science and Engineering A*, vol. 366, no. 2, pp. 310–317, 2004.

[22] N. Luiggi, M. Valera, J. Prin, and M. Linares, "Estudio de la Aleación AA8011 Laminada usando DSC y MET," *Acta Microscópica*, vol. 22, no. 1, pp. 105–110, 2013.

[23] R. Bisbal, T. Torres, S. Camero, and C. Llanos, "Evolución Microestructural de la Aleación de Aluminio AA8011 Sometida a Tratamientos Térmicos," *Acta Microscópica*, vol. 18, pp. 179–181, 2009.

[24] T. Torres, R. Bisbal, S. Camero, and C. Llanos, "Estudio del Efecto del Tratamiento Térmico de Homogeneización en la Microestructura y Propiedades Mecánicas de una Aleación de Aluminio AA8011," *Acta Microscópica*, vol. 20, no. 2, pp. 165–173, 2011.

A Nonmonotone Adaptive Trust Region Method Based on Conic Model for Unconstrained Optimization

Zhaocheng Cui

Department of Mathematics and Physics, Shandong Jiaotong University, Ji'nan, Shandong Province 250023, China

Correspondence should be addressed to Zhaocheng Cui; cuizhaocheng@sina.com

Academic Editor: Adil M. Bagirov

We propose a nonmonotone adaptive trust region method for unconstrained optimization problems which combines a conic model and a new update rule for adjusting the trust region radius. Unlike the traditional adaptive trust region methods, the subproblem of the new method is the conic minimization subproblem. Moreover, at each iteration, we use the last and the current iterative information to define a suitable initial trust region radius. The global and superlinear convergence properties of the proposed method are established under reasonable conditions. Numerical results show that the new method is efficient and attractive for unconstrained optimization problems.

1. Introduction

In this paper, we consider the following unconstrained optimization problem:

$$\min_{x\in R^n} f(x), \tag{1}$$

where $f: R^n \to R$ is a continuously differentiable function.

Trust region method is effective for solving (1). In 1970s, Powell [1] established the convergence result of trust region method. Yuan [2], Nocedal, and Yuan [3] proposed various trust region methods for optimization problems. It calculates a trial step by solving the subproblem

$$\min \phi_k(d) = g_k^T d + \frac{1}{2} d^T B_k d, \quad \text{s.t. } \|d\| \le \Delta_k, \tag{2}$$

where $g_k = \nabla_x f(x_k)$ is the gradient of the objective function at x_k, $B_k \in R^{n\times n}$ is a symmetric matrix which is either the Hessian matrix of f at x_k or an approximation to it, $\|\cdot\|$ refers to the Euclidean norm, and $\Delta_k > 0$ is the radius of an area, called the trust region, where the model is trusted. Let d_k be the solution of (2). The actual reduction of the objective function is defined by

$$\text{Ared}_k = f_k - f(x_k + d_k), \tag{3}$$

and the predicted reduction of $f(x)$ is defined by

$$\text{Pred}_k = \phi_k(0) - \phi_k(d_k). \tag{4}$$

The ratio between these two reductions is defined by

$$r_k = \frac{\text{Ared}_k}{\text{Pred}_k} = \frac{f_k - f(x_k + d_k)}{\phi_k(0) - \phi_k(d_k)}, \tag{5}$$

and it is normally used to test whether the trial step d_k is accepted or the trust region radius needs to be adjusted. The next iterate x_{k+1} is chosen by the following formula:

$$x_{k+1} = \begin{cases} x_k + d_k & \text{if } r_k \ge \eta_0, \\ x_k & \text{otherwise}, \end{cases} \tag{6}$$

where $\eta_0 \in [0, 1)$ is a constant. The next trust region radius is chosen as

$$\Delta_{k+1} = \begin{cases} c_0\Delta_k & \text{if } r_k < \eta_1, \\ c_1\Delta_k & \text{if } r_k > \eta_2, \\ \Delta_k & \text{otherwise}, \end{cases} \tag{7}$$

where $0 \le \eta_1 < \eta_2 < 1$, $0 < c_0 < 1 < c_1$ are constants. In comparison with quasi-Newton methods, trust region methods converge to a point which not only is a stationary point but also satisfies second-order necessary conditions. Trust region method has strong convergence and robustness, so many authors have studied it; see [4–7]. Conn et al. [8], Nocedal and Wright [9], and Yuan and Sun [10] presented the complete introduction.

Recently, Grippo et al. [11] presented the nonmonotone line search method for unconstrained optimization problems. Due to its high efficiency of nonmonotone method, this technique has been combined with trust region methods; we can see [12–16].

Sartenaer [17] presented a strategy for determining automatically an initial trust region radius. Fan and Yuan [18] proposed a trust region method with the trust region radius converging to zero. Recently, Zhang et al. [19] gave an adaptive trust region method, and they solve the subproblem (2) with $\Delta_k = c^p\|g_k\|\|\widehat{B}_k^{-1}\|, 0 < c < 1$, and p is a nonnegative integer. $\widehat{B}_k$ is a safely positive definite matrix based on Schnabal and Eskow modified Cholesky factorization. Therefore, instead of adjusting Δ_k, one adjusts p at each iteration. Based on [19] and a simple subproblem model, Sang and Sun [20] proposed another adaptive trust region method with line search. The numerical results show that the adaptive trust region algorithms [21–24] are more effective than traditional trust region methods for unconstrained optimization problems.

The adaptive trust region methods listed above are based on quadratic model; however, when the objective function has strong nonquadratic behavior, the quadratic model methods often produce a poor prediction of the minimizer of the function. In 1980, Davidon [25] proposed the conic model methods for unconstrained optimization problems. A typical conic trust region subproblem is as follows:

$$\begin{aligned} &\min \varphi_k(s) = \frac{g_k^T s}{1-\alpha_k^T s} + \frac{1}{2}\frac{s^T B_k s}{\left(1-\alpha_k^T s\right)^2}, \\ &\text{s.t. } 1-\alpha_k^T s > 0, \\ &\|s\| \le \Delta_k, \end{aligned} \tag{8}$$

where $\varphi_k(s)$ is called conic model which is an approximation to $f(x_k + s_k) - f(x_k)$. The vector α_k is the associated vector for the collinear scaling in the kth iteration, and it is normally called the horizontal vector. If $\alpha_k = 0$, the conic model reduces to a quadratic model. Also, one can see that $\varphi_k(s)$ is quadratic along any direction $s \in R^n$ satisfying $\alpha_k^T s = 0$. Recently, trust region methods based on conic model have been presented and studied; see [26–32]. Numerical results show the efficiency of the conic trust region methods.

In the traditional trust region methods, the next trust region radius is updated by the current ratio r_k, which is reasonable if the matrix B_k is exactly as the Hessian H_k and if the trust region subproblem (2) is solved exactly. However, in practical computations, the matrix B_k is often obtained approximately and the subproblem is solved approximately. In such a case, it may be more reasonable to adjust the next trust region Δ_{k+1} according to not only r_k but also the previous ratios $\{r_{k-q}, \ldots, r_k\}$, where q is some nonnegative integer.

In order to measure the agreement between the model function and the objective function, we define the following ratio:

$$\widehat{r}_k = \sum_{i=0}^{\min\{k,q\}} \omega_{ki} r_{k-i}, \tag{9}$$

where $\omega_{ki} \in [0,1]$ is the weight of r_{k-i}, such that

$$\sum_{i=0}^{\min\{k,q\}} \omega_{ki} = 1. \tag{10}$$

In this paper, we propose a new adaptive trust region method which combines the conic model and nonmonotone technique which was given in [33] and a new update rule for adjusting the trust region radius. Our method is different from other adaptive trust region methods in three points. First, the subproblem is conic model which can produce a better prediction of the minimizer of the function than quadratic model methods when the objective function has strong nonquadratic behavior. Second, at each iteration, our method generates a suitable trust region radius automatically based on the current and last iterative information. Last, we use a new update rule (8) to adjust the next trust region radius. The new trust region model is more consistent with the objective function at the current iterative point. According to the nonmonotone technique in [33], our method constructs an iterative series $\{x_k\}$ such that the sequence $\{f(x_k)\}$ is not monotonic, which can accelerate the convergence rate of the minimization process, especially in the narrow curved valley. The global and superlinear convergence properties of the proposed method are proved under reasonable conditions. The numerical results show that the new method is more effective than other trust region methods.

The rest of this paper is organized as follows. In next section, we describe a new nonmonotone adaptive trust region method based on conic model for unconstrained optimization problems. In Sections 3 and 4, we prove the global and superlinear convergence properties of the proposed method under reasonable conditions, respectively. Some numerical results are given in Section 5. Conclusions are summarized in Section 6.

2. Algorithm Description

In this section, we describe a new nonmonotone adaptive trust region method based on conic model for unconstrained optimization problems.

We obtain the trial step s_k by solving the following subproblem:

$$\begin{aligned} &\min \varphi_k(s) = \frac{g_k^T s}{1-\alpha_k^T s} + \frac{1}{2}\frac{s^T B_k s}{\left(1-\alpha_k^T s\right)^2}, \\ &\text{s.t. } 1-\alpha_k^T s > 0, \\ &\|s\| \le \Delta_k, \end{aligned} \tag{11}$$

where $\Delta_k = \mu_k(\|s_{k-1}\|^2/s_{k-1}^T\widehat{B}_k s_{k-1})\|g_k\|, \Delta_0 = \mu_0\|g_0\|, \mu_0 > 0$. μ_k is updated according to (8), s_{k-1} is the solution of the last subproblem. $\widehat{B}_k$ is generated by the procedure $s_{k-1}^T \widehat{B}_k s_{k-1} = s_{k-1}^T B_k s_{k-1} + i\|s_{k-1}\|^2$ and i is the smallest nonnegative integer such that

$$s_{k-1}^T \widehat{B}_k s_{k-1} = s_{k-1}^T B_k s_{k-1} + i\|s_{k-1}\|^2 > 0. \tag{12}$$

The new reduction of $f(x)$ is defined by

$$\text{Ared}_k = M_k - f(x_k + s_k), \tag{13}$$

where

$$M_k = \max\left\{f(x_k), \sum_{r=0}^{m(k)-1} \lambda_{kr} f(x_{k-r})\right\}, \tag{14}$$

$\lambda \in (0,1], m \geq 1$ is a positive integer. We define $m(k) = \min\{k+1, m\}$ and choose

$$\lambda_{kr} \geq \lambda, \quad r = 0, 1, \ldots, m(k) - 1, \quad \sum_{r=0}^{m(k)-1} \lambda_{kr} = 1. \tag{15}$$

The predicted reduction of $f(x)$ is

$$\text{Pred}_k = -\frac{g_k^T s_k}{1 - \alpha_k^T s_k} - \frac{1}{2}\frac{s_k^T B_k s_k}{(1 - \alpha_k^T s_k)^2}. \tag{16}$$

The new ratio r_k is defined by

$$r_k = \frac{\text{Ared}_k}{\text{Pred}_k}. \tag{17}$$

In what follows, we describe our new nonmonotone adaptive trust region method based on conic model.

Algorithm 1 (nonmonotone adaptive trust region method based on conic model). Consider the following.

Step 0. Given $x_0 \in R^n$, $B_0 \in R^{n\times n}$ is a symmetric positive definite matrix, $\varepsilon > 0$, $\mu_0 > 0$, $\Delta_0 = \mu_0\|g_0\|$, $0 < \lambda \leq 1$, $0 < \eta_0 < \eta_1 < 1$, $0 < c_0 < 1 < c_1$, and an integer constant $m \geq 1$; set $m(0) = 0$, $k := 0$.

Step 1. Compute g_k. If $\|g_k\| < \varepsilon$, then stop. Otherwise, go to Step 2.

Step 2. Solve the subproblem (11) and let s_k be an approximate solution of the subproblem (11).

Step 3. Compute Ared_k, Pred_k, and r_k.

Step 4. If $r_k < \eta_0$, set $\mu_k := c_0\mu_k$ and go to Step 2. Otherwise, go to Step 5.

Step 5. Choose $\omega_{ki} \in [0,1]$ that satisfying (9). Compute $\widehat{r}_k$ and, by (8), set $x_{k+1} = x_k + s_k$ and

$$\mu_{k+1} = \begin{cases} c_1\mu_k & \text{if } \eta_1 \leq \widehat{r}_k, \\ \mu_k & \text{if } \eta_0 \leq \widehat{r}_k < \eta_1. \end{cases} \tag{18}$$

Step 6. Update $\alpha_{k+1}, m(k+1)$, and the symmetric matrix B_{k+1}. Let $\Delta_{k+1} = \mu_{k+1}(\| d_k\|^2/d_k^T \widehat{B}_{k+1} d_k)\|g_{k+1}\|$ and set $k := k+1$ and go to Step 1.

Remark 2. In Algorithm 1, we define a new reduction of $f(x)$ which is different from the actual reduction $f_k - f(x_k + s_k)$ in other methods. Instead of requiring f_{k+1} to be smaller than f_k, it is only required that f_{k+1} is either less than f_k or less than the weighted mean of the function values at the last $m(k)$ iterates.

Remark 3. How to choose the horizontal vector α_k and the approximate Hessian matrix B_k is one of the crucial issues of a conic model method. More details for generating α_k and B_k can be found in [25]. If $\alpha_k = 0$, Algorithm 1 reduces to the quadratic model trust region method. If we choose $m = 1$, we obtain a monotonic adaptive trust region method. Hence, Algorithm 1 is a generalization and development of quadratic model trust region method and monotonic trust region method.

Remark 4. "Step 2-Step 3-Step 4" are called the internal circulation.

3. Convergence Analysis

In this section we discuss the global convergence of Algorithm 1. Before we address some theoretical issues, we would like to give the following assumptions.

Assumption 5. (*i*) The level set $L(x_0) = \{x \in R^n \mid f(x) \leq f(x_0)\}$ is bounded for any given $x_0 \in R^n$ and $f(x)$ is twice continuously differentiable in $L(x_0)$. (*ii*) The sequences $\{B_k\}$ and $\{\alpha_k\}$ are uniformly bounded; that is, there exist two positive scalars M_1, M_2 such that $\|B_k\| \leq M_1$ and $\|\alpha_k\| \leq M_2$ hold for all k. (*iii*) There exists a positive constant $\sigma \in (0,1)$, such that $\|\alpha_k\|\Delta_k \leq \sigma$.

It is similar to Theorem 3.1 in [28], and we can prove the following lemma.

Lemma 6. *Suppose that Assumption 5 holds. Then, there exists a scalar $\beta \in (0,1)$ such that*

$$\begin{aligned} Pred_k &\geq \beta\|g_k\| \min\left\{\Delta_k, \frac{\|g_k\|}{\|B_k\|}\right\} \\ &\geq \beta\|g_k\| \min\left\{\|s_k\|, \frac{\|g_k\|}{\|B_k\|}\right\} \end{aligned} \tag{19}$$

holds for all k, where s_k is an inexact solution of the subproblem (11).

Lemma 7. *Suppose that Assumption 5 holds. Then,*

$$|f(x_k) - f(x_k + s_k) - Pred_k| \leq O(\|s_k\|^2), \tag{20}$$

where s_k is the solution of (11) *and is sufficiently close to zero.*

Proof. Since $\{\|\alpha_k\|\}$ is bounded, we can have $1/(1-\alpha_k^T s_k) = 1+O(\|s_k\|)$. Then, from the boundedness of $\{\|g_k\|\}$ and $\{\|B_k\|\}$, we obtain

$$\frac{g_k^T s_k}{1-\alpha_k^T s_k} = g_k^T s_k + O\left(\|s_k\|^2\right),$$
$$\frac{s_k^T B_k s_k}{\left(1-\alpha_k^T s_k\right)^2} = s_k^T B_k s_k + O\left(\|s_k\|^2\right). \tag{21}$$

By $f(x)$ which is twice continuously differentiable in $L(x_0)$, there exists $M_3 > 0$ such that $\|H(x)\| \le M_3$ holds for all Hessian $H(x)$ and all $x \in L(x_0)$.

Therefore, we have that

$$\begin{aligned}
&\left|f(x_k) - f(x_k+s_k) - \mathrm{Pred}_k\right| \\
&= \left|f(x_k) - f(x_k+s_k) + \frac{g_k^T s_k}{1-\alpha_k^T s_k} + \frac{1}{2}\frac{s_k^T B_k s_k}{\left(1-\alpha_k^T s_k\right)^2}\right| \\
&= \left|-g_k^T s_k - \frac{1}{2} s_k^T H(x_k+\theta_k s_k) s_k \right.\\
&\quad \left. + g_k^T s_k + \frac{1}{2} s_k^T B_k s_k + O\left(\|s_k\|^2\right)\right| \\
&\le \frac{1}{2}(M_1+M_3)\|s_k\|^2 + O\left(\|s_k\|^2\right) = O\left(\|s_k\|^2\right).
\end{aligned} \tag{22}$$

This completes the proof. □

Lemma 8. *Suppose that Assumption 5 holds. Then, Algorithm 1 is well defined; that is, Algorithm 1 can not cycle infinitely in the internal circulation.*

Proof. Suppose that Algorithm 1 cycles infinitely many times between Steps 2 and 4 at the current point x_k. We denote the solution of the conic model (11) by $s_{k(i)}$ and the corresponding predicted reduction by $\mathrm{Pred}_{k(i)}$ with the cycling index i. Then, we have

$$r_{k(i)} < \eta_0, \quad i = 0,1,2,\ldots. \tag{23}$$

$$\Delta_{k(i)} = c_0^i \Delta_k \longrightarrow 0, \quad i \longrightarrow \infty. \tag{24}$$

From Lemmas 6 and 7, we can have

$$\begin{aligned}
&\left|\frac{f_k - f(x_k+s_{k(i)})}{\mathrm{Pred}_{k(i)}} - 1\right| \\
&= \left|\frac{f_k - f(x_k+s_{k(i)}) - \mathrm{Pred}_{k(i)}}{\mathrm{Pred}_{k(i)}}\right| \\
&\le \frac{O\left(\|s_{k(i)}\|^2\right)}{\mathrm{Pred}_{k(i)}} \\
&\le \frac{O\left(\Delta_{k(i)}^2\right)}{\beta\|g_k\|\min\{\Delta_{k(i)}, \|g_k\|/\|B_k\|\}} \longrightarrow 0, \\
&\qquad \text{as } i \longrightarrow \infty.
\end{aligned} \tag{25}$$

Hence, the following

$$\frac{f_k - f(x_k+s_{k(i)})}{\mathrm{Pred}_{k(i)}} \ge \eta_0 \tag{26}$$

holds for i sufficiently large. From $M_k \ge f_k$, we obtain that

$$r_{k(i)} = \frac{M_k - f(x_k+s_{k(i)})}{\mathrm{Pred}_{k(i)}} \ge \eta_0, \tag{27}$$

holds for i sufficiently large, which contradicts with (23). The proof is completed. □

In the next lemma, an important decrease property of the function value $f(x_k)$ is established.

Lemma 9. *Suppose that Assumption 5 holds and $\{x_k\}$ is generated by Algorithm 1. Then, there exists a constant $\tau > 0$ such that the following*

$$\begin{aligned}
f(x_k) \le f(x_0) &- \tau\lambda\sum_{r=0}^{k-2}\|g_r\|\min\left\{\Delta_r, \frac{\|g_r\|}{\|B_r\|}\right\} \\
&- \tau\|g_{k-1}\|\min\left\{\Delta_{k-1}, \frac{\|g_{k-1}\|}{\|B_{k-1}\|}\right\}
\end{aligned} \tag{28}$$

holds for $k \ge 1$.

Proof. We prove it by induction. For simplicity, we denote

$$\beta_k = \|g_k\|\min\left\{\Delta_k, \frac{\|g_k\|}{\|B_k\|}\right\}. \tag{29}$$

(1) If $k=1$, by $r_0 \ge \eta_0$, $M_0 = f_0$, and Lemma 6, we have

$$\begin{aligned}
f(x_1) &\le f(x_0) - \eta_0(\varphi_0(0) - \varphi_0(s_0)) \\
&\le f(x_0) - \eta_0\beta\|g_0\|\min\left\{\Delta_0, \frac{\|g_0\|}{\|B_0\|}\right\} \\
&= f(x_0) - \tau\|g_0\|\min\left\{\Delta_0, \frac{\|g_0\|}{\|B_0\|}\right\},
\end{aligned} \tag{30}$$

where $\tau = \eta_0\beta$.

(2) Now, we assume that it holds for $1,2,\ldots,k$, and we consider two cases. □

Case 1. If $M_k = f(x_k)$, by $r_{k+1} \ge \eta_0$, $0 < \lambda \le 1$, and Lemma 6, then we have

$$\begin{aligned}
f(x_{k+1}) = f(x_k+s_k) &\le f(x_k) - \eta_0(\varphi_k(0) - \varphi_k(s_k)) \\
&\le f(x_0) - \tau\lambda\sum_{r=0}^{k-2}\beta_r - \tau\beta_{k-1} - \eta_0\beta\beta_k \\
&\le f(x_0) - \tau\lambda\sum_{r=0}^{k-1}\beta_r - \tau\beta_k.
\end{aligned} \tag{31}$$

Case 2. $M_k = \sum_{r=0}^{m(k)-1} \lambda_{kr} f(x_{k-r})$. Let $q = \min[k, m-1]$, by $r_{k+1} \ge \eta_0$, $0 < \lambda \le 1$, and Lemma 6. Then, we have

$$\begin{aligned} f(x_{k+1}) &= f(x_k + s_k) \\ &\le \sum_{p=0}^{q} \lambda_{kp} f(x_{k-p}) - \eta_0 (\varphi_k(0) - \varphi_k(s_k)) \\ &\le \sum_{p=0}^{q} \lambda_{kp} \left(f(x_0) - \tau\lambda \sum_{r=0}^{k-p-2} \beta_r - \tau\beta_{k-p-1} \right) \\ &\quad - \eta_0 (\varphi_k(0) - \varphi_k(s_k)). \end{aligned} \tag{32}$$

Using $(0,1,2,\ldots,q) \times (0,1,2,\ldots,k-p-2) \subset \{(p,r);\ 0 \le p \le q, 0 \le r \le k-p-2\}$, $\lambda_{kp} \ge \lambda$, $\sum_{p=0}^{q} \lambda_{kp} = 1$, we have

$$\begin{aligned} &f(x_{k+1}) \\ &\le f(x_0) - \tau\lambda \sum_{r=0}^{k-q-2} \left(\sum_{p=0}^{q} \lambda_{kp} \right) \beta_r - \tau \sum_{p=0}^{q} \lambda_{kp} \beta_{k-p-1} - \tau\beta_k \\ &\le f(x_0) - \tau\lambda \sum_{r=0}^{k-q-2} \beta_r - \tau\lambda \sum_{r=k-q-1}^{k-1} \beta_r - \tau\beta_k \\ &\le f(x_0) - \tau\lambda \sum_{r=0}^{k-1} \beta_r - \tau\beta_k. \end{aligned} \tag{33}$$

Lemma 10. *Suppose that the conditions in Lemma 9 hold. Then,* $\{x_k\} \subset L(x_0)$.

Proof. By Lemma 9, $f(x_k) \le f(x_0) - \tau\lambda \sum_{r=0}^{k-1} \|g_r\| \min\{\Delta_r, \|g_r\|/\|B_r\|\} \le f(x_0)$, the conclusion holds obviously.

Now, we prove the global convergence of Algorithm 1. □

Theorem 11. *Suppose that Assumption 5 holds. If* $\varepsilon = 0$, *Algorithm 1 either terminates in finite iterations or generates an infinite sequence* $\{x_k\}$ *which satisfies*

$$\liminf_{k\to\infty} \|g_k\| = 0. \tag{34}$$

Proof. If Algorithm 1 terminates in finite iterations, the theorem is obviously true. Assume that Algorithm 1 generates an infinite sequence $\{x_k\}$ in the following proof.

If (34) is not true, then there exists a positive constant ε_0, such that

$$\|g_k\| \ge \varepsilon_0, \quad \forall k. \tag{35}$$

From Lemma 9, $0 < \lambda \le 1$, and $\|g_k\| \ge \varepsilon_0$, we can have

$$f(x_k) \le f(x_0) - \tau\lambda\varepsilon_0 \sum_{r=0}^{k-1} \min\left\{\Delta_r, \frac{\varepsilon_0}{\|B_r\|}\right\}. \tag{36}$$

By Assumption 5(ii), we can obtain

$$\tau\lambda\varepsilon_0 \sum_{r=0}^{k-1} \min\left\{\Delta_r, \frac{\varepsilon_0}{M_1}\right\} \le f(x_0) - f(x_k). \tag{37}$$

Therefore, by Assumption 5(i), we have

$$\sum_{k=0}^{\infty} \Delta_k < \infty. \tag{38}$$

Thus, we can obtain

$$\lim_{k\to\infty} \Delta_k = 0. \tag{39}$$

By Assumption 5(ii), then $\|\widehat{B}_k\| \le 2M_1 + 1$. From Algorithm 1, we know that

$$\begin{aligned} \Delta_k &= \mu_k \frac{\|s_{k-1}\|^2}{s_{k-1}^T \widehat{B}_k s_{k-1}} \|g_k\| \\ &\ge \mu_k \frac{\|s_{k-1}\|^2}{(2M_1+1)\|s_{k-1}\|^2} \varepsilon_0 = \frac{\mu_k \varepsilon_0}{2M_1+1}. \end{aligned} \tag{40}$$

By (39) and (40), we obtain

$$\lim_{k\to\infty} \mu_k = 0. \tag{41}$$

On the other hand, we have from Lemma 6 that

$$\begin{aligned} &\left| \frac{f_k - f(x_k + s_k)}{\mathrm{Pred}_k} - 1 \right| \\ &= \frac{O(\|s_k\|) + O(\|s_k\|^2 \|B_k\|)}{\mathrm{Pred}_k} \\ &\le \frac{O(\|s_k\|) + O(\|s_k\|^2 \|B_k\|)}{\beta \|g_k\| \min\{\Delta_k, \|g_k\|/\|B_k\|\}} \\ &\le \frac{O(\|s_k\|)}{\Delta_k} \longrightarrow 0. \end{aligned} \tag{42}$$

The above inequality implies that

$$r_k = \frac{M_k - f(x_k + s_k)}{\mathrm{Pred}_k} \ge \frac{f_k - f(x_k + s_k)}{\mathrm{Pred}_k} \ge \eta_0 \tag{43}$$

holds for k sufficiently large. Hence, there exists a positive constant μ^* such that

$$\mu_k \ge \mu^*. \tag{44}$$

holds for all sufficiently large k, which contradicts with (41). The proof is completed. □

4. Superlinear Convergence

In general, the superlinear convergence property of Algorithm 1 requires further assumptions, we give the assumptions and the superlinear convergence property in the following theorem.

Theorem 12. *Suppose that Assumption 5 holds and Algorithm 1 generates an infinite sequence* $\{x_k\}$ *which converges to*

x^*, where $H(x^*)$ is positive definite and $H(x)$ is Lipschitz continuous in a neighborhood of x^*. If the following condition holds

$$\lim_{k\to\infty} \frac{\|g_k + H(x^*) s_k\|}{\|s_k\|} = 0, \quad (45)$$

then the rate of convergence is superlinear; that is, $\|x_{k+1}-x^*\| = o(\|x_k - x^*\|)$.

Proof. From $x_k \to x^*$, we can have $\lim_{k\to\infty}\|s_k\| = 0$. By the Taylor expansion, we can obtain

$$\begin{aligned} g_{k+1} &= g_k + \nabla^2 f(x_k + \theta_k s_k) s_k \\ &= g_k + H(x^*) s_k + \left(\nabla^2 f(x_k + \theta_k s_k) - H(x^*)\right) s_k, \end{aligned} \quad (46)$$

where $\theta_k \in (0, 1)$.

So,

$$\begin{aligned} &\|g_{k+1}\| \\ &\leq \|g_k + H(x^*) s_k\| + \left\|\nabla^2 f(x_k + \theta_k s_k) - H(x^*)\right\| \cdot \|s_k\|. \end{aligned} \quad (47)$$

Dividing both sides by $\|s_k\|$, we get

$$\frac{\|g_{k+1}\|}{\|s_k\|} \leq \frac{\|g_k + H(x^*) s_k\|}{\|s_k\|} + \left\|\nabla^2 f(x_k + \theta_k s_k) - H(x^*)\right\|. \quad (48)$$

By (45), $\|s_k\| \to 0$, and Lipschitz continuous property of $H(x)$, we can have

$$\lim_{k\to\infty} \frac{\|g_{k+1}\|}{\|s_k\|} = 0. \quad (49)$$

By $H(x^*)$ which is positive definite, and f which is a twice continuously differentiable function, there exists $\beta > 0$, such that

$$\beta \|x_{k+1} - x^*\| \leq \|g_{k+1}\| \quad (50)$$

for sufficiently large k.

By (49) and (50), it follows that

$$\begin{aligned} \frac{\|g_{k+1}\|}{\|s_k\|} &\geq \frac{\beta \|x_{k+1} - x^*\|}{\|s_k\|} \geq \frac{\beta \|x_{k+1} - x^*\|}{\|x_{k+1} - x_k\|} \\ &\geq \frac{\beta \|x_{k+1} - x^*\|}{\|x_{k+1} - x^*\| + \|x_k - x^*\|} \\ &= \beta \frac{\|x_{k+1} - x^*\| / \|x_k - x^*\|}{1 + (\|x_{k+1} - x^*\| / \|x_k - x^*\|)}, \end{aligned} \quad (51)$$

and thus

$$\lim_{k\to\infty} \frac{\|x_{k+1} - x^*\|}{\|x_k - x^*\|} = 0, \quad (52)$$

which implies that the sequence $\{x_k\}$ converges to x^* superlinear. □

5. Numerical Results

In this section, we compare the performance of Algorithm 1, denoted by NACTR, with NAQTR if $\alpha_k = 0$ for all k in Algorithm 1, MACTR if $m = 1$ in Algorithm 1, CTR method which is a conic trust region algorithm without adaptive technique, and QTR method which is the traditional trust region method without adaptive technique. All programs are written in MATLAB with double precision.

For these methods, the trial step s_k is computed approximately by the algorithm proposed by Lu and Ni in [34] for solving the subproblem (11). B_0 is chosen as the identity matrix. The parameters in our algorithm are chosen as $\eta_0 = 0.25$, $\eta_1 = 0.7$, $c_0 = 0.25$, $c_1 = 1.5$, $\lambda_{kr} = 1/m(k)$, $(r = 0, 1, \ldots, m(k)-1)$, $m = 4$, and $\alpha_0 = 0$. In our implementation, we prefer the following choice

$$\widehat{r}_k = 0.85 r_k + 0.15 \widehat{r}_{k-1}, \quad (53)$$

where $\widehat{r}_0 = r_0$.

The stopping condition is

$$\|g_k\| \leq 10^{-4}. \quad (54)$$

The iteration is also terminated if the number of evaluations exceeds 300.

We provide the results of our tests in Table 1. The numerical results are given in the form of N_i, N_f, where N_i, N_f denote the numbers of iterations and function evaluations, respectively. We chose the test problems from [35]. From Table 1, we can see that, in most of the problems considered, the numbers N_i and N_f for the conic model methods are considerably smaller than those required for the quadratic model methods, especially for those functions with strong nonquadratic behavior. This means that the conic model methods are very effective to many unconstrained optimization. Compared with the methods without adaptive technique, the adaptive trust region methods are more attractive. Hence, our new nonmonotone adaptive conic trust region method is an improvement of the existing trust region methods.

6. Conclusions

In this paper, we propose a new nonmonotone adaptive conic model trust region method for unconstrained optimization. The global and superlinear convergence properties of the new method are proved under reasonable conditions. Our method is attractive in the following aspects. First, the subproblem is conic model which can produce a better prediction of the minimizer of the function than quadratic model methods

TABLE 1: Numerical comparisons.

Problems	QTR N_i/N_f	CTR N_i/N_f	NAQTR N_i/N_f	MACTR N_i/N_f	NACTR N_i/N_f
Helical Valley	62/74	43/53	56/61	32/45	29/42
Gaussian	7/12	3/8	3/5	3/5	3/5
Box 3-D	41/45	33/37	36/41	30/35	30/33
Var. Dimen.	17/26	11/11	18/27	12/18	13/21
Waston	72/78	66/73	62/81	54/98	48/86
Penalty I	18/21	16/17	18/24	15/20	15/18
Penalty II	21/26	21/24	23/31	17/26	16/25
Brown-Dennis	62/66	78/84	62/75	55/67	53/58
Gulf R.D.	88/83	96/98	66/46	38/42	46/47
Trigonometric	46/52	38/39	45/48	32/39	34/36
Ex. Ros.	91/124	Failed	48/62	33/46	35/48

when the objective function has strong nonquadratic behavior. Second, at each iteration, our method generates a suitable trust region radius automatically based on the current and last iterative information. Last, we use a new update rule (8) to adjust the next trust region radius. According to the nonmonotone technique in [33], our method constructs an iterative series $\{x_k\}$ such that the sequence $\{f(x_k)\}$ is not monotonic, which can accelerate the convergence rate of the minimization process, especially in the narrow curved valley. Numerical results show that the new method is efficient for unconstrained optimization problems.

Conflict of Interests

The author declares that there is no conflict of interests regarding the publication of this paper.

Acknowledgments

The author is thankful to the editor and the reviewers for their constructive comments, which greatly improved this paper. This work was supported by the National Science Foundation of China (Grant no. 11226229) and Foundation of Shandong Jiaotong University.

References

[1] M. J. D. Powell, "Convergence properties of a class of minimization algorithms," in *Nonlinear Programming*, J. B. Rosen, O. L. Mangasarian, and K. Ritter, Eds., pp. 1–27, Academic press, New York, NY, USA, 1975.

[2] Y. Yuan, "On the convergence of trust region algorithms," *Mathematica Numerica Sinica*, vol. 16, pp. 333–346, 1996.

[3] J. Nocedal and Y. Yuan, "Combining trust region and line search techniquesAdvances in Nonlinear Programming," Y. Yuan, Ed., pp. 153–175, Kluwer Academic, Dordrecht, The Netherlands, 1996.

[4] M. J. D. Powell, "A new algorithm for unconstrained optimization," in *Nonlinear Programming*, J. B. Rosen, O. L. Mangassarian, and K. Ritter, Eds., pp. 31–65, Academic press, New York, NY, USA, 1970.

[5] M. J. D. Powell, "On the global convergence of trust region algorithms for unconstrained optimization," *Mathematical Programming*, vol. 29, no. 3, pp. 297–303, 1984.

[6] T. Steilhaug, "The conjugate gradient method and trust region in large scale optimization," *SIAM Journal on Numerical Analysis*, vol. 20, pp. 626–637, 1983.

[7] E. M. Gertz, "A quasi-Newton trust-region method," *Mathematical Programming A*, pp. 1–24, 1963.

[8] A. R. Conn, N. I. M. Gould, and P. L. Toint, *Trust-Region Methods*, Society for Industrial and Applied Mathematics, Philadelphia, Pa, USA, 2000.

[9] J. Nocedal and S. J. Wright, *Numerical Optimization*, Springer, New York, NY, USA, 1999.

[10] Y. X. Yuan and W. Y. Sun, *Optimization Theory and Methods*, Science Press, Beijing, China, 1997.

[11] L. Grippo, F. Lampariello, and S. Lucidi, "A nonmonotone line search technique for Newton's method," *SIAM Journal on Numerical Analysis*, vol. 23, no. 4, pp. 707–716, 1986.

[12] N. Y. Deng, Y. Xiao, and F. J. Zhou, "Nonmonotonic trust region algorithm," *Journal of Optimization Theory and Applications*, vol. 76, no. 2, pp. 259–285, 1993.

[13] W. Y. Sun, "Nonmonotone trust region method for solving optimization problems," *Applied Mathematics and Computation*, vol. 156, no. 1, pp. 159–174, 2004.

[14] P. L. Toint, "Non-monotone trust-region algorithms for nonlinear optimization subject to convex constraints," *Mathematical Programming B*, vol. 77, no. 1, pp. 69–94, 1997.

[15] J. T. Mo, C. Y. Liu, and S. C. Yan, "A nonmonotone trust region method based on nonincreasing technique of weighted average of the successive function values," *Journal of Computational and Applied Mathematics*, vol. 209, no. 1, pp. 97–108, 2007.

[16] X. Ke and J. Han, "A class of nonmonotone trust region algorithms for unconstrained optimization problems," *Science in China A*, vol. 41, no. 9, pp. 927–932, 1998.

[17] A. Sartenaer, "Automatic determination of an initial trust region in nonlinear programming," *SIAM Journal on Scientific Computing*, vol. 18, no. 6, pp. 1788–1803, 1997.

[18] J. Y. Fan and Y. X. Yuan, "A new trust region algorithm with trust region radius converging to zero," in *Proceedings of the 5th International Conference on Optimization: Techniques and Applications*, Hong Kong, China, December 2001.

[19] X. S. Zhang, J. L. Zhang, and L. Z. Liao, "An adaptive trust region method and its convergence," *Science in China A*, vol. 45, no. 5, pp. 620–631, 2002.

[20] Z. Y. Sang and Q. Y. Sun, "A self-adaptive trust region method with line search based on a simple subproblem model," *Journal of Computational and Applied Mathematics*, vol. 232, no. 2, pp. 514–522, 2009.

[21] L. Hei, "A self-adaptive trust region algorithm," *Journal of Computational Mathematics*, vol. 21, pp. 229–236, 2003.

[22] G. D. Li, "A trust region method with automatic determination of the trust region radius," *Chinese Journal of Engineering Mathematicss*, vol. 23, pp. 843–848, 2006.

[23] Z. J. Shi and J. H. Guo, "A new trust region method for unconstrained optimization," *Applied Mathematics and Computation*, vol. 213, pp. 509–520, 2008.

[24] J. L. Zhang and X. S. Zhang, "A nonmonotone adapyive trust region method and its convergence," *Computers and Mathematics with Application*, vol. 45, pp. 1469–1477, 2003.

[25] W. C. Davidon, "Conic approximation and collinear scaling for optimizers," *SIAM Journal on Numerical Analysis*, vol. 17, pp. 268–281, 1980.

[26] S. Di and W. Y. Sun, "A trust region method for conic model to solve unconstrained optimization," *Optimization Methods and Software*, vol. 6, no. 4, pp. 237–263, 1996.

[27] Q. Ni, "Optimality conditions for trust-region subproblems involving a conic model," *SIAM Journal on Optimization*, vol. 15, no. 3, pp. 826–837, 2005.

[28] S.-J. Qu, K.-C. Zhang, and J. Zhang, "A nonmonotone trust-region method of conic model for unconstrained optimization," *Journal of Computational and Applied Mathematics*, vol. 220, no. 1-2, pp. 119–128, 2008.

[29] W. Sun, "On nonquadratic model optimization methods," *Asia and Pacific Journal of Operational Research*, vol. 13, pp. 43–63, 1996.

[30] H. Gourgeon and J. Nocedal, "A Conic algorithm for optimization," *SIAM Journal of Scientific Stochastical Computation*, vol. 6, pp. 253–267, 1985.

[31] W. Y. Sun and Y. X. Yuan, "A conic trust-region method for nonlinearly constrained optimization," *Annal of Operational Research*, vol. 103, pp. 175–191, 2001.

[32] S. Sheng, "Interpolation by conic model for unconstrained optimization," *Computing*, vol. 54, pp. 83–98, 1995.

[33] Z. S. Yu and D. G. Pu, "A new nonmonotone line search technique for unconstrained optimization," *Journal of Computational and Applied Mathematics*, vol. 219, no. 1, pp. 134–144, 2008.

[34] X. P. Lu and Q. Ni, "A quasi-Newton trust region method with a new conic model for the unconstrained optimization," *Applied Mathematics and Computation*, vol. 204, no. 1, pp. 373–384, 2008.

[35] J. J. Moré, B. S. Grabow, and K. E. Hillstrom, "Testing unconstrained optimization software," *ACM Transactions on Mathematical Software*, vol. 7, pp. 17–41, 1981.

14

Beamforming with Reduced Complexity in MIMO Cooperative Cognitive Radio Networks

Mehdi Ghamari Adian

Electrical Engineering Department, University of Zanjan, Zanjan 45371-38791, Iran

Correspondence should be addressed to Mehdi Ghamari Adian; mehdi.ghamari@aut.ac.ir

Academic Editor: Kai-Kit Wong

An approach for beamforming with reduced complexity in MIMO cooperative cognitive radio networks (MIMO-CCRN) is presented. Specifically, a suboptimal approach with reduced complexity is proposed to jointly determine the transmit beamforming (TB) and cooperative beamforming (CB) weight vectors along with antenna subset selection in MIMO-CCRN. Two multiantenna secondary users (SU) constitute the desired link, one acting as transmitter (SU TX) and the other as receiver (SU RX) and they coexist with single-antenna primary and secondary users. Some of single antenna secondary users are recruited by desired link as cooperative relay. The maximization of the achievable rates in the desired link is the objective of this work, provided to interference constraints on the primary users are not violated. The objective is achieved by exploiting transmit beamforming at SU TX, cooperation of some secondary users, and cooperative beamforming. Meanwhile, the costs associated with RF chains at the radio front end at SU RX are reduced. Through simulations, it is shown that better performance in the desired link is attained, as a result of cooperation of SUs.

1. Introduction and Related Works

The rigid structure of current spectrum allocation policies creates a bottleneck for rapidly growing wireless users. On the other hand, the Federal Communication Commission (FCC) measurements reveal that most of the licensed frequency bands are either unused or utilized less than 10% of the time [1]. To address the limitations on spectrum usage, the FCC has motivated the use of opportunistic spectrum sharing to make the licensed frequency bands accessible for unlicensed users.

Transmit beamforming with receive combining is one of the simplest approaches to achieve full diversity [2]. Compared with traditional space time codes, beamforming and combining systems provide the same diversity order as well as significantly more array gain [3] at the expense of requiring channel state information at the transmitter.

The issue of transmit beamforming in cognitive radio networks (CRN) and more specifically for secondary users has been investigated from various points of view in [4–7]. In [4], transmit beamforming (TB) is designed for MIMO cognitive radio networks in a single primary user- (PU-) single secondary user (SU) network, to minimize the transmit power of the SU while limiting the interference temperature to PU and achieving the signal-to-interference-plus-noise ratio (SINR) target at SU. The joint problem of TB and power control in CRN have been considered in [5, 6], where the objective is to optimize the SU users sum rate under the interference constraints of PUs. The joint problems of TB at transmitter and antenna subset selection at receiver of a secondary network are considered in [7], where TB is recruited at multiantenna secondary transmitter to maximize the data rates in SU link; meanwhile the interference on PUs is minimized.

The cooperative beamforming (CB) issue in cooperative cognitive radio networks (CCRN) has been discussed in a few papers such as [8–11]. In [8], a number of relays in a dual-hop amplify-and-forward cooperative scheme in cooperative cognitive radio networks (CCRN) are utilized, aiming to maximize the worst signal-to-interference-plus-noise ratio of the destinations. The bursty traffic in CRN has been considered in [9], where CB is exploited to access busy time slots or spatial spectrum holes. In [10], the relays comprise a distributed beamformer to beamform the signal towards its destination while maintaining the QoS in the primary user. A cooperative beamforming aided incremental relaying scheme in CRN has been proposed in [11], in which the source and relays can exploit CB to activate packet retransmission in busy time slots or spatial spectrum holes without causing interference to primary users. To the best knowledge of authors, the application of joint TB and CB in CCRN has not ever been investigated. In addition to complexity of the CCRN with joint TB and CB, another reason for not taking advantage of joint TB and CB can be attributed to unavailability of complete channel state information (CSI) at transmitter and relay side. In [12], based on Grassmannian line packing technique (GLP), a method has been presented which does not need CSI at transmitter for TB and works when a limited feedback is available from receiver to transmitter. The beamforming codebook is generated using GLP technique. The transmitter and the receiver preserve the same codebook, which contains, A, beamforming weight vectors for the Grassmannian beamforming (GB). In the GB, the index for the optimal beamforming weight vector, not the vector itself, is fed back from the receiver to the transmitter. Thus, the amount of feedback information can be reduced to $[\log_2 A]$ bits.

A promising way of capturing a large portion of the channel capacity in MIMO systems at reduced hardware costs and computational complexity is to select optimally a small number of the best antennas from the larger set of antennas available. The performance of systems that use such selection techniques has been shown to be significantly higher than that of the systems using the same number of antennas without any subset selection [13]. However, the number of computations required for such optimal selection grows exponentially with the total number of the antennas available. In [7], an alternative approach to receive antenna selection for capacity maximization is presented that offers near optimal performance at a complexity significantly lower than the schemes in [14]. In this paper a similar method is employed to reduce the complexity of antenna subset selection process.

The motivation of this work is to determine the optimum TB weight vector at the transmitter side, the optimum CB weight vector at relay side, and antenna subset selection in receiver side of a CCRN with reduced complexity. Meanwhile, the proposed system can be considered an effective system to be implemented in one of related standards to CRN (e.g., IEEE 802.22). Moreover, in order to incorporate the CR concept into the commercial wireless communication system, a popular proposal for physical layer is orthogonal frequency-division multiplexing- (OFDM-) based techniques, among which is multicarrier code division multiplexing (MC-CDM). This is because the MC-CDM is more suitable than any other techniques to provide flexible spectrum management, which makes the CRN effectively reuse the frequency resources.

The contributions of this work can be summarized as follows.

(i) Proposing a method with reduced complexity to determine the optimum weight vectors of transmit beamformer, cooperative beamformer, and the best set of antennas in CCRN.

(ii) Proposing an iterative method to obtain the optimum TB weight vector and optimal subset of antennas at receiver.

(iii) Determining the optimal CB weight vector in the virtual array comprised of cooperating relays to remove the interference on PUs.

The rest of the paper is organized as follows. In Section 2, system model is described and the problem is formulated. Section 3 proposes a solution for the problem formulated in Section 2. In Section 4, simulation results are presented and Section 5 concludes the paper.

Notation. Boldface uppercase is used for matrices and boldface lowercase for vectors. $\det(\cdot)$, $\mathrm{Tr}(\cdot)$, and $(\cdot)^H$ denote the determinant, the trace, and the conjugate transpose operators, respectively. $\mathbf{I}_M$ denotes an $M \times M$ identity matrix. $\mathbb{C}^{t\times u}$ denotes the space of $t \times u$ matrices with complex entries and $CN(\mathbf{0}, \mathbf{I})$ represents the distribution of a zero mean circularly symmetric complex Gaussian (ZMCSCG) vector with covariance matrix $\mathbf{I}$. $\mathbf{H}(i, :)$ is used to represent the ith row of a matrix, $\mathbf{H}$, respectively. $\mathrm{Diag}(\cdot)$ gives the diagonal elements of a matrix.

2. System Model

2.1. System Description. The system model is depicted in Figure 1. In the considered system, a secondary network consisting of $N_{SU} + 2$ secondary users (SUs) coexists with a primary network consisting of N_{PU} single-antenna primary users (PUs). In the secondary network, there exist two multiantenna and N_{SU} single-antenna SUs. All users use the same frequency band. However, PUs own the higher priority in using the frequency band. Two multiantenna secondary users (SU TX and SU RX) constitute the desired SU link. The core aim of this work is to maximize the data rates of the desired SU link, using TB at SU TX and CB at cooperating SUs.

The strategy of cooperation of single-antenna SUs with the desired multiantenna SU link is decode-and-forward. Other cooperation strategies can be analyzed in a similar fashion.

A few issues may arise regarding the proposed system model. First of all, the considered scenario is not an artificial one, since there may exist some multiantenna SUs in any practical cognitive radio networks (for e.g., two cognitive base stations) and also it is wise to assume the existence of ubiquitous single-antenna SUs in any cognitive radio network. Therefore, the motivation for providing an applicable solution with reduced complexity for a practical cognitive radio system is assured. Secondly, the cooperation of single-antenna SUs with the desired multiantenna SU link can be rewarded by the mutual cooperation of the multiantenna SU link with them to resolve the issue of the cooperation

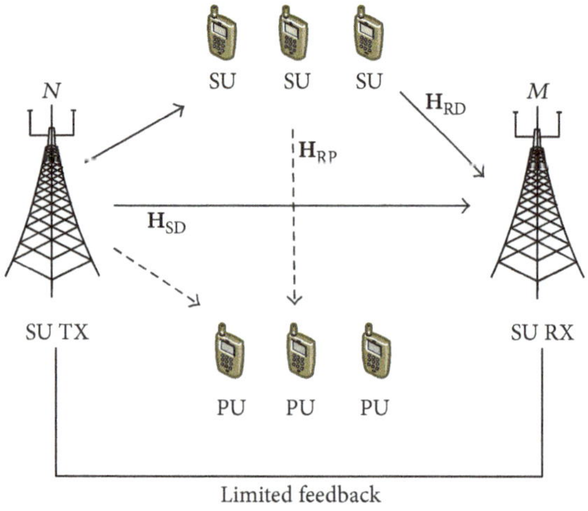

Figure 1: System model.

incentives of the single-antenna SUs. In order to simplify the analysis in this paper, it is presumed that single-antenna SUs are motivated enough to cooperate with SU TX; that is, a kind of compensation has been provided by the MIMO SU link to keep the single-antenna SUs motivated to cooperate with the MIMO SU link.

SU TX is equipped with N antennas that are preceded by a beamformer. It is assumed that channel state information (CSI) is not available at SU TX. Since it is cumbersome to provide the SU TX with the CSI, here we assume that a kind of quantized beamforming, that is, Grassmannian beamforming, is utilized as transmit beamforming scheme at SU TX. As previously stated, the SU TX and the SU RX preserve the same codebook, which contains a number of beamforming weight vectors for the Grassmannian beamforming (GB). The beamforming codebook is generated using Grassmannian line packing technique (GLP) [12]. The GB method works when a limited feedback is available from SU RX to SU TX. Therefore, the optimum TB weight vector, $\mathbf{w}_{\mathrm{TB}}^{\mathrm{opt}}$, is determined at SU RX. Then the index for the optimal TB weight vector, not the vector itself, is fed back, using a few bits of limited feedback, from SU RX to SU TX. SU RX is equipped with M antennas. At SU RX m out of M antennas are selected. Also, the optimum TB weight vector is the one which maximizes the achievable rates, on condition that interference on PUs does not exceed the threshold. Moreover, zero-forcing beamforming is utilized in the virtual MIMO of cooperating SUs, in order not to disturb the existing single-antenna PUs as a result of cooperation. The channels between all nodes are assumed to experience frequency flat Rayleigh fading.

2.2. Structure of SU TX and SU RX. The structure of SU TX and SU RX is depicted in Figure 2. L data symbols on the mth antenna, $[d_1^m \cdots d_L^m]^T$, are code division multiplexed, where L denotes the number of subcarriers of the MC-CDM modulation. Orthogonal Walsh codes are used for spreading and each symbol is multiplied with a length L Walsh code before all spread replicas of the modulated symbols are summed up to form the frequency domain length L vector $x^m = [X_1^m \cdots X_L^m]^T$; that is, $\mathbf{x}^m = \mathbf{C}_L \mathbf{d}^{(m)}$, where

$$\mathbf{C}_L = \frac{1}{\sqrt{L}} \begin{bmatrix} \mathbf{C}_{L/2} & \mathbf{C}_{L/2} \\ \mathbf{C}_{L/2} & -\mathbf{C}_{L/2} \end{bmatrix} \quad (1)$$

is the normalized Walsh-Hadamard transformation matrix. Note that $L = 2^n$, $n \geq 1$, and $\mathbf{C}_1 = 1$. Consequently on each subcarrier we have a MIMO channel with the following input vector: $\mathbf{x}_l = \begin{bmatrix} X_l^1 & X_l^2 & \cdots & X_l^M \end{bmatrix}^T$ and $1 \leq l \leq L$. The OFDM modulator on each antenna of SU TX consists of an IFFT block, parallel to serial converter followed by a cyclic prefix insertion unit and a digital to analog converter (DAC). The length of the cyclic prefix is assumed to be greater than the largest maximum excess delay over all spatial channels.

2.3. Problem Formulation. The received signal at SU RX can be written as

$$\mathbf{y} = \mathbf{H}_{\mathrm{SD}} \mathbf{w}_{\mathrm{TB}} x + \mathbf{H}_{\mathrm{RD}} \mathbf{w}_{\mathrm{ZFBF}} \widehat{x} + \mathbf{i} + \mathbf{z}, \quad (2)$$

where $\mathbf{w}_{\mathrm{TB}}$ and $\mathbf{w}_{\mathrm{ZFBF}}$ represent the transmit beamforming vector at the SU-TX and zero-forcing beamforming weight vector at the cooperating SU, respectively. $\widehat{x}$ stands for the decoded signal at the cooperating SUs. $\mathbf{H}_{\mathrm{SD}} \in \mathbb{C}^{M \times N}$ and $\mathbf{H}_{\mathrm{RD}} \in \mathbb{C}^{M \times N_{\mathrm{SU}}}$ represent the channel coefficients matrix with ZMCSCG entries from SU TX to SU RX and from single-antenna SUs (virtual MIMO) to SU RX, respectively; $\mathbf{i}$ is interference due to primary users and $\mathbf{z}$ represents the white noise (where $\mathbf{z} \sim CN(0, \mathbf{I}_M)$ and $\mathbf{z} \in \mathbb{C}^{N \times 1}$). Note that all channel matrices must be considered for each subcarrier, to be more accurate. However, as we aim to find the optimal set of antennas at SU RX, the optimum transmit and cooperative beamformer weight vectors are applied to all of the subcarriers regardless of different channel characteristics of different subcarriers and the dependence of all system parameters on the subcarrier index can be dropped. For simplicity we also assume that the detection process at cooperating SUs is error-free and as a result (2) can be written as

$$\mathbf{y} = \left(\mathbf{H}_{\mathrm{SD}} \mathbf{w}_{\mathrm{TB}} + \mathbf{H}_{\mathrm{RD}} \mathbf{w}_{\mathrm{ZFBF}} \right) x + \mathbf{i} + \mathbf{z}. \quad (3)$$

We assume that the total transmit power is limited to P_{SU}; that is,

$$\mathbf{Q}_{\mathrm{SU}} = E\left\{ \mathbf{w}_{\mathrm{TB,opt}} \mathbf{w}_{\mathrm{TB,opt}}^H x x^* \right\} = \mathbf{w}_{\mathrm{TB,opt}} \mathbf{w}_{\mathrm{TB,opt}}^H P_x \leq P_{\mathrm{SU}}, \quad (4)$$

where $E\{xx^*\} = P_x$ and $\mathbf{Q}_{\mathrm{SU}}$ stands for the transmit covariance matrix at SU TX. The covariance matrix of the noise and interference at SU RX is given by

$$\mathbf{U} = E\left\{ \mathbf{i}\mathbf{i}^H + \mathbf{z}_T \mathbf{z}_T^H \right\} = \mathbf{I}_M + \mathbf{H}_{\mathrm{PS}} \mathbf{H}_{\mathrm{PS}}^H, \quad (5)$$

in which $\mathbf{I}_M$ is $M \times M$ identity matrix and $\mathbf{H}_{\mathrm{PS}}$ denotes the channel matrix from PUs to SU RX ($\mathbf{H}_{\mathrm{PS}} \in \mathbb{C}^{M \times N_{\mathrm{PU}}}$). Note that, in (5), it is assumed that $\mathbf{i}$ and $\mathbf{z}$ are independent random processes. For satisfactory operation of the incumbent PUs in the presence of the SU TX, interference seen at the

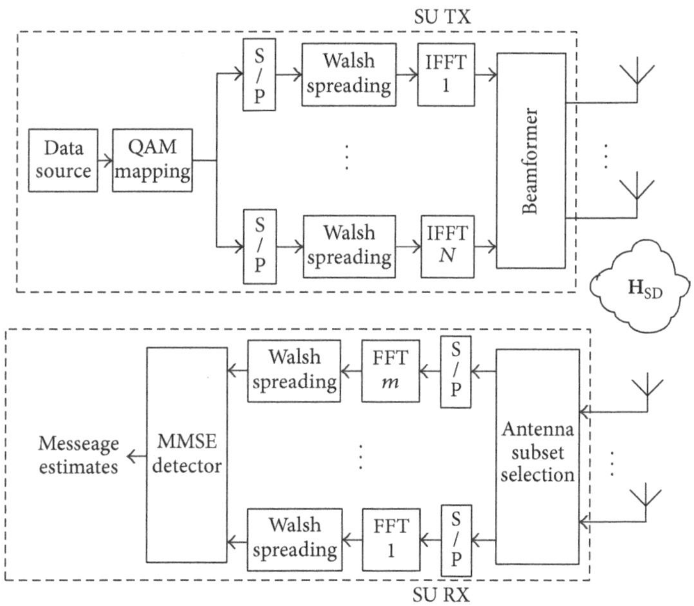

FIGURE 2: Structure of SU TX and SU RX.

PU RX should not exceed a particular threshold, P_i ($i = 1, 2, \ldots, N_{PU}$):

$$\mathbf{h}_{SP,i}\mathbf{Q}_{SU}\mathbf{h}_{SP,i}^{H} \leq P_i, \quad i = 1, 2, \ldots, N_{PU}, \tag{6}$$

where $\mathbf{H}_{SP}$ is channel matrix from secondary user transmitter (SU TX) to primary users (PUs) and $\mathbf{h}_{SP,i} = \mathbf{H}_{SP}(i, :)$ and $\mathbf{h}_{SP,i} \in \mathbb{C}^{1\times N}$ denotes the ith row of $\mathbf{H}_{SP}$. The achievable rates of the multiantenna SU link, without cooperation of single-antenna SUs and using all antennas at SU RX, is given by [15]

$$R^{(1)}\left(\mathbf{S}, \mathbf{w}_{TB,opt}, P_x\right) = \log_2 \det\left(\mathbf{I}_M + \mathbf{H}_{SD}\mathbf{Q}_{SU}\mathbf{H}_{SD}^{H}\mathbf{U}^{-1}\right). \tag{7}$$

The achievable rates of the desired SU link at the output of the maximum ratio combiner at SU RX, using the cooperation of single-antenna SUs and also all available antennas at SU RX, can be written as

$$R^{(2)}\left(\mathbf{S}, \mathbf{w}_{TB,opt}, \mathbf{w}_{ZFBF,opt}, P_x\right) = \frac{1}{2}\log_2 \det\left(\mathbf{I}_M + \mathbf{H}_{SD}\mathbf{Q}_{SU}\mathbf{H}_{SD}^{H}\mathbf{U}^{-1} + \mathbf{H}_{RD}\mathbf{Q}'_{SU}\mathbf{H}_{RD}^{H}\mathbf{U}^{-1}\right), \tag{8}$$

where $\mathbf{Q}'_{SU} = E\{\mathbf{w}_{ZFBF}\mathbf{w}_{ZFBF}^{H}xx^*\} = \mathbf{w}_{ZFBF}\mathbf{w}_{ZFBF}^{H}P_x \leq P_{SU}$. The coefficient 1/2 is due to the fact that cooperative transmission only uses half of the resources (e.g., time slots, frequency bands, etc.). For simplicity, the same transmit power limit, P_{SU}, is assumed at cooperating SUs. Similar to the method presented in [7], we define a diagonal matrix $\mathbf{S}$ (where $\mathbf{S} \in \mathbb{R}^{M\times M}$):

$$(\mathbf{S})_{ii} = \begin{cases} 1 & i\text{th receive antenna selected} \\ 0 & \text{otherwise.} \end{cases} \tag{9}$$

Antennas (in SU RX) that maximize the achievable data rates in the system are selected. The diagonal matrix provides us with the index of selected antennas in the SU RX. Hence, if m antennas are chosen ($m \leq M$), new channel matrices, $\widehat{\mathbf{H}}_{SD}$ and $\widehat{\mathbf{H}}_{RD}$, with the same dimension as $\mathbf{H}_{SD}$ and $\mathbf{H}_{RD}$, will have $M - m$ all-zero rows. Thus, the rate expression at SU RX becomes

$$R^{(2)}\left(\mathbf{S}, \mathbf{w}_{TB,opt}, \mathbf{w}_{ZFBF,opt}, P_x\right) = \frac{1}{2}\log_2 \det\left(\mathbf{I}_M + \widehat{\mathbf{H}}_{SD}\mathbf{Q}_{SU}\widehat{\mathbf{H}}_{SD}^{H} + \widehat{\mathbf{H}}_{RD}\mathbf{Q}'_{SU}\widehat{\mathbf{H}}_{RD}^{H}\right), \tag{10}$$

where $\widehat{\mathbf{H}}_{SD} = \widehat{\mathbf{U}}^{-1/2}\mathbf{S}\mathbf{H}_{SD}$, $\widehat{\mathbf{H}}_{RD} = \widehat{\mathbf{U}}^{-1/2}\mathbf{S}\mathbf{H}_{RD}$, and $\widehat{\mathbf{U}}$ is defined as follows. With the selected receive antennas we have reduced $m \times 1$ interference and noise vectors which give a new interference and noise covariance matrix, $\mathbf{U}_{reduced}$, of dimension $m\times m$. This matrix is inflated to form $\widehat{\mathbf{U}}$, an $M\times M$ matrix, by adding rows and columns of zeros corresponding to the receive antennas not selected.

The PUs are vulnerable to cooperation of single-antenna SUs with SU TX and a relaying scheme must be devised to minimize the alleged interference to other PUs. We propose to use zero-forcing beamforming in virtual antenna array, formed by single-antenna SUs, before their relaying of PU BS's signals. More specifically, this distributed beamforming scheme is aimed not only to remove the interference to other PUs, but also to maximize the data rates from single-antenna SUs (relays) to SU RX. Therefore, it can also be considered as cooperative beamforming. The distributed beamforming can be realized by a virtual antenna array, which can be created by a set of relays in cooperative relaying networks [16].

With this notation, the problem of joint transmit beamforming, cooperative beamforming, and antenna selection can be mathematically explained by

$$
\begin{aligned}
\mathbf{P}: \max \quad & \frac{1}{2}\log_2 \det\left(\mathbf{I}_M + \widehat{\mathbf{H}}_{\mathrm{SD}}\mathbf{Q}_{\mathrm{SU}}\widehat{\mathbf{H}}_{\mathrm{SD}}^H\mathbf{U}^{-1}\right. \\
& \left. + \widehat{\mathbf{H}}_{\mathrm{RD}}\mathbf{Q}'_{\mathrm{SU}}\widehat{\mathbf{H}}_{\mathrm{RD}}^H\mathbf{U}^{-1}\right) \\
\text{s.t.} \quad & (\mathbf{S})_{ii} \in \{0,1\}, \quad i = 1,2,\ldots,M \\
& \mathrm{Tr}\left(\mathbf{Q}_{\mathrm{SU}}\right) \le P_{\mathrm{SU}} \Longrightarrow \mathrm{Tr}\left(\mathbf{w}_{\mathrm{TB}}\mathbf{w}_{\mathrm{TB}}^H\right) \le \frac{P_{\mathrm{SU}}}{P_x} \\
& \mathrm{Tr}\left(\mathbf{Q}'_{\mathrm{SU}}\right) \le P_{\mathrm{SU}} \Longrightarrow \mathrm{Tr}\left(\mathbf{w}_{\mathrm{ZFBF}}\mathbf{w}_{\mathrm{ZFBF}}^H\right) \le \frac{P_{\mathrm{SU}}}{P_x} \\
& \mathrm{Tr}\left(\mathbf{S}\right) = m \\
& P_x\mathbf{h}_{\mathrm{SP},i}\mathbf{w}_{\mathrm{TB}}\mathbf{w}_{\mathrm{TB}}^H\mathbf{h}_{\mathrm{SP},i}^H \le P_i, \quad i = 1,2,\ldots,N_{\mathrm{PU}} \\
& \left|\mathbf{h}_{\mathrm{RP},i}\mathbf{w}_{\mathrm{ZFBF}}\right| = 0, \quad i = 1,2,\ldots,N_{\mathrm{PU}} \\
\text{variables} \quad & \mathbf{w}_{\mathrm{TB}}, \mathbf{w}_{\mathrm{ZFBF}}, \mathbf{S}, P_x,
\end{aligned}
\tag{11}
$$

where $\mathbf{h}_{\mathrm{RP},i}$, $i = 1,2,\ldots,N_{\mathrm{PU}}$, denotes the channel coefficients from single-antenna SUs to other single-antenna PUs.

3. A Suboptimal Method to Determine the Optimum TB and CB Weight Vectors and Best Antennas

In problem **P**, four unknown variables must be determined jointly. Due to excessive complexity, in this section we propose a suboptimum solution with reduced complexity to solve **P**. In this way, problem **P** is decomposed into two problems, **P1** and **P2**, according to the following:

$$
\begin{aligned}
\mathbf{P1}: \max \quad & \log_2 \det\left(\mathbf{I}_M + \widehat{\mathbf{H}}_{\mathrm{SD}}\mathbf{Q}_{\mathrm{SU}}\widehat{\mathbf{H}}_{\mathrm{SD}}^H\mathbf{U}^{-1}\right) \\
\text{s.t.} \quad & (\mathbf{S})_{ii} \in \{0,1\}, \quad i = 1,2,\ldots,M \\
& \mathrm{Tr}\left(\mathbf{Q}_{\mathrm{SU}}\right) \le P_{\mathrm{SU}} \Longrightarrow \mathrm{Tr}\left(\mathbf{w}_{\mathrm{TB}}\mathbf{w}_{\mathrm{TB}}^H\right) \le \frac{P_{\mathrm{SU}}}{P_x} \\
& \mathrm{Tr}\left(\mathbf{S}\right) = m \\
& P_x\mathbf{h}_{\mathrm{SP},i}\mathbf{w}_{\mathrm{TB}}\mathbf{w}_{\mathrm{TB}}^H\mathbf{h}_{\mathrm{SP},i}^H \le P_i, \quad i = 1,2,\ldots,N_{\mathrm{PU}} \\
\text{variables} \quad & \mathbf{w}_{\mathrm{TB}}, \mathbf{S}, P_x, \\
\mathbf{P2}: \max \quad & \left\|\mathbf{H}_{\mathrm{RD}}\mathbf{w}_{\mathrm{ZFBF}}\right\|^2 \\
\text{s.t.} \quad & \mathrm{Tr}\left(\mathbf{Q}'_{\mathrm{SU}}\right) \le P_{\mathrm{SU}} \Longrightarrow \mathrm{Tr}\left(\mathbf{w}_{\mathrm{ZFBF}}\mathbf{w}_{\mathrm{ZFBF}}^H\right) \le \frac{P_{\mathrm{SU}}}{P_x} \\
& \left|\mathbf{h}_{\mathrm{RP},i}\mathbf{w}_{\mathrm{ZFBF}}\right| = 0, \quad i = 1,2,\ldots,N_{\mathrm{PU}} \\
\text{variable} \quad & \mathbf{w}_{\mathrm{ZFBF}}.
\end{aligned}
\tag{12}
$$

Note that, prior to solving problem **P2**, problem **P1** has to be solved to determine the optimum value of P_x and **S**. Moreover, in problem **P2**, the received signal power at SU RX due to cooperation of single-antenna SUs is aimed to be maximized, instead of achievable data rates at SU RX. This facilitates the finding the optimal zero-forcing beamforming weight vector, as will be discussed soon. In the sequel, possible ways of solving the above problems are discussed.

3.1. Solving **P1**

3.1.1. Exhaustive Search. A straightforward way to solve **P1** is to perform an exhaustive search (ES) over all possible combinations of antenna elements and transmit beamforming weight vectors and optimize over P_x. Hence, ES amounts to optimizing P_x, $\binom{M}{m} \times \binom{K}{1}$ times subject to interference and total transmit power constraints, where K denotes the number codewords in the transmit beamforming codebook matrix. Each optimization of P_x can be considered a convex problem. However, the need to iterate through all the possible combinations gives a complexity which explodes for higher dimensional systems.

3.1.2. Convex Approximation. Problem **P1** is highly nonconvex and can be classified as an example of an integer programming problem, since matrix **S** has only binary elements [17]. The nonconvexity of the problem arises due to the nature of the objective function, interference, and binary constraints. Further, the binary variable renders the problem NP-hard problem. In order to obtain a more computationally efficient approach, we modify the problem in the following way. The binary structure of **S** can be relaxed so that the antenna selection variable takes on values in the interval 0 to 1. This makes the problem far easier to solve than the original integer program. Finally we note that in this approach the effect of the **U** matrix cannot be included. This limitation is discussed below. With these changes, **P1** can be written as

$$
\begin{aligned}
\mathbf{P3}: \max \quad & \log_2 \det\left(\mathbf{I}_M + P_x\mathbf{S}\mathbf{H}_{\mathrm{SD}}\mathbf{w}_{\mathrm{TB}}\mathbf{w}_{\mathrm{TB}}^H\mathbf{H}_{\mathrm{SD}}^H\mathbf{S}^H\right) \\
\text{s.t.} \quad & 0 \le (\mathbf{S})_{ii} \le 1, \quad i = 1,2,\ldots,M \\
& \mathrm{Tr}\left(\mathbf{Q}_{\mathrm{SU}}\right) \le P_{\mathrm{SU}} \Longrightarrow \mathrm{Tr}\left(\mathbf{w}_{\mathrm{TB}}\mathbf{w}_{\mathrm{TB}}^H\right) \le \frac{P_{\mathrm{SU}}}{P_x} \\
& \mathrm{Tr}\left(\mathbf{S}\right) = m \\
& P_x\mathbf{h}_{\mathrm{SP},i}\mathbf{w}_{\mathrm{TB}}\mathbf{w}_{\mathrm{TB}}^H\mathbf{h}_{\mathrm{SP},i}^H \le P_i, \quad i = 1,2,\ldots,N_{\mathrm{PU}} \\
\text{variables} \quad & \mathbf{w}_{\mathrm{TB}}, \mathbf{S}, P_x.
\end{aligned}
\tag{13}
$$

Note that problem **P3** is still nonconvex, due to nonconcavity of the objective function. Thus, we seek a convex approximation (CA) to this problem.

Proposition 1. *With two of the three variables known, the cost function in problem* **P3** *is concave in the third one and this renders the problem convex in this variable.*

Proof. Three different cases must be investigated as follows.

(i) When **S** and P_x are known and $\mathbf{w}_{\mathrm{TB}}$ is unknown.

(ii) When $\mathbf{w}_{\mathrm{TB}}$ and P_x are known and $\mathbf{S}$ must be specified.

(iii) When $\mathbf{w}_{\mathrm{TB}}$ and $\mathbf{S}$ are known and P_x must be found.

Case 1 (when $\mathbf{S}$ and P_x are known and $\mathbf{w}_{\mathrm{TB}}$ is unknown). In this case, the objective function can be written as

$$g(\mathbf{W}_{\mathrm{TB}}) = \log_2 \det\left(\mathbf{I}_M + P_x \mathbf{S}\mathbf{H}_{\mathrm{SU}}\mathbf{W}_{\mathrm{TB}}\mathbf{H}_{\mathrm{SU}}^H\mathbf{S}^H\right), \quad (14)$$

where $\mathbf{W}_{\mathrm{TB}} = \mathbf{w}_{\mathrm{TB}}\mathbf{w}_{\mathrm{TB}}^H$ and, considering the constraints, there is no difference between maximizing $\mathbf{w}_{\mathrm{TB}}$ and $\mathbf{W}_{\mathrm{TB}}$. According to [17], $g(\mathbf{W}_{\mathrm{TB}})$ is concave, if and only if $f(t) = g(t\mathbf{W}_{\mathrm{TB}}^{(1)} + (1-t)\mathbf{W}_{\mathrm{TB}}^{(2)})$ is concave for every feasible $\mathbf{W}_{\mathrm{TB}}^{(1)}$, $\mathbf{W}_{\mathrm{TB}}^{(2)}$, and $0 \le t \le 1$. Therefore $f(t)$ can be written as

$$\begin{aligned} f(t) &= \log_2 \det\Big(\mathbf{I}_M + P_x \mathbf{S}\mathbf{H}_{\mathrm{SU}} \\ &\qquad \times \left(\mathbf{W}_{\mathrm{TB}}^{(2)} + t\left(\mathbf{W}_{\mathrm{TB}}^{(1)} - \mathbf{W}_{\mathrm{TB}}^{(2)}\right)\right)\mathbf{H}_{\mathrm{SU}}^H\mathbf{S}^H\Big) \\ &= \log_2 \det(\mathbf{A} + t\mathbf{B}), \end{aligned} \quad (15)$$

where $\mathbf{A} = \mathbf{I}_M + P_x\mathbf{S}\mathbf{H}_{\mathrm{SU}}\mathbf{W}_{\mathrm{TB}}^{(2)}\mathbf{H}_{\mathrm{SU}}^H\mathbf{S}^H$ and $\mathbf{B} = P_x\mathbf{S}\mathbf{H}_{\mathrm{SU}}(\mathbf{W}_{\mathrm{TB}}^{(1)} - \mathbf{W}_{\mathrm{TB}}^{(2)})\mathbf{H}_{\mathrm{SU}}^H\mathbf{S}^H$. $f(t)$ can be further manipulated as

$$\begin{aligned} f(t) &= \log_2 \det\left(\mathbf{A}^{1/2}\left(\mathbf{I}_M + t\mathbf{A}^{-1/2}\mathbf{B}\mathbf{A}^{-1/2}\right)\mathbf{A}^{1/2}\right) \\ &= \sum_{i=1}^{n}\log_2(1 + t\lambda_i) + \log_2 \det \mathbf{A}. \end{aligned} \quad (16)$$

In which $\lambda_1, \ldots, \lambda_n$ denote the eigenvalues of $\mathbf{A}^{-1/2}\mathbf{B}\mathbf{A}^{-1/2}$. Hence,

$$\begin{aligned} \frac{df(t)}{dt} &= \frac{1}{\ln 2}\sum_{i=1}^{n}\frac{\lambda_i}{1 + t\lambda_i}, \\ \frac{d^2 f(t)}{dt^2} &= \frac{-1}{\ln 2}\sum_{i=1}^{n}\frac{\lambda_i^2}{(1 + t\lambda_i)^2}. \end{aligned} \quad (17)$$

Negativity of $d^2 f(t)/dt^2$ results in concavity of $f(t)$ and concavity of $g(\mathbf{W})$ is concluded.

Case 2 (P_x and $\mathbf{w}_{\mathrm{TB}}$ known and $\mathbf{S}$ unknown). First of all, the objective function must be manipulated as

$$\begin{aligned} &\log_2 \det\left(\mathbf{I}_M + P_x\mathbf{S}\mathbf{H}_{\mathrm{SU}}\mathbf{w}_{\mathrm{TB}}\mathbf{w}_{\mathrm{TB}}^H\mathbf{H}_{\mathrm{SU}}^H\mathbf{S}^H\right) \\ &\quad = \log_2 \det\left(1 + P_x\mathbf{w}_{\mathrm{TB}}^H\mathbf{H}_{\mathrm{SU}}^H\mathbf{S}^H\mathbf{S}\mathbf{H}_{\mathrm{SU}}\mathbf{w}_{\mathrm{TB}}\right). \end{aligned} \quad (18)$$

And, by introducing = $\mathbf{S}^H\mathbf{S}$, the concavity of the objective function is proved in a similar method employed in Case 1.

Case 3 ($\mathbf{S}$ and $\mathbf{w}_{\mathrm{TB}}$ known and P_x unknown). In this case, the objective function can be written as

$$g(P_x) = \log_2 \det(\mathbf{I}_M + P_x\mathbf{A}), \quad (19)$$

where $\mathbf{A} = \mathbf{S}\mathbf{H}_{\mathrm{SU}}\mathbf{w}_{\mathrm{TB}}\mathbf{w}_{\mathrm{TB}}^H\mathbf{H}_{\mathrm{SU}}^H\mathbf{S}^H$. Once again, $f(t)$ is defined as $g(tP_x^{(1)} + (1-t)P_x^{(2)})$, where $P_x^{(1)}$ and $P_x^{(2)}$ are any feasible amounts for P_x and $0 \le t \le 1$:

$$\begin{aligned} f(t) &= \log_2 \det\left(\mathbf{I}_M + \left(P_x^{(2)} + t\left(P_x^{(1)} - P_x^{(2)}\right)\right)\mathbf{A}\right) \\ &= \log_2 \det(\mathbf{B} + t\mathbf{C}), \end{aligned} \quad (20)$$

where $\mathbf{B} = \mathbf{I}_M + P_x^{(2)}\mathbf{A}$ and $\mathbf{C} = (P_x^{(1)} - P_x^{(2)})\mathbf{A}$ and the concavity of the $g(P_x)$ is derived in a similar manner as employed in previous cases. □

Thus, to solve **P3**, we initialize P_x and $\mathbf{w}_{\mathrm{TB}}$ and optimize over $\mathbf{S}$. Then, using the optimum $\mathbf{S}$ and the initial value for P_x, optimum transmit beamforming weight vector is chosen and ultimately, with $\mathbf{S}$ and $\mathbf{w}_{\mathrm{TB}}$ known, optimum value for transmit power, P_x, is obtained. Since elements of $\mathbf{S}$ are nonbinary, index of chosen antennas are the m-largest diagonal elements of $\mathbf{S}$. In the following steps, the proposed procedure for solving the problem is summarized.

A comment on the convergence of the proposed iterative algorithm is in order here. Using a similar approach to [18], we argue that during the $(k+1)$th iteration we calculate $\mathbf{S}^{(k+1)} = \arg\max_{\mathbf{S}} \mathbf{P3}(\mathbf{S}, \mathbf{w}_{\mathrm{TB}}^{(k)}, P_x^{(k)})$ and obtain achievable data rate r_1.

Then we calculate $\mathbf{w}_{\mathrm{TB}}^{(k+1)} = \arg\max_{\mathbf{w}_{\mathrm{TB}}} \mathbf{P3}(\mathbf{w}_{\mathrm{TB}}, P_x^{(k)}, \mathbf{S}^{(k+1)})$ giving rate r_2. Finally, we evaluate $P_x^{(k+1)} = \arg\max_{P_x} \mathbf{P3}(P_x, \mathbf{S}^{(k+1)}, \mathbf{w}_{\mathrm{TB}}^{(k+1)})$ and the corresponding achievable data rate r_3. Since $r_1 \le r_2 \le r_3$ forms a monotonically increasing sequence which is bounded above (due to transmit power constraint) we conclude that the sequence of achievable data rates converges to a limit. Our simulations indicate that iterating 6 times is almost sufficient to attain an optimum value of **P3**. Since the problem **P3** is not convex in nature, the maximum achievable rates obtained from the proposed iterative rate may not be globally optimum. However, results suggest that the values obtained are robust and are globally optimal most of the time for the parameters and scenario discussed in Section 2.

Proposed Iterative Method to Solve ***P1***

Step 1 (initializations).

(1) Select an initial value for P_x and P_{SU}.

(2) Select an Initial value for $\mathbf{w}_{\mathrm{TB}}$ (columns of codebook matrix) which satisfies $\mathrm{Tr}(\mathbf{w}_{\mathrm{TB}}\mathbf{w}_{\mathrm{TB}}^H) \le P_{\mathrm{SU}}/P_x$.

Step 2 (calculation step).

(1) Solve the convex optimization problem and find $\mathbf{S}$.

(2) Find the optimum $\mathbf{w}_{\mathrm{TB}}$ which maximizes achievable rates or SNR (depending on the scenario selected) using exhaustive search method.

(3) Solve the resultant convex optimization problem, knowing $\mathbf{S}$ and $\mathbf{w}_{\mathrm{TB}}$, to calculate the optimum P_x.

Step 3 (iterative step).

(1) Repeat the Steps 1 and 2 until convergence. The achievable rate at any SNR will be the average of results at that specific SNR.

(2) Using the optimum values of transmit power in each SNR, P_x, find the symbol error rate (using Monte Carlo method).

3.2. Solving ***P2***. After obtaining the optimum value of P_x and **S**, it is time to calculate the optimum zero-forcing beamforming weight vector. The cooperating SUs are assumed to be cognitive in the sense that they can obtain the channel state information (CSI) on the channels from themselves to PUs. The objective function of **P2** can be written as

$$\left\|\mathbf{H}_{\mathrm{RD}}\mathbf{w}_{\mathrm{ZFBF}}\right\|^2 = \sum_{i=1}^{m}\left|\mathbf{h}_{\mathrm{RD},i}\mathbf{w}_{\mathrm{ZFBF}}\right|^2, \quad (21)$$

where $\mathbf{h}_{\mathrm{RD},i} = \mathbf{H}_{\mathrm{RD}}(i,:)$ denotes the ith row of $\mathbf{H}_{\mathrm{RD}}$ matrix and m represents the number of selected antennas at SU RX.

Theorem 2. *The optimal zero-forcing (ZF) beamforming weight vector which maximizes* $\|\boldsymbol{H}_{RD}\boldsymbol{w}_{ZFBF}\|^2$ *and satisfies the constraints of the problem* ***P2*** *is one of the orthogonal projections of rows of* $\boldsymbol{H}_{RD}$ *onto the orthogonal complementary* $\gamma^{\perp}$ *of the subspace* $\gamma = \mathrm{span}\{\boldsymbol{h}_{CP}^{(2)},\ldots,\boldsymbol{h}_{CP}^{(N_{PU})}\}$ *which maximizes* $\sum_{i=1}^{m}|\boldsymbol{h}_{RD,i}\boldsymbol{w}_{ZFBF}|^2$*. To satisfy the cooperative transmit power constraint, the elements of the optimum ZFBF weight vector have to be multiplied to* $\sqrt{(P_{SU})/P_x\,\mathrm{Tr}(\boldsymbol{w}_{ZFBF}\boldsymbol{w}_{ZFBF}^H)}$.

Proof. The channel vector $\mathbf{h}_{\mathrm{RD},i} \in \mathbb{C}^{1\times N_{\mathrm{SU}}}$, $i = 1,\ldots,m$, can be written as $\mathbf{h}_{\mathrm{RD},i} = a_1^{(i)}\mathbf{e}_1 + a_2^{(i)}\mathbf{e}_2 + \cdots + a_{N_{\mathrm{SU}}}^{(i)}\mathbf{e}_{N_{\mathrm{SU}}}$, using the N_{SU} basis vectors $\{\mathbf{e}_1,\mathbf{e}_2,\ldots,\mathbf{e}_{N_{\mathrm{SU}}}\}$ (where $a_k^{(i)} \sim CN(0,1)$, $k = 1,2,\ldots,N_{\mathrm{SU}}$, $i = 1,\ldots,m$) [19]. Hence, the matrix $= [\mathbf{e}_1,\mathbf{e}_2,\ldots,\mathbf{e}_{N_{\mathrm{SU}}}] = \mathbf{I}_{N_{\mathrm{SU}}}$ is $N_{\mathrm{SU}}\times N_{\mathrm{SU}}$ identity matrix. Then, we consider the set of basis vectors $\{\mathbf{e}'_1,\mathbf{e}'_2,\ldots,\mathbf{e}'_{N_{\mathrm{PU}}}\}$, which is the orthogonal basis for the subspace $\gamma = \mathrm{span}\{\mathbf{h}_{\mathrm{RP},1},\ldots,\mathbf{h}_{\mathrm{RP},N_{\mathrm{PU}}}\}$. Actually, γ is a N_{PU}-dimensional subspace, for the reason that the probability of the realizations of the independent and continuous random vectors $\mathbf{h}_{\mathrm{RP},1},\ldots,\mathbf{h}_{\mathrm{RP},N_{\mathrm{PU}}}$ being interrelated is very much small and thus can be ignored. If $\{\mathbf{e}'_{N_{\mathrm{PU}}+1},\mathbf{e}'_{N_{\mathrm{PU}}+2},\ldots,\mathbf{e}'_{N_{\mathrm{SU}}}\}$ is an orthogonal basis for $\gamma^{\perp}$, the orthogonal set $\{\mathbf{e}'_1,\mathbf{e}'_2,\ldots,\mathbf{e}'_{N_{\mathrm{SU}}}\} \cup \{\mathbf{e}'_{N_{\mathrm{PU}}+1},\mathbf{e}'_{N_{\mathrm{PU}}+2},\ldots,\mathbf{e}'_{N_{\mathrm{SU}}}\}$ is another orthogonal basis for $\mathbb{C}^{1\times N_{\mathrm{SU}}}$. Similarly, $\mathbf{h}_{\mathrm{RD},i}$ can be represented by $\mathbf{h}_{\mathrm{RD},i} = b_1^{(i)}\mathbf{e}'_1 + b_2^{(i)}\mathbf{e}'_2 + \cdots + b_{N_{\mathrm{SU}}}^{(i)}\mathbf{e}'_{N_{\mathrm{SU}}}$. Clearly, $\mathbf{L} = [\mathbf{e}'_1,\mathbf{e}'_1,\ldots,\mathbf{e}'_{N_{\mathrm{SU}}}]$ is also a unitary matrix. Moreover, by matrix manipulation we have

$$\left[b_1^{(i)},b_2^{(i)},\ldots,b_{N_{\mathrm{SU}}}^{(i)}\right]^T = \mathbf{L}^H\left[a_1^{(i)},a_2^{(i)},\ldots,a_{N_{\mathrm{SU}}}^{(i)}\right]^T. \quad (22)$$

Since the random matrix $\mathbf{L}^H$ is unitary and independent with $\mathbf{h}_{\mathrm{RP},i}$, $[a_1^{(i)},a_2^{(i)},\ldots,a_{N_{\mathrm{SU}}}^{(i)}]$ has the same distribution as $[b_1^{(i)},b_2^{(i)},\ldots,b_{N_{\mathrm{SU}}}^{(i)}]$, from (A.22) in [20]. As a result, $b_k^{(i)} \sim CN(0,1)$, $k = 1,2,\ldots,N_{\mathrm{SU}}$, $i = 1,\ldots,m$. The ZF beamforming weight vector $\mathbf{w}_{\mathrm{ZFBF}}$ is orthogonal to each $\mathbf{h}_{\mathrm{RP},i}$ ($i = 1,\ldots,N_{\mathrm{PU}}$). Hence, it is perpendicular to each vector in γ and belongs to $\gamma^{\perp}$. In order to maximize $\sum_{i=1}^{m}|\mathbf{h}_{\mathrm{RD},i}\mathbf{w}_{\mathrm{ZFBF}}|^2$, we need to find the vector $\mathbf{w}_{\mathrm{ZFBF}}^{(i)} \in \gamma^{\perp}$ which is the closest to $\mathbf{h}_{\mathrm{RP},i}$ ($i = 1,\ldots,m$). From closest point theorem, $\mathbf{w}_{\mathrm{ZFBF}}^{(i)}$ is the orthogonal projection of $\mathbf{h}_{\mathrm{RP},i}$ onto the subspace $\gamma^{\perp}$. The optimum ZFBF weight vector is the one which maximizes $\sum_{i=1}^{m}|\mathbf{h}_{\mathrm{RD},i}\mathbf{w}_{\mathrm{ZFBF}}|^2$:

$$\mathbf{w}_{\mathrm{ZFBF,opt}} = \arg\max_{\mathbf{w}_{\mathrm{ZFBF}}^{(k)}}\sum_{i=1}^{m}\left|\mathbf{h}_{\mathrm{RD},i}\mathbf{w}_{\mathrm{ZFBF}}\right|^2, \quad k = 1,\ldots,m. \quad (23)$$

The cooperative transmit power constraint in **P2** makes such $\mathbf{w}_{\mathrm{ZFBF,opt}}$ unique and Theorem 2 is proved. □

A comparison between the computational complexity of the optimal and proposed suboptimal methods is presented here. If you choose to solve the problem in (11) using the dual method as an optimal algorithm, $(2N_{\mathrm{PU}}+3)$ dual variables are updated in every iteration. Using these values, M^2 function evaluations are performed to find the power allocation. Then, the optimal solution has a complexity of $O(T(M^2N_{\mathrm{PU}}))$, where T is the number of iterations required to converge, which is usually high [17]. In the proposed scheme, every subcarrier in the source side requires not more than MN_{PU} function evaluations; therefore the complexity of the proposed algorithm is $O(T'(MN_{\mathrm{PU}}))$, where T' is the number of

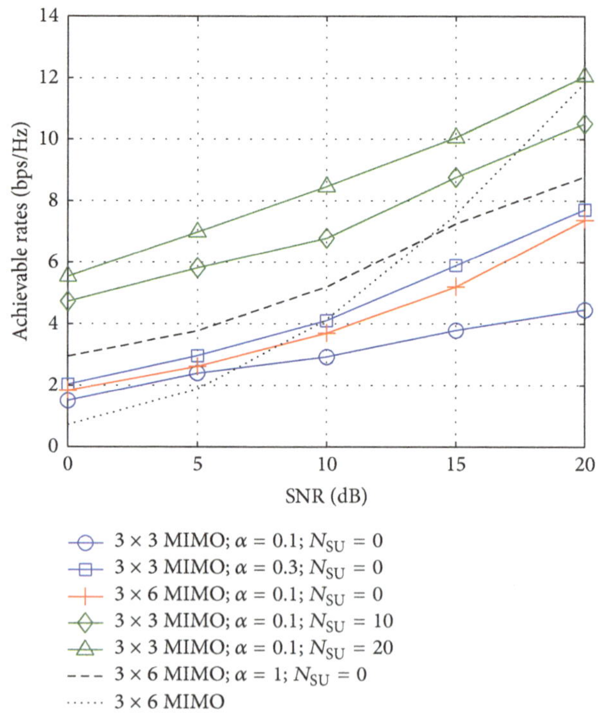

FIGURE 3: Achievable data rates versus SNR for different numbers of cooperating SUs, different values of α, and different numbers of selected antennas at SU RX.

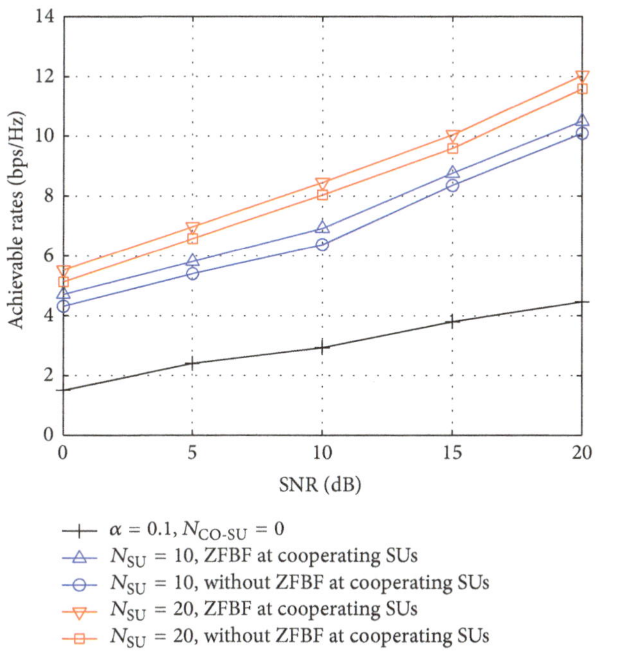

FIGURE 4: The effect of cooperative beamforming (zero-forcing beamforming) on the achievable rates of the system.

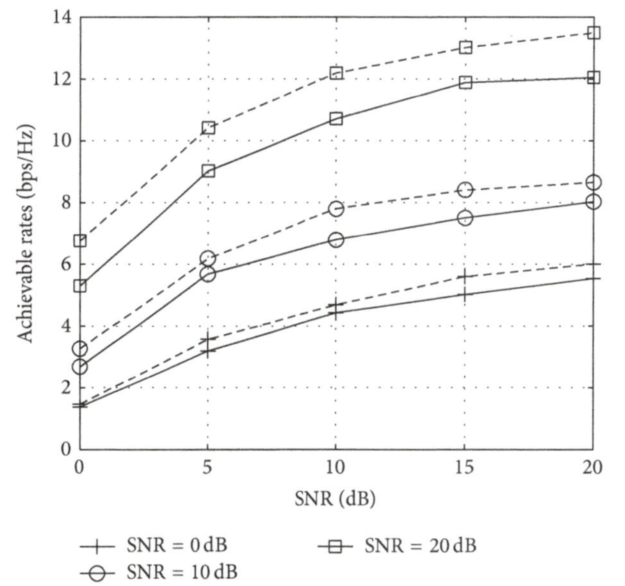

FIGURE 6: Achievable data rates versus N_{SU} in the multiantenna SU link for different values of α and SNR, when 3 out of 6 antennas are selected at SU RX and CB is employed at cooperating SUs. For dashed lines, $\alpha = 0.2$ and, for solid lines, $\alpha = 0.1$.

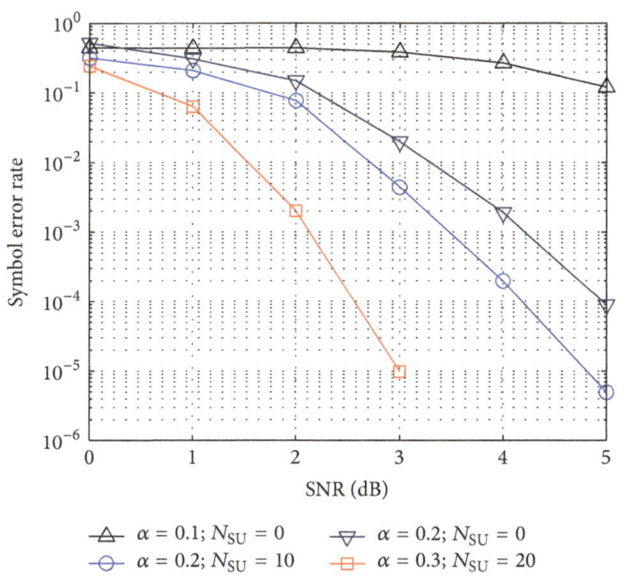

FIGURE 5: Symbol error rates versus SNR for different values of α and N_{SU}.

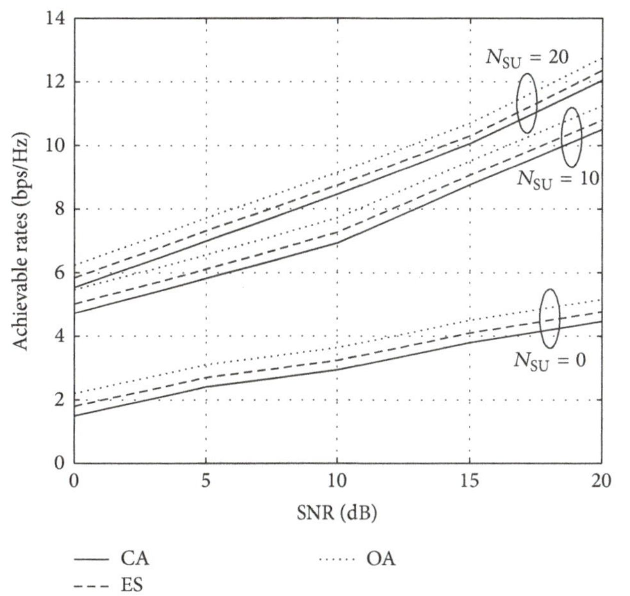

FIGURE 7: Comparison between the performance of convex approximation and exhaustive search method.

iterations required to converge for the suboptimal algorithm and is much smaller than T, according to simulations.

4. Simulation Results

In this section we explain the simulation results based on the proposed solutions for problems **P1**, **P2**, and **P3**. However, before we elaborate the results, we introduce a parameter, α, which controls the interference threshold at the PUs. α is chosen so that allowable interference at the PUs is a fraction of PU SNR; that is, $P_i = \alpha \text{SNR}_{\text{PU}}$ at the PUs. To compare the different approaches, we use the measures of achievable rates and symbol error rates. These are the assumptions for the simulations:

(i) achievable rates are determined by averaging over the results obtained from 1000 i.i.d. channels realizations;

(ii) the number of subcarriers of the MC-CDM modulation is assumed to be 16;

(iii) CVX package is used along with MATLAB for simulations [21];

(iv) the SU RX is equipped with 6 antennas, while 3 antennas exist at SU TX. The codebook matrix, whose columns are potential transmit beamformer weight vectors, for three antennas at SU TX and three limited feedback bits, that is, eight possible TB weight vectors, is given by [12]:

$$\mathbf{K} = \begin{bmatrix} \frac{1}{\sqrt{2}} & \frac{1}{\sqrt{2}} & 0 & \frac{1}{\sqrt{2}}e^{2\pi j/3} & \frac{1}{\sqrt{2}}e^{2\pi j/3} & \frac{1}{\sqrt{2}}e^{4\pi j/3} & \frac{1}{\sqrt{2}}e^{4\pi j/3} & 0 \\ \frac{1}{\sqrt{2}} & 0 & \frac{1}{\sqrt{2}} & \frac{1}{\sqrt{2}}e^{4\pi j/3} & 0 & 0 & \frac{1}{\sqrt{2}}e^{2\pi j/3} & \frac{1}{\sqrt{2}}e^{4\pi j/3} \\ 0 & \frac{1}{\sqrt{2}} & \frac{1}{\sqrt{2}} & 0 & \frac{1}{\sqrt{2}}e^{4\pi j/3} & \frac{1}{\sqrt{2}}e^{2\pi j/3} & 0 & \frac{1}{\sqrt{2}}e^{2\pi j/3} \end{bmatrix}. \quad (24)$$

Achievable data rates for different values of α, different numbers of cooperating SUs, and different numbers of antennas at SU RX are presented in Figure 3. CA method is recruited to solve problem **P1** and also cooperative beamforming is implemented on virtual MIMO of cooperating SUs. It can be implied from Figure 3 that utilizing the cooperation of single-antenna SUs, larger data rates can be attained. For example, by recruiting 10 SUs as relay and for $\alpha = 0.1$, achievable data rate is larger than the case with no cooperative SUs and $\alpha = 0.3$. Note that in both cases, 3 out of 6 available antennas are selected at SU RX. As depicted in Figure 3, even if PUs are able to endure an amount of interference as equal as their receive SNR, that is, $\alpha = 1$, and also all antennas at SU RX are employed (3×6 MIMO), data rates of the multiple-antenna SU link are far less than the data rates achieved using 20 cooperating SUs. Moreover, the achievable data rates in a 3×6 MIMO system, that is, there are no PUs and thus no power constraint exists, has also been depicted.

In Figure 4, we demonstrate that CB not only removes the interference on PUs, due to cooperation of single-antenna SU with desired SU link, but also leads to an increase in the achievable rates of the desired link. Note that, for all graphs in Figure 4, $\alpha = 0.1$.

In Figure 5 we consider the effects of TB, CB, and antenna selection on the symbol error rates (SER) of the multiantenna SU link. CA method is employed to solve problem **P1** and 3 out of 6 antennas at SU RX are selected. Clearly, the SER in the multiantenna SU link diminishes as a result of cooperation of single-antenna SUs. In Figure 6, we plot the achievable data rates versus N_{SU} in the multiantenna SU link for different values of and SNR. Also, 3 out of 6 antennas are selected at SU RX and CB is employed at cooperating SUs. It can be implied from Figure 6 that recruiting many cooperative SUs would not result in expectable increase in data rates.

A comparison between the achievable rates of the multiantenna SU link, using the CA and ES as suboptimal methods and the optimal approach (OA) (using the dual method) has been performed in Figure 7. Evidently, the proposed suboptimal method has near-optimal performance, while the proposed CA method performs very close to ES method, which is very promising.

5. Conclusions

Jointly determining the optimum TB and CB weight vectors and antenna selection in the MIMO-CCRN was considered in this work. The scenario consists of 2 multiantenna SUs and a number of single-antenna SUs and PUs. The problem was formulated and, to achieve a computationally efficient solution with much less complexity, we utilized convex approximation method. It was shown through simulations that using the proposed method, a rather complex problem can be solved with reduced complexity. Furthermore, taking advantage of cooperation of single-antenna secondary users, better data rates are achieved, without imposing intolerable interference on primary users.

Conflict of Interests

The authors declare that there is no conflict of interests regarding the publication of this paper.

References

[1] "FCC report of the spectrumefficiency working group," Tech. Rep., FCC Spectrum Policy Task Force, 2002, http://transition.fcc.gov/sptf/files/SEWGFinalReport_1.pdf.

[2] D. J. Love and R. W. Heath Jr., "Equal gain transmission in multiple-input multiple-output wireless systems," in *Proceedings of the IEEE Global Telecommunications Conference (GLOBECOM '02)*, vol. 2, pp. 1124–1128, November 2002.

[3] B. Hassibi and B. M. Hochwald, "High-rate codes that are linear in space and time," *IEEE Transactions on Information Theory*, vol. 48, no. 7, pp. 1804–1824, 2002.

[4] H. Du and T. Ratnarajah, "Transmit beamforming in MIMO cognitive radio network via semidefinite programming," in *Proceedings of the IEEE 73rd Vehicular Technology Conference (VTC '11)*, pp. 1–5, Yokohama, Japan, May 2011.

[5] S. You, G. Noh, J. Lee, H. Wang, and D. Hong, "Joint beamforming and power control algorithm for cognitive radio network with the multi-antenna base station," in *Proceedings of the IEEE Wireless Communications and Networking Conference (WCNC '10)*, pp. 1–6, Sydney, Australia, April 2010.

[6] F. Wang and W. Wang, "Robust beamforming and power control for multiuser cognitive radio network," in *Proceedings of the*

IEEE Global Telecommunications Conference (GLOBECOM '10), pp. 1–5, Miami, Fla, USA, December 2010.

[7] M. G. Adian and H. Aghaeinia, "Joint transmit beamforming and antenna selection in cognitive radio networks," in *Proceedings of the 5th International Symposium on Telecommunications (IST '10)*, pp. 33–37, Tehran, Iran, December 2010.

[8] K. Zarifi, S. Affes, and A. Ghrayeb, "Joint source power control and relay beamforming in amplify-and-forward cognitive networks with multiple source-destination pairs," in *Proceedings of the IEEE International Conference on Communications (ICC '11)*, pp. 1–6, Kyoto, Japan, June 2011.

[9] J. Liu, W. Chen, Z. Cao, and Y. J. Zhang, "Delay optimal scheduling for cognitive radio networks with cooperative beamforming," in *Proceedings of the IEEE International Conference on Communications (ICC '11)*, pp. 1–5, Kyoto, Japan, June 2011.

[10] M. A. Beigi and S. M. Razavizadeh, "Cooperative beamforming in cognitive radio networks," in *Proceedings of the 2nd IFIP Wireless Days (WD '09)*, pp. 1–5, Paris, France, December 2009.

[11] J. Liu, W. Chen, Z. Cao, and Y. J. Zhang, "Cooperative beamforming aided incremental relaying in cognitive radios," in *Proceedings of the IEEE International Conference on Communications (ICC '11)*, pp. 1–5, Kyoto, Japan, June 2011.

[12] D. J. Love, R. W. Heath, and T. Strohmer, "Grassmannian beamforming for multiple-input multiple-output wireless systems," *IEEE Transactions on Information Theory*, vol. 49, no. 10, pp. 2735–2747, 2003.

[13] A. F. Molisch, M. Z. Win, and J. H. Winters, "Capacity of MIMO systems with antenna selection," in *Proceedings of the International Conference on Communications (ICC '01)*, pp. 570–574, Helsinki, Finland, June 2000.

[14] A. Gorokhov, D. A. Gore, and A. J. Paulraj, "Receive antenna selection for MIMO spatial multiplexing: theory and algorithms," *IEEE Transactions on Signal Processing*, vol. 51, no. 11, pp. 2796–2807, 2003.

[15] M. Kang, L. Yang, and M. S. Alouini, "Capacity of MIMO Rician channels with multiple correlated Rayleigh co-channel interferers," in *Proceedings of the IEEE Global Telecommunications Conference (GLOBECOM '03)*, vol. 2, pp. 1119–1123, December 2003.

[16] J. N. Laneman and G. W. Wornell, "Distributed space-time-coded protocols for exploiting cooperative diversity in wireless networks," *IEEE Transactions on Information Theory*, vol. 49, no. 10, pp. 2415–2425, 2003.

[17] W. Yu and R. Lui, "Dual methods for nonconvex spectrum optimization of multicarrier systems," *IEEE Transactions on Communications*, vol. 54, no. 7, pp. 1310–1322, 2006.

[18] S. Boyd and L. Vandenberghe, *Convex Optimization*, Cambridge University Press, New York, NY, USA, 2004.

[19] P. Ubaidulla and A. Chockalingam, "Non-linear transceiver designs with imperfect CSIT using convex optimization," in *Proceedings of the IEEE Wireless Communications and Networking Conference (WCNC '09)*, pp. 1–6, Budapest, Hungary, April 2009.

[20] C. D. Meyer, *Matrix Analysis and Applied Linear Algebra*, SIAM, Philadelphia, Pa, USA, 2000.

[21] D. N. C. Tse and P. Viswanath, *Fundamentals of Wireless Communication*, Cambridge University Press, Cambridge, UK, 2005.

15

Determining Bounds on Assumption Errors in Operational Analysis

Neal M. Bengtson

School of Business, Barton College, Wilson, NC 27893, USA

Correspondence should be addressed to Neal M. Bengtson; bengtson@barton.edu

Academic Editor: Zne-Jung Lee

The technique of operational analysis (OA) is used in the study of systems performance, mainly for estimating mean values of various measures of interest, such as, number of jobs at a device and response times. The basic principles of operational analysis allow errors in assumptions to be quantified over a time period. The assumptions which are used to derive the operational analysis relationships are studied. Using Karush-Kuhn-Tucker (KKT) conditions bounds on error measures of these OA relationships are found. Examples of these bounds are used for representative performance measures to show limits on the difference between true performance values and those estimated by operational analysis relationships. A technique for finding tolerance limits on the bounds is demonstrated with a simulation example.

1. Introduction

The analysis of the performance of a network of devices is important in many areas. Computer systems and industrial manufacturing systems are two examples. The types of networks considered in this paper are operationally connected, queue and server devices. That is, each device is connected in some way with every other device in the network and each device may have a queue assigned to it. Certain information about these types of networks may be obtained using a technique known as operational analysis (OA). Relationships used to estimate performance measures (PMs) of networks may be derived in operational analysis under a few restrictive assumptions. OA is a technique which was originally defined as an aid in computer system performance analysis [1–6]. It can be an aid in the understanding of system performance in general [7] and is a complementary approach to stochastic analysis used in many networks of servers performance analyzes and in computer programs [8–14]. Other used or suggested applications for the OA approach include telecommunications [15], E-commerce [16, 17], flexible manufacturing systems [18], and Petri nets [19–21]. The performance measures derived are such things as average number of units at a device, average response time, and throughput. The behavior of a single, arbitrary device in a network will be considered.

Two basic principles define the OA approach [2].

(1) All assumptions that are made in analyzing the performance of a real system should be subject to direct verification.

(2) All variables that appear in any equation which characterize the performance of a real system should be verifiable by direct measurement.

The validity of PM equations developed using these principles can be shown for a particular set of data because they are based on assumptions which can be directly tested by the observation of data produced by the system of interest over a finite period of time.

The most widely used assumption about the data is that of *job flow balance*, that is, the number of arrivals to a network (for global flow balance) or to a device (for local flow balance) must be equal to the number of departures from that network or device. Also assumed is *one step behavior*: only one unit may arrive or depart the network or device at a time. Arrivals and completions do not occur simultaneously. The OA approach assumes that devices must have *homogeneous service*. That is, the service time of a device in a network is independent of the queue length at any device. *Homogeneous arrivals* are the corresponding condition for the arrival times. *Homogeneous routing* holds when the routing frequencies of jobs leaving a device are independent of the queue lengths at other devices in the network. *Device homogeneity* exists when the rate of output from a device is determined only by its queue length. Other assumptions can be invoked as the need arises to derive OA relations [22]. The only requirement is that these assumptions meet the two basic principles of testability given above.

OA assumptions allow for the development of relationships which enable us to determine PMs by collecting only a few types of data, namely, the number of arrivals and departures for each device state and the total time spent in each state [23]. These PMs will be accurate only if the OA assumptions are met and only for the finite time period observed. The accuracy of the assumptions can be measured. A device state is the number of items (customers, jobs, entities, etc.) both waiting and in service at a device.

While OA research was originally proposed as an aid in computer performance analysis, it is more general in that developments can be applicable to any system that generates time series data. This would include computer simulation.

The Abbreviations section gives definitions of variable used in this paper. Error measures of various OA assumptions have been defined and are summarized in Table 1 for job flow balance, homogeneous service, and homogeneous arrival [24]. The limit over time of the expected value and variance of the job flow balance error is zero [25] so that over time this error is not significant for data runs of reasonable length. The expected value over time of other error measures, such as, for the assumptions of homogeneous service and arrival, may not, in general, tend to zero [25].

Different error measures, in the form of relative errors, have been defined by Brumfield [22]. By presenting a set of new assumptions, formulas for the calculation of response time and average queue length in terms of the average and coefficient of variation of service times are developed. Two examples of these new assumptions are homogeneity of queueing and service and homogeneity of residuals. For arrival j "forward residual is either the time remaining in the service period during which j arrives or zero if arrival j begins a service period ... similarly backward residual is either the time since the beginning of the service period during which j arrives or zero if arrival j begins a service period" [22]. Relative error formulas for response time are determined with these new assumptions in addition to the old assumptions of homogeneous service and homogeneous arrivals. Unfortunately, the error terms are quite complex since there are more assumptions with which to deal.

If we are using a relationship to determine a PM derived under OA assumptions, then the resulting value of the PM is in error if the founding assumptions do not hold. This can be checked from the data because of the way OA assumptions are defined. The degree of error in the PM calculated is a function of the assumption error measures. Correction terms have been developed using the assumption error measures [24]. When added to the PMs these correction terms produce exact results. It is these correction terms which are studied in this paper.

As an example, assume we are interested in obtaining a value for the average number of jobs in a computer system that a new job sees upon arrival. If we make the homogeneous service assumption, then this average may be estimated by [4]

$$\overline{n}_A^S = \frac{\overline{n}}{U} - 1, \tag{1}$$

where $\overline{n}_A^S$ is average number of jobs at a device seen by an arriver, assuming homogeneous service, $\overline{n}$ is average number of jobs at the device, and U is device utilization.

We are interested in finding a correction term, such that

$$\overline{n}_A = \overline{n}_A^S + C_{\overline{n}_A}. \tag{2}$$

The correction term is equal to

$$C_{\overline{n}_A} = e_F - e_S^*, \tag{3}$$

where

$$e_F = \sum_{n=0}^{N-1} (n+1)\left(p_A(n) - p_C(n)\right),$$
$$e_S^* = \sum_{n=1}^{N} n p_C(n-1)\, e_S(n). \tag{4}$$

Check the Abbreviations section for all symbol definitions. e_F is the error measure for the job flow balance assumption and e_S^* is a weak form of the homogeneous service error measure because it may be equal to 0 even when some or all the individual $e_S(n)$ values are not.

There are a couple of problems with using the error measures to derive correction terms. One problem is that the amount of data needed to calculate an exact value for an assumption error measure is the same as to find the performance measure of interest directly. The process of determining the error measure for each assumption used to derive a relation, the correction term, and the PM estimate to which the correction term is added is a way of getting something which may be observed more directly. Finding exact values for performance measures in this indirect way over a finite time period may be worthwhile only if a number of PMs are desired. In this case, a single assumption error measure is determined for each assumption and applied to all the PM correction terms of interest.

Another problem with the error measure technique is that these measures apply only to the data observed. For another run of data, new error measure values need to be found. This limitation may be acceptable if PMs cannot be measured directly without changing the nature of the system, for example, in a complex computer system. We would like a way to extend assumption error measures over longer sets of data and, thereby, say something about the system that generated the data. As Sevcik and Klawe [26] stated shortly after OA was introduced "Because operational analysis is based on assumptions that can be tested but that are very unlikely to be satisfied exactly in any finite time period, it is very important

TABLE 1: Operational analysis error measures.

Assumption	State dependent error	Overall error
Job-flow balance	$e_F(n) = (n+1)(p_A(n) - p_C(n)), \quad n \in I$	$e_F = \sum_{n=0}^{N-1} e_F(n)$
Homogeneous service	$e_S(n) = \frac{1}{p_C(n-1)}\left(\frac{T(n)}{T - T(0)}\right) - 1, \quad n \in I^+$	$e_S^* = \sum_{n=1}^{N} np_c(n-1)e_S(n)$
Homogeneous arrival	$e_A(n) = p_A(n)\left(\frac{T - T(N)}{T(n)}\right) - 1, \quad n \in I$	$e_A^* = \sum_{n=1}^{N-1} e_A(n)\frac{nT(n)}{T - T(N)}$

to develop a means of dealing with 'fuzzy homogeneity' or situations in which the various independence assumptions are satisfied within some tolerance." This paper addresses this need to define these assumption bounds.

The next section will illustrate how OA relations may be used to reduce data collection while estimating performance measures. This will be followed by a discussion of the determination of bounds on the OA assumption measurement errors for homogeneous service and homogeneous arrival. Sample calculations of these bounds will be presented afterward. An illustration of the use of bounds in a simulation will then be given.

2. Simplifying Data Collection

Calculating performance measures with OA relationships that are derived under one or more of the system behavior assumptions is usually simpler than using more direct relationships. This is because by making the assumptions a model has been created which reduces, perhaps artificially, the complexity of the behavior of the system. The result is that less information is needed to make an estimate of the PM than would be needed for a direct measurement.

There are situations where it is impractical, if not impossible, to collect sufficient data to determine exact values for PMs over a finite period of time. In some cases, only an estimate of a PM is needed and it is not worthwhile to go to the trouble of determining the precise PM values. Any PM value obtained for a behavior sequence is only an estimate of the underlying system PM. With this realization in mind, it may seem unwise to spend a great deal of effort to obtain an exact value for a sequence which is, in turn, only an estimate of some other value. A good approximation of the sequence estimate may be sufficient.

If we want the average response time, R, of a behavior sequence, we could accumulate the response times of all the jobs that go through the system and get the exact R by dividing by the number of jobs. A simpler procedure would be to say that response time is

$$R = \frac{S}{1 - U}, \tag{5}$$

where S is mean time between completions during busy periods and U is utilization.

This equation will give the exact R value if we have a behavior sequence for a single server queue which is in flow balance and has homogeneous arrivals and services. If these conditions do not hold, the equation will not give R exactly, but an estimate of R, call it R'. If we collect only the idle time, $T(0)$, and the number of completions, C, we can use the same equation to find R'. If the behavior sequence lasts for time T, then

$$U = 1 - \frac{T(0)}{T}, \qquad S = \frac{T - T(0)}{C}. \tag{6}$$

Another example calculates the average number of jobs at a device. With the same assumptions as for estimating response time, the average number of jobs in the queue/server system is

$$\overline{n} = \frac{U}{1 - U}. \tag{7}$$

This value takes even less data to calculate than does R'. The direct calculation of $\overline{n}$ requires accumulating data every time there is an arrival or completion or requires keeping track of the total amount of time spent at each of the states.

Using these equations for predicting future values for R and $\overline{n}$ of a system presents certain problems. For example, over future time, will the assumptions of the system behave in the same way? Since we can determine and use error measures of the assumptions in order to correct assumption derived PM estimate, it is not necessary that assumptions hold in the future if they have not in the past. With the determination of correction terms all that is really necessary is for the correction terms to remain relatively constant, that is, for the system's violations of assumptions to remain the same over future time periods.

Without knowledge of the assumption error measures and through them the correction terms, the performance measure estimates may be quite bad for any particular behavior sequence [23]. As stated before, in a stable system the job flow balance assumption error measure will go to zero as time increases, but, as shown in [24], this is not necessarily true for other assumption error measures. For any behavior sequence it is important to make some assessment, if possible, of the behavior of the PM correction terms.

3. Performance Measure Bounds

One approach to use the simplified OA formulas for PMs is to determine bounds on the maximum PM error. That is, we are interested in defining bounds on the difference between true values of various PMs at a device for particular state sequences and those PMs estimated by using relationships derived under operational analysis assumptions. We will assume the network is in steady state.

In the following, bounds are found for the assumptions of homogeneous services and homogeneous arrivals. In the case of the job-flow balance assumption, we know that the expected value of the error measure and its variance go to zero [25]:

$$\lim_{t\to\infty} E\left[e_F\right] = 0, \qquad \lim_{t\to\infty} V\left[e_F\right] = 0. \tag{8}$$

Therefore, as the length of the sequence increases the e_F can be expected to become insignificant.

We will need to assume that a maximum value of the error for services and arrivals for any state is known or can be set. Call these values δ_S and δ_A for the maximum service error and maximum arrival error, respectively.

3.1. Bounds on Homogeneous Service Assumption Error. If δ_S is the maximum error for the homogeneous service assumption, then

$$\left|e_S(n)\right| \le \delta_S \quad n \in I^+. \tag{9}$$

A more useful bound would be on the weak overall homogeneous service error:

$$e_S^* \le \delta_S^*. \tag{10}$$

But, this limit may be harder to know beforehand. Equation (9) may be used to find an upper bound on e_S^* by using the definition

$$e_S^* = \sum_{n=1}^{N} n p_C(n-1)\, e_S(n). \tag{11}$$

Substituting (9) yields

$$e_S^* = \sum_{n=1}^{N} n p_C(n-1)\left|e_S(n)\right| \le \delta_S \sum_{n=1}^{N} n p_C(n-1). \tag{12}$$

The term

$$\sum_{m=1}^{N} n p_C(n-1) = \overline{n}_C \tag{13}$$

is the average number at the device seen by a completer. Therefore,

$$e_S^* \le \delta_S \overline{n}_C. \tag{14}$$

The bound given by (14) does not take into consideration the fact that the $e_S(n)$ values are not independent. In fact, they are related by the expression

$$\sum_{n=1}^{N} p_C(n-1)\, e_S(n) = 0. \tag{15}$$

We can get a tighter bound of the e_S^* values by taking this dependence into consideration. Equation (15) can be shown by substituting the definition

$$e_S(n) = \frac{S(n)}{S} = \frac{1}{p_C(n-1)}\left[\frac{T(n)}{T-T(0)}\right] - 1 \tag{16}$$

as follows:

$$\begin{aligned}
&\sum_{n=1}^{N} p_C(n-1)\, e_S(n)\\
&\quad = \sum_{n=1}^{N} p_C(n-1)\left[\frac{1}{p_C(n-1)}\frac{T(n)}{T-T(0)} - 1\right]\\
&\quad = \sum_{n=1}^{N} \frac{T(n)}{T-T(0)} - \sum_{n=1}^{N} p_C(n-1)\\
&\quad = \frac{T-T(0)}{T-T(0)} - 1 = 0.
\end{aligned} \tag{17}$$

Since what is desired is an upper bound on e_S^* a solution to the optimization problem below will give the desired result:

$$\begin{aligned}
\text{Max} \quad & e_S^* = \sum_{m=1}^{N} n p_C(n-1)\, e_S(n)\\
\text{s.t.} \quad & \sum_{n=1}^{N} p_C(n-1)\, e_S(n) = 0\\
& \left|e_S(n)\right| \le \delta_S \quad n \in I^+.
\end{aligned} \tag{18}$$

In order to show the optimal solution, first put this problem in primal and dual forms:

primal:

$$\begin{aligned}
\text{Max} \quad & \sum_{n=1}^{N} n p_C(n-1)\, e_S(n)\\
\text{s.t.} \quad & \sum_{n=1}^{N} p_C(n-1)\, e_S(n) = 0 \quad : u\\
& -e_S(n) \le \delta_S \quad n \in I^+ : y(n)\\
& e_S(n) \le \delta_S \quad n \in I^+ \quad : w(n);
\end{aligned} \tag{19}$$

dual:

$$\begin{aligned}
\text{Min.} \quad & \delta \sum_{n=1}^{N} \left(y(n) + w(n)\right)\\
\text{s.t.} \quad & p_C(n-1)\, u - y(n) + w(n)\\
& = n p_C(n-1) \quad n \in I^+\\
& y(n) \ge 0 \quad n \in I^+\\
& w(n) \ge 0 \quad n \in I^+.
\end{aligned} \tag{20}$$

The optimal solution to the problem will have to satisfy the Karush-Kuhn-Tucker (KKT) conditions, that is, feasibility of the Primal and Dual, as well as complementary slackness [27]. The KKT conditions give the necessary conditions for optimality of the general constrained problem.

Checking dual feasibility, for any u the constraints can be satisfied by construction such that if $(n-u) > 0$, then $y(n) = 0$, $w(n) = (n-u)p_C(n-1)$ and if $(n-u) < 0$, then $y(n) = -(n-u)p_C(n-1)$, $w(n) = 0$. This is because the main dual constraint is

$$(n-u)\, p_C(n-1) = w(n) - y(n). \tag{21}$$

In order to satisfy complementary slackness, if $e_S(n) > -\delta_S$, then it must be true that $y(n) = 0$, and if $e_S < \delta_S$, then it must be true that $w(n) = 0$. In terms of the above primal-dual construction, if $e_S(n) > -\delta_S$, then $(n-u) > 0$ and if $e_S(n) < \delta_S$, then $(n-u) < 0$. Any solution $e_S(n), n \in I^+$ and u satisfying the primal constraints and the above two conditions is optimal.

Consider the solution

$$\begin{aligned} e_S(n) &= -\delta_S \quad \text{if } n < \tilde{n}_C, \\ e_S(n) &= \delta_S \quad \text{if } n > \tilde{n}_C, \\ u &= \tilde{n}_C, \end{aligned} \tag{22}$$

where $\tilde{n}_C$ is the median state at completions.

Assume for simplicity that there is an even number of states so that $\tilde{n}_C \neq n$ for any n. This solution is dual feasible since we showed above that any n is a solution to the dual.

Checking primal feasibility, the solution satisfies $-e_S(n) \le \delta_S$ and $e_S(n) \le \delta_S$.

The main constraint is

$$\begin{aligned} &\sum_{n=1}^{N} p_C(n-1)\, e_S(n) \\ &\quad = -\delta_S \sum_{n<\tilde{n}_C} p_C(n-1) + \delta_S \sum_{n>\tilde{n}_C} p_C(n-1) \\ &\quad = \delta_S \left[-\sum_{n<\tilde{n}_C} p_C(n-1) + \sum_{n>\tilde{n}_C} p_C(n-1) \right] \\ &\quad = \delta_S [0] \end{aligned} \tag{23}$$

since $\tilde{n}_C$ is median.

Lastly, we check for complementary slackness. Now, $e_S(n) > -\delta_S$ when $n > \tilde{n}_C$; then $n - y > 0$ and $e_S(n) < \delta_S$ when $n < \tilde{n}_C$; then $n - y < 0$. Therefore, complementary slackness holds and the solution is an optimal one.

The solution value is

$$\begin{aligned} &\sum_{n=1}^{N} np_C(n-1)\, e_S(n) \\ &\quad = \sum_{n<\tilde{n}_C} np_C(n-1)(-\delta_S) + \sum_{n>\tilde{n}_C} np_C(n-1)(\delta_S). \end{aligned} \tag{24}$$

Set this value equal to Δ_S, which is the overall completer's average minus the average of the set truncated at the median. This can be shown by first taking

$$\Delta_S = \sum_{n<\tilde{n}_C} np_C(n-1)(-\delta_S) + \sum_{n>\tilde{n}_C} np_C(n-1)(\delta_S). \tag{25}$$

δ_S times the completer's average, $\overline{n}_C$, is

$$\delta_S \overline{n}_C = \sum_{n<\tilde{n}_C} np_C(n-1)(\delta_S) + \sum_{n>\tilde{n}_C} np_C(n-1)(\delta_S). \tag{26}$$

Subtracting Δ_S yields

$$\begin{aligned} \delta_S \overline{n}_C - \Delta_S &= 2 \sum_{n<\tilde{n}_C} np_C(n-1)(\delta_S), \\ \Delta_S &= \delta_S \left[\overline{n}_C - 2 \sum_{n<\tilde{n}_C} np_C(n-1) \right]. \end{aligned} \tag{27}$$

We know that

$$\sum_{n<\tilde{n}_C} p_C(n-1) = \frac{1}{2} \tag{28}$$

since $\tilde{n}_C$ is a median. Therefore,

$$\begin{aligned} \sum_{n<\tilde{n}_C} 2p_C(n-1) &= 1, \\ \sum_{n<\tilde{n}_C} n2p_C(n-1) &= \overline{n}_{CT} \end{aligned} \tag{29}$$

is the average of the set of states truncated at the median. So the bound on e_S^* is

$$\Delta_S = \delta_S [\overline{n}_C - \overline{n}_{CT}]. \tag{30}$$

As an example, take the behavior sequence in Figure 1. If we want to use the OA equation [24]

$$\overline{n}^{AS} = \frac{U}{1 - U - p(N)} (1 - (N+1)\, p(N)) \tag{31}$$

to calculate $\overline{n}^{AS}$, which is the average number assuming flow balance, homogeneous arrival, and services, we would be interested in the bound of the difference between $\overline{n}$ and $\overline{n}^{AS}$. We can calculate

$$U = \frac{T - T(0)}{T} = \frac{5}{6} \tag{32}$$

p(n)=1/3, and $n_C = 7/4$. Assume the maximum error is $\delta_S = 3/5$. If the other assumption errors are zero, then the difference, $\overline{n} - \overline{n}^{AS}$ is equal to the correction term:

$$C_{\overline{n}} = \frac{U}{1 - U - p(N)} (1 - p(N))\, e_S^*. \tag{33}$$

The upper bound on this correction term, using (14), is

$$\frac{U}{1 - U - p(N)} (1 - p(N))\, \delta_S \overline{n}_C = -3.5. \tag{34}$$

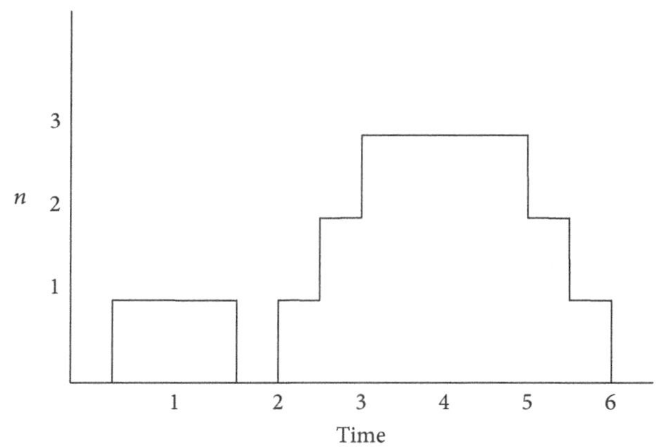

FIGURE 1: Example Behavior Sequence for estimating. Homogeneous service error bound.

Using the tighter bound, Δ_S, we get

$$\begin{aligned}\frac{U}{1-U-p(N)}(1-p(N))\Delta_S &= -\frac{10}{3}(\delta_S(\overline{n}_C-\overline{n}_{CT})) \\ &= -\frac{10}{3}\left(\frac{3}{5}\left(\frac{7}{4}-1\right)\right)=-1.5.\end{aligned} \tag{35}$$

This is a reduction of 57.14%, for this example of the difference bound.

3.2. Bounds on Homogeneous Arrival Assumption Error. As in the previous section, we can assume that a maximum error, δ_A, for any state is known beforehand. That is, we assume

$$|e_A(n)| \le \delta_A, \quad n \in I. \tag{36}$$

Then, the weak overall homogeneous arrival error is bounded by

$$e_A^* \le \sum_{n=0}^{N-1} n\frac{T(n)}{T-T(N)}|e_A(n)| \le \delta_A \sum_{n=0}^{N-1}\frac{nT(n)}{T-T(N)} \le \delta_A\overline{n}_N, \tag{37}$$

where $\overline{n}_N$ is the average number at the device excluding the maximum state.

As with the service errors, $e_S(n)$, the $e_A(n)$ are not independent. The dependency is

$$\sum_{n=0}^{N-1}\frac{T(n)}{T-T(N)}e_A(n)=0. \tag{38}$$

This can be shown by substituting the definition of $e_A(n)$ into the equation. As with the homogeneous service assumptions, we set up the following optimization problem:

$$\begin{aligned}\text{Max} \quad & e_A^* = \sum_{n=0}^{N-1} n\frac{T(n)}{T-T(N)}e_A(n) \\ \text{s.t.} \quad & \sum_{n=0}^{N-1}\frac{T(n)}{T-T(N)}e_A(n)=0 \\ & |e_A(n)| \le \delta_A \quad n \in I.\end{aligned} \tag{39}$$

Using the Karush-Kuhn-Tucker conditions as before, we can show that with $\tilde{n}$ as the median state

$$\begin{aligned} e_A(n) &= -\delta_A \quad \text{if } n<\tilde{n}, \\ e_A(n) &= \delta_A \quad \text{if } n>\tilde{n}, \\ u &= \tilde{n}\end{aligned} \tag{40}$$

is the solution to the optimization problem. The value of the solution is found by substituting into the primal objective function to get

$$\begin{aligned}\Delta_A &= \max e_A^* \\ &= \sum_{n<\tilde{n}} n\frac{T(n)}{T-T(N)}(-\delta_A) + \sum_{n>\tilde{n}}^{N-1} n\frac{T(n)}{T-T(N)}(\delta_A).\end{aligned} \tag{41}$$

δ_A times the average excluding the maximum, $\overline{n}_N$, is

$$\delta_A\overline{n}_N = \sum_{n<\tilde{n}}\frac{nT(n)}{T-T(N)}(\delta_A) + \sum_{n>\tilde{n}}^{N-1}\frac{nT(n)}{T-T(N)}(\delta_A). \tag{42}$$

Subtracting

$$\delta_A\overline{n}_N - \Delta_A = 2\sum_{n<\tilde{n}}\frac{nT(n)}{T-T(N)}\delta_A \tag{43}$$

therefore,

$$\begin{aligned}\Delta_A &= \delta_A\left[\overline{n}_N - 2\sum_{n<\tilde{n}}\frac{nT(n)}{T-T(N)}\right] \\ &= \delta_A\left[\overline{n}_N - 2\sum_{n<\tilde{n}}\frac{np(n)}{1-p(N)}\right].\end{aligned} \tag{44}$$

The expression

$$2\sum_{n<\tilde{n}} np(n) \tag{45}$$

is the average of the values truncated at the median. Call this value $\overline{n}_T$. Then, substitution yields an upper bound on the error measure due to violations in homogeneous arrivals of

$$\Delta_A = \delta_A\left[\overline{n}_N - \frac{\overline{n}_T}{1-p(N)}\right]. \tag{46}$$

This bound may not be as useful as the bound on the homogeneous service assumption error because the value of Δ_A is based on knowing the $T(n)$ values, whereas, for Δ_S only completion counts are necessary.

4. Example Performance Measure Error Bounds

Some examples are given next of how Δ_A and Δ_S may be used to determine bounds on the difference between exact values of various PMs and the OA estimated values for particular behavior sequences.

4.1. Arriver's Average Queue Length. Using the example from the introduction, for a behavior sequence the average queue length seen by an arriving job may be calculated by

$$\overline{n}_A = \overline{n}_A^S + C_{\overline{n}_A^S}, \tag{47}$$

where $C_{\overline{n}_A^S} = e_F - e_S^*$ and $\overline{n}_A^S = (\overline{n}/U) - 1$ is the average arriver's queue length assuming homogeneous services.

Rearranging and using the Δ_A bound give

$$\begin{aligned} \overline{n}_A - \overline{n}_A^S &= e_F - e_S^*, \\ \overline{n}_A^S - \overline{n}_A &= e_S^* - e_F \le \Delta_S - e_F. \end{aligned} \tag{48}$$

Since $e_F \rightarrow 0$ for any sequence of data in steady state, we assume flow balance holds. Then,

$$\overline{n}_A^S - \overline{n}_A \le \Delta_S. \tag{49}$$

This expression shows us that the difference between estimating the average length seen by arrivers with the relationship that assumes homogeneous servers and the true value of $\overline{n}_A$ is less than or equal to the bound on e_S^*.

4.2. Response Time. The exact response time for a behavior sequence can be found by

$$R = R^S + C_{R^S}, \tag{50}$$

where $C_{R^S} = -SC_{\overline{n}_A^S} = -S(e_F - e_S^*)$ and $R^S = S(\overline{n}_A + 1)$.

Since, as before, we are interested in the difference between a possible observed value (R) and a calculated value (R^S) we should do some rearranging and get $R - R^S = S(e_S^* - e_F)$. Again, assuming e_F is small and substituting $e_S^* \le \Delta_S$ yield

$$\begin{aligned} R - R^S &\le S\Delta_S, \\ R - R^S &\le S\left(\delta_S\left(\overline{n}_C - \overline{n}_{CT}\right)\right). \end{aligned} \tag{51}$$

4.3. Average Number at Device. If homogeneous arrivals, homogeneous services, and job flow balance hold then the average number at a device (i.e., those both in queue and in service) can be calculated by

$$\overline{n}^{AS} = \frac{U}{1 - U - p(N)}\left(1 - (N+1)\,p(N)\right). \tag{52}$$

The exact average number if these assumptions do not hold is

$$\overline{n} = \overline{n}^{AS} + C_{\overline{n}^{AS}}, \tag{53}$$

where

$$C_{\overline{n}^{AS}} = \frac{U}{1 - U - p(N)}\left(1 - p(N)\right)\left(e_A^* + e_S^* - e_F\right). \tag{54}$$

Rearranging again,

$$\begin{aligned} \overline{n} - \overline{n}^{AS} &= C_{\overline{n}^{AS}} \\ &= \frac{U}{1 - U - p(N)}\left(1 - p(N)\right)\left(e_A^* + e_S^* - e_F\right) \end{aligned} \tag{55}$$

and substituting $e_A^* \le \Delta_A$, $e_S^* \le \Delta_S$, and $e_F = 0$ we get

$$\overline{n} - \overline{n}^{AS} \le \frac{U}{1 - U - p(N)}\left(1 - p(N)\right)\left(\Delta_A + \Delta_S\right) \tag{56}$$

if $1 > U + p(N)$ or

$$\overline{n} - \overline{n}^{AS} \ge \frac{U}{1 - U - p(N)}\left(1 - p(N)\right)\left(\Delta_A + \Delta_S\right) \tag{57}$$

if $1 < U + p(N)$.

Because the calculation of Δ_A requires knowledge of $T(n)$ values, there may be little benefit in using the right hand side of the above relationship to find this error bound. If, however, the behavior sequence can be assumed to have homogeneous arrivals, then we can get a bound on the average number at a device error without the knowledge of the individual $T(n)$ values. In that case, the only time statistics we need are utilization, U, and the fraction of time spent at the maximum state, $p(N)$.

4.4. Throughput. Using the OA version of Little's Law [1] we can say that the difference between a behavior sequence's actual throughput and that calculated assuming both homogeneous arrivals and services will be

$$X - X^{AS} = \frac{\overline{n}}{R} - \frac{\overline{n}^{AS}}{R^{AS}}. \tag{58}$$

For R^{AS}, we can substitute R^S which is the more general expression since it does not need the homogeneous arrival assumption. From previous developments we know the following relations hold:

$$\overline{n} = \overline{n}^{AS} + C_{\overline{n}^{AS}}, \tag{59}$$

where

$$C_{\overline{n}^{AS}} \le \left|\frac{U}{1 - U - p(N)}\left(1 - p(N)\right)\left(\Delta_A + \Delta_S\right)\right| \tag{60}$$

and $R = R^S + C_{R^S}$, where

$$C_{R^S} \le S\Delta_S. \tag{61}$$

Substituting into (58) gives

$$X - X^{AS} = \frac{\overline{n}^{AS} + C_{\overline{n}^{AS}}}{R^S + C_{R^S}} - \frac{\overline{n}^{AS}}{R^S} = \frac{R^S C_{\overline{n}^{AS}} - \overline{n}^{AS} C_{R^S}}{R^S\left(R^S + C_{R^S}\right)}. \tag{62}$$

If the bound of (60) is substituted into (62), then it must be that

$$\frac{R^S\left|(U/(1-U-p(N)))(1-p(N))(\Delta_A+\Delta_S)\right|-\overline{n}^{AS}C_{R^S}}{R^S(R^S+C_{R^S})} \geq \frac{R^S C_{\overline{n}^{AS}}-\overline{n}^{AS}C_{R^S}}{R^S(R^S+C_{R^S})}. \quad (63)$$

If instead of bound (60), bound (61) is substituted, the numerator can only decrease and the denominator only increase; therefore,

$$\frac{R^S C_{\overline{n}^{AS}}-\overline{n}^{AS}S\Delta_S}{R^S(R^S+S\Delta_S)} \leq \frac{R^S C_{\overline{n}^{AS}}-\overline{n}^{AS}C_{R^S}}{R^S(R^S+C_{R^S})}. \quad (64)$$

If both (60) and (61) are substituted, then the value of the expression must fall within the limits defined by (63) and (64). Therefore, the bounds on our throughput error are

$$\begin{aligned}&\frac{R^S C_{\overline{n}^{AS}}-\overline{n}^{AS}S\Delta_S}{R^S(R^S+S\Delta_S)}\\ &\leq X-X^{AS}\\ &\leq \frac{R^S\left|(U/(1-U-p(N)))(1-p(N))(\Delta_A+\Delta_S)\right|}{R^S(R^S+C_{R^S})}\\ &\quad-\frac{\overline{n}^{AS}C_{R^S}}{R^S(R^S+C_{R^S})}.\end{aligned} \quad (65)$$

5. Using the PM Bounds

The bounds derived in the previous sections are actually limits on PM correction terms. Assuming we know the δ_A and δ_S values *a priori* and that we can say something about the homogeneous arrival assumption, then these bounds can be found without knowledge of all the $T(n)$ values for each n. This is the same simplification of data collection that we have in using the OA formulas instead of direct calculations.

As an example of using the bounds in a simulation study, assume we are interested in finding an estimate for average number at a device. A series of 10 runs is made and the bound for the correction term, $C_{\overline{n}}$, is calculated. If it is positive, we can call this an upper bound. If it is negative, let this be a lower bound. In both cases the other bound is 0. Assume in the simulation runs these bounds always fall between −6 and 3.3. The correction term for each run is approximated by taking the average value between the upper and lower bounds. We would like to be able to say something about the probability that future runs will fall within the [−6, 3.3] limits that have already appeared. We can do this using tolerance limit calculations.

Assume the 10 runs produced the results given in Table 2. The average and standard deviation for these observations are −1.34 and 1.55, respectively.

Table 2: Example estimated correction terms for average number at a device.

Run	$C_{\overline{n}}$
1	−1.75
2	−3.00
3	−1.25
4	1.65
5	−2.50
6	−1.90
7	−1.75
8	1.15
9	−2.60
10	−1.45

Since the tolerance limits are going to be set at the observed limits, we can say

$$\begin{aligned}[-6, 3.3] &= \left[C_{\overline{n}}-Ks, C_{\overline{n}}+Ks\right],\\ K &= \frac{C_{\overline{n}}-(-6)}{s}=\frac{(3.3)-C_{\overline{n}}}{s}=3.0.\end{aligned} \quad (66)$$

From tolerance limit tables [28] we can say with 95% confidence that at least 91% of future observations of $C_{\overline{n}}$ will fall within the interval [−6, 3.3]. That is, if we use $\overline{n}^{AS}$ values, we have 95% confidence that the correct $\overline{n}$ values for each run will fall within these limits 91% of the time.

6. Conclusion

In this paper, bounds were developed for the operational analysis error measures of homogeneous service and arrival assumptions. These bounds allow us to take advantage of the simplified data collection made possible by the use of operational analysis relationships, even when the assumptions used to derive those relationships are violated. A tolerance limit based method was given in order to be able to say something about the confidence that future correction term values in a time series would be within certain limits.

Abbreviations

- C: Total number of job completions at a device
- $C(n)$: Total number of completions when n jobs are at a device
- C_x: Correction term makes the OA term x equal to the directly calculated performance measure
- I: The set of states $\{0, 1, 2, \ldots, N-1\}$
- I^+: The set of states $\{1, 2, \ldots, N\}$
- K: Tolerance limit multiplier
- N: The maximum number of jobs seen at a device, both in queue and in service

n: State of a device, number of jobs both in queue and in service
$\tilde{n}$: The median state seen by a device
$\tilde{n}_C$: The median state at a device at job completions
$\overline{n}$: Average number of jobs at a device
$\overline{n}^{AS}$: The average number of jobs at a device assuming flow balance, homogeneous arrivals, and homogeneous services
$\overline{n}_C$: The average number at the device seen by a completing job
$\overline{n}_{CT}$: The average of the set of states seen by completing job truncated at the median state value
$\overline{n}_N$: The average state at a device, excluding the maximum state
$\overline{n}_T$: Time-average of the set of states truncated at the median
$\overline{n}_A^S$: Average number of jobs at a device seen by an arriver, assuming homogeneous service
$p(n)$: Proportion of time spent in state n
$p_A(n)$: The proportion of arrivals when n jobs are at a device, $A(n)/A$
$p_C(n)$: The proportion of completions that leave n jobs at a device, $C(n+1)/C$
R: Average response time
R^S: Average response time assuming homogeneous services
R^{AS}: Average response time assuming homogeneous arrivals and homogeneous services
S: Mean time between completions during busy periods
s: Standard deviation
T: Total time of the period of observation
$T(n)$: Total time device was in state n
U: Utilization, $1-p(0)$
X: Throughput
X^{AS}: Throughput assuming homogeneous arrivals and homogeneous services
δ_A: Maximum error in homogeneous arrival assumption error measure
δ_S: Maximum error in homogeneous service assumption error measure
Δ_A: A bound on the homogeneous arrival assumption error measure, $\leq \delta_A$
Δ_S: A bound on the homogeneous service assumption error measure, $\leq \delta_S$.

Conflict of Interests

The author declares that there is no conflict of interests regarding the publication of this paper.

References

[1] J. P. Buzen, "Fundamental operational laws of computer system performance," *Acta Informatica*, vol. 7, no. 2, pp. 167–182, 1976.

[2] J. P. Buzen, "Operational analysis: an alternative to stochastic modeling," in *Performance of Computer Installations*, pp. 175–194, North-Holland, Amsterdam, The Netherlands, 1978.

[3] J. P. Buzen and P. J. Denning, "Operational treatment of queue distributions and mean value analysis," Tech. Rep. CSD-TR 309, Department of Computer Science, Purdue University, 1979.

[4] J. P. Buzen and P. J. Denning, "Measuring and calculating queue length distributions," *Computer*, vol. 13, no. 4, pp. 33–44, 1980.

[5] P. J. Denning and J. P. Buzen, "The operational analysis of queueing network models," *Computer Surveys*, vol. 10, no. 3, pp. 225–261, 1978.

[6] J. P. Buzen, "From stochastic modeling to operational analysis: the journey begins," in *Fundamental Concepts in Computer Science*, E. Gelenbe and J. Kahane, Eds., pp. 141–149, Imperial College Press, London, UK, 2009.

[7] E. D. Lazowska, J. Zahorjan, G. S. Graham, and K. C. Sevcik, *Quantitative System Performance*, Prentice-Hall, Englewood Cliffs, NJ, USA, 1984.

[8] R. Suri, "Robustness of queueing network formulas," *Journal of the ACM*, vol. 30, no. 3, pp. 564–594, 1983.

[9] E. D. Lazowska, J. Zahorjan, and E. C. Sevcik, "Computer system performance evaluation using queueing network models," *Annual Reviews in Computer Science*, vol. 1, pp. 107–137, 1986.

[10] B. V. Ivanovskii, "Operational analysis of communication networks with lockups," *Automatic Control and Computer Sciences*, vol. 22, no. 3, pp. 26–32, 1988.

[11] P. J. Denning, "Queueing in networks of computers," *American Scientist*, vol. 79, pp. 206–209, 1991.

[12] P. J. Denning, "In the queue: mean values," *American Scientist*, vol. 79, pp. 402–403, 1991.

[13] Y. Dallery and X.-R. Cao, "Operational analysis of stochastic closed queueing networks," *Performance Evaluation*, vol. 14, no. 1, pp. 43–61, 1992.

[14] M. El-Taha and S. Stidham Jr., "Deterministic analysis of queueing systems with heterogeneous servers," *Theoretical Computer Science*, vol. 106, no. 2, pp. 243–264, 1992.

[15] J. P. Buzen, "An overview of performance prediction in MVS systems and SNA networks," in *Proceedings of the ACM Fall Joint Computer Conference*, pp. 751–759, Dallas, Tex, USA, 1986.

[16] D. A. Menasce and V. A. Almeida, *Scaling for E-Business: Technologies, Models, Performance, and Capacity Planning*, Prentice-Hall, Englewood Cliffs, NJ, USA, 2000.

[17] D. K. Buch and V. M. Pentkovski, "Performance characterization experience of multi-tier E-business systems using queuing operational analysis," in *Proceedings of the IEEE International Symposium on Performance Analysis of Systems and Software*, Tucson, Ariz, USA, 2001.

[18] M. P. Fanti, B. Maione, G. Piscitelli, and B. Turchiano, "Two method for real-time routing selection in flexible manufacturing systems," in *Proceedings of the IEEE International Conference on Robotics and Automation*, pp. 1158–1166, May 1992.

[19] W. M. Zuberek, "Throughput analysis in timed Petri nets," in *Proceedings of the 35th Midwest Symposium on Circuits and Systems*, pp. 1576–1580, Washington, DC, USA, 1992.

[20] W. M. Zuberek, "Throughput analysis of simple closed timed Petri net models," in *Proceedings of the 36th Midwest Symposium on Circuits and Systems*, pp. 930–933, Detroit, Mich, USA, August 1993.

[21] G. Chiola, C. Anglano, J. Campos, J. M. Colom, and M. Silva, "Operational analysis of timed Petri nets and application to the

computation of performance bounds," in *Quantitative Methods in Parallel Systems*, F. Baccelli, A. Jean-Marie, and I. Mitrani, Eds., Esprit Basic Research Series, pp. 161–174, Springer, Berlin, Germany, 1995.

[22] J. A. Brumfield, *Operational Analysis of Queuing Phenomena [Ph.D. thesis]*, Department of Computer Science, Purdue University, 1982.

[23] N. M. Bengtson, "Using operational analysis in simulation: a queuing network example," *Journal of the Operational Research Society*, vol. 39, no. 12, pp. 1125–1136, 1988.

[24] N. M. Bengtson, "Measuring errors in operational analysis assumptions," *IEEE Transactions on Software Engineering*, vol. 13, no. 7, pp. 767–776, 1987.

[25] N. M. Bengtson, "Operational analysis revisited: error measure limits of assumptions," *GSTF International Journal on Computing*, vol. 3, no. 2, 2013.

[26] K. C. Sevcik and M. Klawe, "Operational analysis versus stochastic modeling of computer systems," in *Proceedings of the Computer Science and Statistics: 12th Annual Symposium on the Interface*, pp. 177–184, Waterloo, Canada, 1979.

[27] F. S. Hillier and G. J. Lieberman, *Introduction to Operations Research*, McGraw-Hill, New York, NY, USA, 8th edition, 2005.

[28] C. Eisenhart, M. W. Hastay, and W. A. Wallis, Eds., *Selected Techniques of Statistical Analysis*, McGraw-Hill, New York, NY, USA, 1947.

16

Memetic Algorithm with Local Search as Modified Swine Influenza Model-Based Optimization and Its Use in ECG Filtering

Devidas G. Jadhav,[1] Shyam S. Pattnaik,[1] and Sanjoy Das[2]

[1] *National Institute of Technical Teachers' Training & Research (NITTTR), Chandigarh 160019, India*
[2] *Kansas State University, Manhattan, KS 66506, USA*

Correspondence should be addressed to Shyam S. Pattnaik; shyampattnaik@yahoo.com

Academic Editor: Zne-Jung Lee

The Swine Influenza Model Based Optimization (SIMBO) family is a newly introduced speedy optimization technique having the adaptive features in its mechanism. In this paper, the authors modified the SIMBO to make the algorithm further quicker. As the SIMBO family is faster, it is a better option for searching the basin. Thus, it is utilized in local searches in developing the proposed memetic algorithms (MAs). The MA has a faster speed compared to SIMBO with the balance in exploration and exploitation. So, MAs have small tradeoffs in convergence velocity for comprehensively optimizing the numerical standard benchmark test bed having functions with different properties. The utilization of SIMBO in the local searching is inherently the exploitation of better characteristics of the algorithms employed for the hybridization. The developed MA is applied to eliminate the power line interference (PLI) from the biomedical signal ECG with the use of adaptive filter whose weights are optimized by the MA. The inference signal required for adaptive filter is obtained using the selective reconstruction of ECG from the intrinsic mode functions (IMFs) of empirical mode decomposition (EMD).

1. Introduction

Genetic algorithm, particle swarm optimization, bacterial foraging optimization, differential evolution, evolutionary programming, and so forth are the stochastic optimizers that have drawn the attentions in recent time [1–7]. In these, a population of the solutions is utilized in the search process. These algorithms are capable of exploring and exploiting the promising regions in the search space but take relatively longer time [1]. Hence, algorithms are combined for required properties like faster exploration and exploitation capabilities making the combination faster and accurate [8–11].

Since 1960, genetic algorithm (GA) is a potential optimizer in the field of optimization [2]. In GA, chromosomes are used to encode problem parameters and to search for solution with iterations or generations in a natural selection process using genetic operations such as selection, crossover, and mutation. GA offers a robust search mechanism in complex spaces applicable to problems in various fields like science, engineering, commerce, and so forth.

The hybrid algorithm tries to keep a balance in processes of exploration and exploitation achieving global optimization. Hybrid methods like least-square-fuzzy BFO [8], GA-BFO [9], and BFO-Nelder-Mead [10] are developed for taking advantage of such combinations. In developing some of the memetic algorithms, GA is used, in which gradient-based information is utilized for local search. But gradient-based methods fail in the cases of multimodal and nondifferentiable functions [12]. Population-based search algorithms have advantages over the gradient-type searches as the former is capable of overcoming the local optima issue. But some of the population-based search algorithms like particle swarm optimization (PSO) sometimes have tendency of premature convergence. So, to overcome that, stochastic local search is utilized to take out the solution from local trapping [13]. This property of heuristic search methods having the capability to overcome premature convergence makes them suitable for local refinement in the memetic algorithm [14]. One such combination is known as memetic algorithm (MA). It is inspired by both Darwinian's principle of natural evolution and Dawkins' notion of a meme as a unit of information or idea transmission from brain to brain or cultural evolution capable of individual learning [1, 15].

This paper proposal is in two ways, that is, modification of the newly introduced algorithms, Swine Influenza Model-Based Optimization (SIMBO) [16], and development of the new variants of memetic algorithm (MA) having genetic algorithm (GA) as the main algorithm with modified SIMBOs as local search mechanisms.

The paper is organized in seven folds. Section 2 briefs the SIMBO. In Section 3, modifications of SIMBOs are discussed, and in Section 4, development of memetic algorithms with the use of modified SIMBOs as local searches is discussed. The experimental results of the proposed modified SIMBO and the proposed MAs on benchmark functions are presented in Section 5. The application of ECG signal processing is discussed in Section 6, and Section 7 is the conclusion.

2. Swine Influenza Model-Based Optimization (SIMBO)

The Swine Influenza Model-Based Optimization (SIMBO) developed by Pattnaik et al. [16] introduces a novel optimization scheme. The SIMBO is imitated the concept in the models of swine flu that is, Susceptible-Infectious-Recovered (SIR) [17]. This is similar to the development of algorithm family based on natural phenomena in physics [18]. The initial population is divided into three classes: susceptible (S), infectious (I), and recovered (R) [17]. The population with single infectious individual goes susceptible (S). The infectious individual (I) transmits influenza towards all susceptible individuals (S). Afterwards, researchers modified the SIR model by incorporating seasonality, vaccination, treatment, quarantine, isolation, and so forth [17]. The susceptible individuals (S) can be vaccinated, and infected individuals (I) can be treated with antiviral drugs (T) [17]. SIR model was further extended by adding two behavioural modifications: quarantine and isolation measures (Q) [17]. The expansion of SIMBO is with the help of treatment (SIMBO-T), vaccination (SIMBO-V), and quarantine (SIMBO-Q) depending on the probabilities. These SIMBO variants are employed in optimization of the multimodal functions and this shows the enhancement in convergence as well as accuracy. In the process, first, swine flu test depending on the dynamic threshold is recognizing confirmed cases of swine flu. Then the susceptible cases in the community population are recommended the vaccination for gaining immunity against swine flu. The confirmed cases of swine flu in the population are quarantined. Also, the suspected cases are administered the antiviral drugs and are treated. The antiviral drugs quantity administered to the individual is based on the current health of the individual. The states of the individuals are updated with direct vaccination/quarantine and indirect treatment in SIMBO-V and SIMBO-Q. Details on SIMBO can be obtained from [16].

2.1. Comparison with Artificial Immune System. Artificial immune system (AIS) is inspired by the immunology of the body's immune system that categorizes all cells (or molecules) within the body as self-cells or nonself-cells [19, 20]. In implementing a preliminary artificial immune system, four steps are present, namely, encoding, similarity measure, selection, and mutation [19, 20]. There are a variety of the AIS techniques developed like negative selection algorithms, artificial immune networks, clonal selection algorithms, danger theory, and dendritic cell algorithms [19, 20]. These are based on harmful entity suppression mechanism inside the body, whereas the SIMBO is based on the mechanism that suppresses the viral spread in the human populations by adopting some techniques like treatment, vaccination, and quarantine. The biological immune system is highly distributed, highly adaptive, self-organising in features, preserving a memory of earlier period came across, and has the ability of continuous learning about new experiences [19, 20] that is tried to adopt in AIS. In the SIMBO, also the adaptability and memory in terms of best health information are present [16]. The AIS is made for optimization and learning, whereas SIMBO is purely an optimization technique [16, 19, 20].

3. Modifications in Swine Influenza Model-Based Optimization (SIMBO)

The modifications in the SIMBO are incorporated in the proposed work as (i) constraining the individual range in the process of optimization, and (ii) new process of state change, having population diversity in the vicinity of the current individual, in SIMBO-V and SIMBO-Q.

3.1. Modified Swine Influenza Model-Based Optimization with Treatment (Modified SIMBO-T). The key terms and definitions used in SIMBO have been used in the modified SIMBO. These are listed below.

Key Terms and Definitions

Day (D): current generation or iteration.

TD: total number of days or generations.

State (S): individual position.

Health (H): fitness.

Pandemic state (PS): pandemic (global) best state in all the individuals.

Pandemic health (PH): fitness value corresponding to pandemic state.

Primary symptoms of swine flu: swine flu symptoms are fever, runny nose, cough, sore throat, headache, body aches, fatigue, and chills.

Swine flu test: the laboratory test-based verification of an infected individual with swine flu.

Table 1: List of the parameters in SIMBO used in mSIMBO.

Parameters of SIMBO used in mSIMBO	Value/range
Amount of vaccine (Vc): the vaccine dose supplied to the individual	Randomly generated in the range $[-0.5, 1.0)$
Quarantine factor (Vq): the state change factor to change state from best individual to new one	Randomly generated in the range $[-0.5, 1.0)$
μ: Vaccination probability	$\mu = 0.8$
α: Recovery probability	$\alpha = 0.2$
β: Quarantine probability	$\beta = 0.8$
Momentum factor of dose (Md): it is for controlling the individual dose	Randomly generated in the range $[0.0, 2.0)$
Momentum factor of state (MS): it is for controlling the individual state	MS = 0.2
Fe: fever, Co: cough, fathead: fatigue and headache, NV: nausea and Vomiting, Dai: diarrhea All are combined and are called primary symptoms, and the product of combining all of them is taken as a random number	(Fe × Co × fathead × NV × Dai) = random number in the range [0,1)

Vaccination (V): it vaccinates against the swine flu to the susceptible/unexposed population.

Amount of vaccine (Vc): the vaccine dose supplied to the individual.

Dose: antiviral drugs supplied to individual for curing the swine flu.

μ: vaccination probability.

α: recovery probability.

β: quarantine probability.

Momentum factor of dose (Md): it is for controlling the individual dose.

Momentum factor of state (MS): it is for controlling the individual state.

The list of the parameters in SIMBO used in mSIMBO is shown in Table 1 with their values and ranges. The same tuning of the parameters is used in cases of all problems and parameters are not tuned differently for every problem.

The modified SIMBO-T, which is named as mSIMBO-T, also does optimization by treatment mechanism using probability of treatment similar to that of basic SIMBO-T. Initially, all individuals are susceptible due to the infected individual and the treatment is given to all susceptible cases by antiviral drug dose. The amount of dose is dependent on primary and secondary symptoms as well as current health and pandemic health.

The mSIMBO-T is also doing optimization through two steps as given below.

Steps of mSIMBO-T

Step 1: evaluate health.

Step 2: treatment within limits.

3.1.1. Evaluate Health. The health of the individual depends upon given fitness function. Initially, the health of all individuals is evaluated for checking susceptible patient to swine flu as diagnostic confirmation.

3.1.2. Treatment. Treatment is based on symptoms and is often based on a trial and error. A physician generally begins the treatment with a typical dose and monitors for a response as well as side effects. The standardization of a dose is based on population dose response characteristics and not on individual optimal outcomes [16]. When the dose is accustomed eventually depending on the individual feedback, then optimal result is reached faster.

Similar to the basic SIMBO algorithms (i.e., SIMBO-T, SIMBO-V, and SIMBO-Q), the modified SIMBO algorithms, which are named as mSIMBO-T, mSIMBO-V, and mSIMBO-Q, also have the amount of dose based on primary symptoms, secondary symptoms, current health, and pandemic health, where the primary symptoms and secondary symptoms are shown in (1) and (2).

$$\text{Primary}(\text{Day}) = (\text{Fe} \times \text{Co} \times \text{fathead} \times \text{NV} \times \text{Dai}) \times e^{(-\text{TD}/\text{Day})}, \quad (1)$$

where Fe: fever, Co: cough, fathead: fatigue and headache, NV: Nausea and Vomiting, and Dai: diarrhea. The first term in (1) is total influence of primary symptom and second term increases the primary symptom per day:

$$R0(\text{Day}) = 1 - e^{(-\text{Primary}(\text{Day}))}. \quad (2)$$

The dose administered to the individual is given by (3). After the progress in the health of the individual, the dose is reduced due to the last two terms in (3) as the current health of the individual is closer to the pandemic health. The momentum factors are utilized in controlling the dose and state of individual for current iteration those are found based on their previous iteration status. For applying the treatment, the probability of recovery used is $\alpha = 0.2$. As the part of proposed work, the modification is incorporated as: after calculating the dose it is checked for its value with the help of range and limited it to the feasible range. With this controlled dose, the individual state is changed as indicated in (4):

$$\begin{aligned}\text{Dose}(m+1) = {} & \text{Dose}(m) \times \text{Md} + \text{Primary}(\text{Day}) \\ & \times \text{rand} \times \left(1 - \frac{\text{Current\_health}(m)}{\text{rand} \times \text{PH}}\right) \\ & + R0(\text{Day}) \times \text{rand} \\ & \times (\text{Current\_health}(m) + \text{PH}),\end{aligned} \tag{3}$$

$$S(m+1) = S(m) \times \text{Ms} + \text{Dose}(m+1). \tag{4}$$

Here, one of the modifications is incorporated in the SIMBO family of the algorithms as applying boundary constrain on the dose keeps it within the search range. In the modified SIMBO algorithms, for such constrain, the dose is checked for its value and it is put at one of the extremum of range that is nearer if it is out of range. As the dose is within the range, the individual will also eventually return to the search range. When the feasible range is (range1, range2) then repair is incorporated with application of this feasibility constrain on the dose as shown in (5):

$$\text{Dose}(m+1) = \begin{cases} \max(\text{Dose}(m+1), \text{range1}), \\ \text{if Dose}(m+1) < \text{range2} \\ \min(\text{Dose}(m+1), \text{range2}), \\ \text{if Dose}(m+1) > \text{range1}. \end{cases} \tag{5}$$

3.2. Modified Swine Influenza Model-Based Optimization with Vaccination (mSIMBO-V). It does the optimization by utilizing the vaccination and treatment mechanism. The susceptible individuals go through the swine flu test. The susceptible class is presented for swine flu vaccination and acquiring immunity. The SIMBO-V mechanism of optimization has four steps as shown below and mSIMBO-V also utilizes these.

Steps of mSIMBO-V

Step 1: evaluate health.

Step 2: swine flu test.

Step 3: modified Vaccination.

Step 4: treatment within limits.

The explanation of Steps 2 and 3 is as follows.

3.2.1. Swine Flu Test. Dynamic threshold is used for evaluation of the patient states due to affect of swine flu virus. When current health values of individuals are more than the dynamic threshold (*D_Threshold*), they are susceptible and the remaining other are recovered cases. Dynamic threshold is calculated by utilizing the health information of better half of population, quarantine or vaccination probability (β or μ), and primary symptoms as given in (6):

$$\begin{aligned}D\_\text{Threshold} = {} & \frac{\sum_1^{\text{Sr}} \text{Current\_health}}{\text{Sr}} \times \mu \\ & \times \text{rand} \times \text{Primary}(\text{Day}).\end{aligned} \tag{6}$$

The following is the algorithm showing the application of swine flu test:

for $k = 1 : TI$
if *current_health (k)* > *D_Threshold*
Susceptible case
else
Recovered case
end
end

3.2.2. Vaccination. With the application of a vaccine, the vaccination is carried out for the protection from disease with the help of immunity [21]. In the basic SIMBO-V, the vaccination is done by the amount of vaccine by multiplying the state of individual with random value in the range of 0 to 1 [16].

In mSIMBO-V, this process of the state change of the individual is also modified and multiplicand Vc modified as in (7) that randomly perturbs the original state value in the range from 100% to −50% of original state value to get all directional diversities for changing the health of individual:

$$\text{Vc} = (1.5 \times \text{rand} - 0.5). \tag{7}$$

The probability of vaccination (μ) is selected high (0.8) for having higher exploitation in the neighbourhood of the individual by direct state change like the local refinement.

The following is the algorithm showing the application of modified vaccination:

for $k = 1 : TI$
if *rand* > μ
if *current_health (k)* > *D_Threshold*
if *flag (k)*= =0
$S(k) = S(k) * Vc$
flag $(k) = 1$
end
end
end
end

3.3. Modified Swine Influenza Model-Based Optimization with Quarantine (mSIMBO-Q). It does the optimization with quarantine and treatment mechanism. Out of all susceptible individuals who undergo the swine flu test, the confirmed cases have quarantine from population. The treatment is applied to all individuals by dose amount based on current health. Similar to the basic SIMBO-Q, mSIMBO-Q also does optimization in four steps as given below. Steps 1 and 4 of mSIMBO-Q are the same as those of mSIMBO-T. Step 2 is the same as the above explanation, whereas Step 3 is explained below. Steps 1, 2, and 4 of mSIMBO-Q are similar to those of mSIMBO-V.

Steps of mSIMBO-Q

Step 1: evaluate health.

Step 2: swine flu test.

Step 3: modified Quarantine.

Step 4: treatment within limits.

3.3.1. Quarantine. Quarantine is the separation and isolation of individuals or restriction forced on unbound reposition to prevent contagious disease spread. The confirmed cases in swine flu tests are isolated or quarantined. The original algorithm quarantines the individual by multiplying the best state of the individual with Vq that is a random number in the range from 0 to 1.

The modification introduced in calculating the value of Vq is used for quarantining the individuals. This is able to generate new individual in the vicinity to the original value of the individual by perturbation ranging from 100% to −50%. This introduces the diversity in all directions in the population and also keeps the individual in the vicinity to the original cluster of the population. This is achieved with the help of (8) shown below:

$$\mathrm{Vq} = (1.5 \times \mathrm{rand} - 0.5). \quad (8)$$

The best state is chosen because quarantine generates the individual with probably having immunity and best health for faster convergence. If β is greater than rand and current health of individual is greater than *D_Threshold*, then individual is quarantined. The probability of quarantine (β) is kept high (0.8) for having more exploitation in the neighbourhood of the best individual having local refinement.

The following is the algorithm that shows the application of modified quarantine:

for $k = 1 : TI$

if $rand > \beta$

if *current_health (k)* > *D_Threshold*

$S(k) = PS * Vq$

end

end

end

4. Memetic Algorithm Using Local Search as Modified SIMBO Algorithms

Hybridization of genetic algorithm (GA) with local search algorithm is memetic algorithm (MA) [1, 2, 14, 15, 22, 23]. In general, MA is a synergy of evolution and individual learning [1, 22], which improves the capability of evolutionary algorithms like GA to find optimal solutions accurately in function optimization problems with higher speed of convergence. Genetic algorithm (GA) is nature-inspired optimization technique that represents an intelligent exploration having a random search confined within a defined search space for solving a problem optimally with the help of population [24]. Elitism helps prevent the loss of good solutions once they are found. This also can speed up the performance of the GA significantly [25, 26]. Elitism is used to keep the best solution intact without changing it due to the evolution in the evolutionary algorithms like genetic algorithm [24–26]. These best solutions are used for the reproduction, which helps the algorithm to converge fast with better accuracy.

The unique aspect of the MA is that the chromosomes and offspring are facilitated to gain some experience with a local search process in between regular evolutionary process [15, 27]. The local search process is inspired by individual learning using meme, which is a unit of information or idea that gets transmitted. Similar to the GA, the MA also creates an initial population randomly and searches the fitness landscape. The local search process drags solutions in the direction of local optima. These improvements go on accumulating over all generations, resulting in a larger improvement in the total performance [27]. The meta-heuristic search mechanism in the memetic algorithms offers it with the speed and quality of convergence [28].

Evolutionary algorithms follow heuristic search procedures that incorporate random selection and variation. Therefore, Swine Influenza Model-based Optimization Algorithms (SIMBOs) are modified and used as local searches in the main algorithm GA to make MAs. In this work, the variants of MAs are named like memetic algorithm with BLX-α and SBX crossover operators, having mSIMBO-T for local search, as MA-BLX-T and MA-SBX-T, respectively. Similarly, MAs with mSIMBO-V as local search are named as MA-BLX-V and MA-SBX-V; also with mSIMBO-Q as local search they are named as MA-BLX-Q and MA-SBX-Q. The modified SIMBOs are population-based heuristic algorithms derived from SIR model of swine influenza and are utilized for local search in development of the new variants of memetic algorithm that converges in cases of unimodal, multimodal, shifted, rotated, and hybrid composite functions.

The GA is used for the purpose of exploration and the modified SIMBOs as local search for exploitation in the meta-heuristic combination of the proposed MAs. So, this gives a better converging capability to the algorithm for global convergence with better accuracy in convergence besides speed along with higher success rate. The inherent properties of basic SIMBO family are present in mSIMBO-T, mSIMBO-V, and mSIMBO-Q and they are having the dynamic adaptation for every individual with the help of learning from its neighbours facilitating the mSIMBO-T, mSIMBO-V, and mSIMBO-Q in dealing with complex, multimodal search landscapes efficiently. When these mechanisms are used in searching the search landscapes under limited exposure of local areas of landscapes with local boundary constraints in memetic framework, these give the capability of finding solutions in the complex local areas also. The exploration-exploitation balancing adaptation in the evolutionary algorithm like modified SIMBO family inherently has its local refining characteristics and again it is utilized in refining the smaller area of landscape in MA framework making it sort of meta-meme in nature. This is observed in the memetic framework under consideration.

The GAs and the MAs based on real number representation are called real-coded genetic algorithms (RCGAs) and real-coded memetic algorithms (RCMAs), respectively [2, 23]. The population is real coded for both evolving algorithm and local search algorithm. The evolving algorithm has the population that is obtained by generation process. The generation process generates the uniformly distributed random population in the specified range in [7] for particular function shown in Table 2. The fitness of the population is determined and current best individual is opted for the elite preservation. Then this population goes under the GA operations such as crossover and mutation.

The crossovers used are BLX-α crossover [30] and SBX crossover [31]. The offsprings $O^1 = \{o_1^1, o_2^1, \ldots, o_i^1, \ldots, o_d^1\}$ and $O^2 = \{o_1^2, o_2^2, \ldots, o_i^2, \ldots, o_d^2\}$ are generated from parents $P^1 = \{p_1^1, p_2^1, \ldots, p_i^1, \ldots, p_d^1\}$ and $P^2 = \{p_1^2, p_2^2, \ldots, p_i^2, \ldots, p_d^2\}$ with "d" dimension. In BLX-α crossover, O is chosen randomly between the interval $[(P^1 - I \cdot \alpha), (P^2 + I \cdot \alpha)]$ with the condition $P^1 < P^2$ and $I = \max(P^1, P^2) - \min(P^1, P^2)$. In SBX crossover, the effect of the one-point crossover of the binary representation is tried to emulate. The crossover generates two offspring $O^1 = 1/2[(1 + B)P^1 + (1 - B)P^2]$ and $O^2 = 1/2[(1 - B)P^1 + (1 + B)P^2]$, where $B \geq 0$ is a sample from random generator having density function as shown in (9):

$$p(B) = \begin{cases} \dfrac{1}{2}(\eta + 1) B^{\eta}, & \text{if } 0 \leq B \leq 1, \\ \dfrac{1}{2}(\eta + 1) \dfrac{1}{B^{\eta+2}}, & \text{if } B > 1. \end{cases} \qquad (9)$$

This distribution is obtained by transforming with uniform random number source $u(0, 1)$ in (10) which is expressed as

$$B(u) = \begin{cases} (2u)^{1/(\eta+1)}, & \text{if } u \leq \dfrac{1}{2}, \\ \left(\dfrac{1}{2(1 - u)}\right)^{1/(\eta+1)}, & \text{if } u > \dfrac{1}{2}, \end{cases} \qquad (10)$$

where $\eta = 2$ is taken for better results. Exploitation capacity of operator increases for higher values of η. The mutation mutates the individual in the specified range of the variables to change it randomly under uniform normal distribution. Elitism speeds up the performance of the GA largely, which also prevents the loss of good solutions once they are found [25, 26]. The elitism is also used in the MA for enhancing its convergence process.

The local search algorithm is evoked after a decided number of iterations or generations of evolving algorithm (known as Glocal) which is selected as 2 for better result. In the MA, local searching is of the search space around best candidate solution obtained due to the previous process of exploration. The smaller population for the refinement is generated around the current best individual by little perturbations.

Then the individuals undergo treatment, quarantine, and vaccination processes according to the threshold, state of the individual, quarantine and vaccination probabilities, and so forth. The calculated dose used for treatment depends on primary and secondary symptoms, current health and best health of individual, and previous dose for the same individual. Then the controlled dose is applied to the individual for recovery. The best individual is found out in all iterations during the refinement process. After the local refinement, the better result due to the refinement process is taken back in the evolving algorithmic process. It strengthens the algorithm for convergence of shifted and/or rotated unimodal/multimodal single/composite well-known standard benchmark functions [7] and the functions in the CEC 2005 [29], which are listed in Table 2. The performance of the algorithms is checked with 38 standard benchmark functions shown in Table 2. The

TABLE 2: Standard numerical benchmark functions (basic and CEC2005) used for experimental study [7, 29].

Function	Names
	Basic standard benchmark functions
Fsph	Sphere model (NFFE = 150000)
Fs2.22	Schwefel's problem 2.22 (NFFE = 200000)
Fs1.2	Schwefel's problem 1.2 (NFFE = 500000)
Fs2.21	Schwefel's problem 2.21 (NFFE = 500000)
Fros	Generalized Rosenbrock's function (NFFE = 500000)
Fste	Step function (NFFE = 150000)
Fqua	Quartic function (i.e., noise) (NFFE = 300000)
Fs2.26	Generalized Schwefel's problem 2.26 (NFFE = 300000)
Fras	Generalized Rastrigin problem (NFFE = 300000)
Fack	Ackley's function (NFFE = 150000)
Fgri	Generalized Griewanks function (NFFE = 300000)
Fpen1	Generalized penalized function no. 1 (NFFE = 150000)
Fpen2	Generalized penalized function no. 2 (NFFE = 150000)
	CEC 2005 Standard benchmark function
F1	Shifted sphere function
F2	Shifted Schwefel's problem 1.2
F3	Shifted rotated high conditioned elliptic function
F4	Shifted Schwefel's problem 1.2 with noise in fitness
F5	Schwefel's problem 2.6 with global optimum on bounds
F6	Shifted Rosenbrock's function
F7	Shifted rotated Griewank's function without bounds
F8	Shifted rotated Ackley's function with global optimum on bounds
F9	Shifted Rastrigin's function
F10	Shifted rotated Rastrigin's function
F11	Shifted rotated weierstrass function
F12	Schwefel's problem 2.13
F13	Expanded extended Griewank's plus Rosenbrock's function (F8F2)
F14	Shifted rotated expanded Scaffer's F6
F15	Hybrid composition function
F16	Rotated hybrid composition function
F17	Rotated hybrid composition function with noise in fitness
F18	Rotated hybrid composition function
F19	Rotated hybrid composition function with a narrow basin for the global optimum
F20	Rotated hybrid composition function with the global optimum on the bounds
F21	Rotated hybrid composition function
F22	Rotated hybrid composition function with high condition number matrix
F23	Noncontinuous rotated hybrid composition function
F24	Rotated hybrid composition function
F25	Rotated hybrid composition function without bounds

proposed algorithms are proved on the standard benchmark functions used and the performance is noted by determining accuracy with the help of error and convergence speed by the number of fitness function evaluations (NFFEs).

Pseudo-Code of the Proposed MA

Initialization: Generate a random initial population

Number of generation: $G = 0$, Glocal $= 0$

while Stopping conditions are not satisfied do

Evaluate all individuals in the population

Find current best individual

If Glocal = 2 for best individual do (*refinement by modified SIMBO*)

Obtain the local population around the current best individual by small perturbation

While stopping conditions are not satisfied do

Evaluate and sort the population

Classify the individuals in classes S, I and R

Update current_best_individual, current_best_health

Calculate dynamic threshold (DT)

Classify and tag individuals into S, I, R classes

Quarantine or vaccination S & I individuals based probability (β or μ), DT, health

Calculate Dose for applying to individual

Provide treatment to S based on probability α, health and PH

end while

Glocal = 0

Return best individual to evolving algorithm

end if

Select individuals for crossover based on crossover probability P_c

Crossover the parents by BLX-α crossover or SBX crossover operators

Correct the feasibility of the produced individuals

Mutate some of the descendant population based on mutation probability P_m

Replace the old population by new preserving the elite

G++, Glocal++

end while

where G is generations, Glocal is counter for application of local refinement, current_best_individual is current best individual from local population, current_best_health is current best health of the individual which is among the local population, P_c is crossover probability, and P_m is mutation probability.

5. Experimentation

The experimental results on benchmark functions using modified SIMBO variants and modified SIMBO-based memetic algorithms are presented here in this section.

5.1. Parameter Setting and Experimentation. In the genetic as well as in memetic algorithms, a population of 100 individuals of real-valued representation is used. The crossover operator used is the BLX-α crossover or SBX crossover with crossover probability set to 0.8. For BLX-α crossover, $\alpha = 0.3$ and for SBX crossover, $\eta = 2$ are used. The mutation probability is set to 0.05. The algorithm uses generational replacement of individuals preserving one elite. Two stopping criteria are used either the number of fitness function evaluations (NFFE as in Table 2 for basic functions; and 20000 or 100000 for 02 or 10 dimensions, resp., in case of CEC 2005 functions) or an error value of 10^{-8}. In case of basic SIMBO family, modified SIMBO family for its individual performance validity and for combining modified SIMBO variants as local search with GA for making MA, all similar parameters are used, like real-coded population size 100, the treatment probability of recovery $\alpha = 0.2$, probability for quarantine $\beta = 0.8$, and probability for vaccination $\mu = 0.8$. The ensemble strategy to tune parameters and operators can be useful to develop further the more adaptive techniques that collectively may take care of parameters [32]. All algorithms with various dimensions, namely, 02, 10, 30, and 50 are executed for 25 independent trials and the best errors in trails (runs) are used for averaging and calculating standard deviation (Std Dev) presented in Tables 3, 4, and 5 in cases of basic standard benchmark functions and in Tables 7 and 8 in cases of CEC 2005 functions. Also, the algorithms are compared with statistical testing (Student's t-test) for statistical validity of the obtained results in Table 6 in cases of basic standard benchmark functions and in Tables 9 and 10 in cases of CEC 2005 functions [33].

5.2. Discussion. The algorithms are tested using two sets of standard benchmark functions, namely, (1) well-known basic benchmark functions [7] and (2) benchmark problems in CEC 2005 [29].

5.2.1. Basic Standard Benchmark Functions [7]. The algorithms are tested on well-known basic standard benchmark functions having unimodal, multimodal properties with noise and discontinuity. Functions Fsph (sphere model), Fs2.22 (Schwefel's problem 2.22), Fs1.2 (Schwefel's problem 1.2), Fs2.21 (Schwefel's problem 2.21), and Fros (generalized Rosenbrock's function) are functions having unimodal property. Function Fste (step function) is step function, which has one minimum and which is discontinuous function. Function Fqua (quartic function) is a noisy quartic function comprising of random [0, 1) which is a uniformly distributed random variable in [0, 1). Functions Fs2.26 (generalized Schwefel's problem 2.26), Fras (generalized Rastrigin problem), Fack (Ackley's function), Fgri (generalized Griewank function), Fpenl (generalized penalized function no. 1), and

TABLE 3: Comparison of the experimental results of algorithms used for dimension 10.

10D	F	Fsph	Fs2.22	Fsl.2	Fs2.21	Fros	Fste	Fqua	Fs2.26	Fras	Fack	Fgri	Fpenl	Fpen2
GA-BLX	Mean	1.67e − 03	7.96e − 04	4.09e + 00	1.07e − 01	2.87e + 00	0.00e + 00	2.07e − 02	−1.30e − 05	8.61e − 05	6.32e − 03	6.53e − 02	2.96e − 05	1.04e − 04
	STD	1.54e − 03	5.77e − 04	2.10e + 00	2.34e − 02	2.55e + 00	0.00e + 00	1.44e − 02	6.25e − 05	1.04e − 04	3.98e − 03	2.20e − 02	8.24e − 05	1.80e − 04
GA-SBX	Mean	1.35e − 03	7.03e − 04	2.07e − 02	3.59e − 02	4.59e + 00	0.00e + 00	1.83e − 02	−1.00e − 10	1.22e − 04	3.37e − 01	6.61e − 02	3.76e − 06	6.19e − 05
	STD	1.56e − 03	5.92e − 04	2.08e − 02	1.20e − 02	4.40e + 00	0.00e + 00	1.51e − 02	5.03e − 09	1.87e − 04	1.18e + 00	2.15e − 02	6.28e − 06	9.32e − 05
SIMBO-T	Mean	2.77e − 09	4.02e − 09	2.66e − 09	4.50e − 09	9.00e + 00	0.00e + 00	8.54e − 12	4.18e + 03	2.19e − 09	4.22e − 09	3.27e − 09	8.83e − 01	1.00e + 00
	STD	2.88e − 09	2.75e − 09	3.08e − 09	2.88e − 09	0.00e + 00	0.00e + 00	4.26e − 11	9.28e − 13	2.58e − 09	3.14e − 09	3.18e − 09	3.40e − 16	0.00e + 00
SIMBO-V	Mean	2.33e − 09	3.97e − 09	3.54e − 09	3.78e − 09	9.00e + 00	0.00e + 00	2.18e − 13	4.18e + 03	2.96e − 09	3.96e − 09	2.72e − 09	8.83e − 01	1.00e + 00
	STD	2.52e − 09	3.12e − 09	3.25e − 09	2.75e − 09	0.00e + 00	0.00e + 00	1.09e − 12	9.28e − 13	3.06e − 09	2.59e − 09	3.21e − 09	3.40e − 16	0.00e + 00
SIMBO-Q	Mean	2.21e − 09	3.21e − 09	9.72e − 10	2.75e − 09	9.00e + 00	0.00e + 00	6.09e − 20	4.18e + 03	1.95e − 09	2.48e − 09	9.17e − 10	8.83e − 01	1.00e + 00
	STD	2.54e − 09	2.86e − 09	1.35e − 09	2.95e − 09	0.00e + 00	0.00e + 00	2.90e − 19	9.28e − 13	2.98e − 09	2.67e − 09	2.07e − 09	3.40e − 16	0.00e + 00
mSIMBO-T	Mean	2.85e − 09	3.41e − 09	2.70e − 09	4.23e − 09	9.00e + 00	0.00e + 00	8.30e − 11	4.18e + 03	1.79e − 09	4.97e − 09	2.08e − 09	8.83e − 01	1.00e + 00
	STD	3.06e − 09	2.34e − 09	2.82e − 09	3.13e − 09	0.00e + 00	0.00e + 00	4.15e − 10	9.28e − 13	2.24e − 09	2.67e − 09	2.53e − 09	3.40e − 16	0.00e + 00
mSIMBO-V	Mean	2.82e − 09	3.88e − 09	3.50e − 09	4.86e − 09	9.00e + 00	0.00e + 00	5.02e − 26	4.18e + 03	2.99e − 09	3.40e − 09	3.20e − 09	8.83e − 01	1.00e + 00
	STD	2.91e − 09	2.53e − 09	2.61e − 09	3.19e − 09	0.00e + 00	0.00e + 00	2.51e − 25	9.28e − 13	3.18e − 09	2.61e − 09	2.80e − 09	3.40e − 16	0.00e + 00
mSIMBO-Q	Mean	1.29e − 09	2.22e − 09	8.08e − 10	1.74e − 09	9.00e + 00	0.00e + 00	4.09e − 11	4.18e + 03	3.92e − 10	1.84e − 09	8.51e − 10	8.83e − 01	1.00e + 00
	STD	2.45e − 09	2.59e − 09	1.27e − 09	2.51e − 09	0.00e + 00	0.00e + 00	1.29e − 10	9.28e − 13	6.26e − 10	2.23e − 09	1.42e − 09	3.40e − 16	0.00e + 00
MA-BLX-T	Mean	2.80e − 09	4.73e − 09	3.32e − 09	5.58e − 09	0.00e + 00	0.00e + 00	1.42e − 02	1.80e + 02	3.38e − 09	3.87e − 09	3.08e − 09	3.34e − 06	1.21e − 04
	STD	2.98e − 09	3.24e − 09	3.54e − 09	2.90e − 09	0.00e + 00	0.00e + 00	1.59e − 02	2.58e + 02	2.90e − 09	2.68e − 09	2.90e − 09	5.15e − 06	2.58e − 04
MA-BLX-V	Mean	3.60e − 09	4.96e − 09	3.78e − 09	5.16e − 09	0.00e + 00	0.00e + 00	1.34e − 02	1.47e + 02	3.30e − 09	4.72e − 09	3.62e − 09	7.47e − 06	6.81e − 05
	STD	3.59e − 09	2.97e − 09	3.22e − 09	2.72e − 09	0.00e + 00	0.00e + 00	1.76e − 02	2.03e + 02	3.25e − 09	2.71e − 09	3.39e − 09	1.75e − 05	7.78e − 05
MA-BLX-Q	Mean	2.33e − 09	4.89e − 09	3.04e − 09	4.74e − 09	0.00e + 00	0.00e + 00	1.53e − 02	2.27e + 02	3.28e − 09	4.84e − 09	3.46e − 09	1.44e − 05	6.13e − 05
	STD	2.80e − 09	2.87e − 09	2.43e − 09	2.67e − 09	0.00e + 00	0.00e + 00	1.80e − 02	2.53e + 02	2.91e − 09	2.78e − 09	3.09e − 09	2.85e − 05	7.76e − 05
MA-SBX-T	Mean	2.86e − 09	3.52e − 09	1.80e − 09	3.20e − 09	0.00e + 00	0.00e + 00	2.21e − 02	1.00e − 09	2.31e − 09	3.81e − 09	2.51e − 09	3.73e − 06	9.57e − 04
	STD	3.06e − 09	2.82e − 09	2.35e − 09	3.25e − 09	0.00e + 00	0.00e + 00	1.58e − 02	4.93e − 09	2.64e − 09	2.94e − 09	2.91e − 09	5.36e − 06	3.02e − 03
MA-SBX-V	Mean	1.94e − 09	4.10e − 09	1.59e − 09	3.29e − 09	0.00e + 00	0.00e + 00	2.69e − 02	1.27e − 09	3.13e − 09	4.17e − 09	3.52e − 09	2.60e − 06	1.02e − 03
	STD	2.41e − 09	3.19e − 09	2.59e − 09	2.32e − 09	0.00e + 00	0.00e + 00	2.10e − 02	5.03e − 09	3.27e − 09	2.88e − 09	3.07e − 09	2.77e − 06	3.03e − 03
MA-SBX-Q	Mean	3.92e − 09	3.94e − 09	2.09e − 09	3.79e − 09	0.00e + 00	0.00e + 00	2.59e − 02	3.50e − 10	2.96e − 09	4.19e − 09	1.97e − 09	1.10e − 05	5.47e − 04
	STD	3.19e − 09	1.89e − 09	2.22e − 09	3.19e − 09	0.00e + 00	0.00e + 00	2.10e − 02	5.07e − 09	2.90e − 09	2.77e − 09	2.39e − 09	2.56e − 05	2.17e − 03

Table 4: Comparison of the experimental results of algorithms used for dimension 30.

30D	F	Fsph	Fs2.22	Fs1.2	Fs2.21	Fros	Fste	Fqua	Fs2.26	Fras	Fack	Fgri	Fpenl	Fpen2
GA-BLX	Mean	4.06e − 02	9.10e − 02	1.29e + 02	3.97e − 01	4.72e + 01	0.00e + 00	6.95e − 02	−1.60e − 04	2.94e + 00	5.47e + 00	2.54e − 02	3.24e − 04	2.96e − 03
	STD	1.72e − 02	1.11e − 01	3.89e + 01	5.91e − 02	3.56e + 01	0.00e + 00	4.29e − 02	7.78e − 04	3.42e + 00	6.03e − 01	1.68e − 02	4.74e − 04	3.70e − 03
GA-SBX	Mean	5.98e − 02	1.64e − 02	3.12e + 01	4.52e − 01	5.78e + 01	0.00e + 00	1.29e − 01	1.81e − 09	8.86e − 01	1.05e + 01	1.96e − 02	2.31e − 04	3.38e − 03
	STD	2.99e − 02	5.16e − 03	1.32e + 01	6.06e − 02	3.40e + 01	0.00e + 00	5.48e − 02	6.21e − 09	2.67e + 00	1.68e + 00	2.14e − 02	4.53e − 04	4.17e − 03
SIMBO-T	Mean	1.82e − 09	4.71e − 09	2.05e − 09	3.24e − 09	2.90e + 01	0.00e + 00	4.29e − 14	1.25e + 04	2.72e − 09	4.42e − 09	2.89e − 09	1.67e + 00	3.00e + 00
	STD	2.02e − 09	3.2e − 09	2.48e − 09	3.00e − 09	0.00e + 00	0.00e + 00	2.14e − 13	0.00e + 00	2.56e − 09	2.73e − 09	2.90e − 09	2.27e − 16	9.06e − 16
SIMBO-V	Mean	2.13e − 09	4.81e − 09	2.07e − 09	4.00e − 09	2.90e + 01	0.00e + 00	9.62e − 20	1.26e + 04	3.18e − 09	5.01e − 09	1.92e − 09	1.67e + 00	3.00e + 00
	STD	2.38e − 09	3.05e − 09	2.65e − 09	2.84e − 09	0.00e + 00	0.00e + 00	3.58e − 19	0.00e + 00	3.33e − 09	3.42e − 09	2.92e − 09	2.27e − 16	9.06e − 16
SIMBO-Q	Mean	1.29e − 09	2.14e − 09	1.57e − 09	2.09e − 09	2.90e + 01	0.00e + 00	6.36e − 27	1.26e + 04	1.51e − 09	1.63e − 09	9.21e − 10	1.67e + 00	3.00e + 00
	STD	1.73e − 09	2.58e − 09	2.64e − 09	2.01e − 09	0.00e + 00	0.00e + 00	3.18e − 26	0.00e + 00	2.70e − 09	1.55e − 09	1.47e − 09	2.27e − 16	9.06e − 16
mSIMBO-T	Mean	3.10e − 09	4.96e − 09	1.30e − 09	4.65e − 09	2.90e + 01	0.00e + 00	1.73e − 12	1.25e + 04	1.98e − 09	4.52e − 09	2.00e − 09	1.66e + 00	3.00e + 00
	STD	3.14e − 09	2.51e − 09	1.52e − 09	3.09e − 09	0.00e + 00	0.00e + 00	8.64e − 12	0.00e + 00	2.57e − 09	3.22e − 09	2.51e − 09	2.27e − 16	9.06e − 16
mSIMBO-V	Mean	2.16e − 09	4.44e − 09	2.61e − 09	5.59e − 09	2.90e + 01	0.00e + 00	3.46e − 21	1.25e + 04	2.63e − 09	3.21e − 09	2.68e − 09	1.67e + 00	3.00e + 00
	STD	2.53e − 09	2.51e − 09	2.48e − 09	2.98e − 09	0.00e + 00	0.00e + 00	1.73e − 20	0.00e + 00	2.72e − 09	2.21e − 09	2.75e − 09	2.27e − 16	9.06e − 16
mSIMBO-Q	Mean	1.24e − 09	2.40e − 09	4.80e − 10	2.03e − 09	2.90e + 01	0.00e + 00	7.48e − 26	1.25e + 04	1.53e − 09	1.70e − 09	4.62e − 10	1.67e + 00	3.00e + 00
	STD	1.90e − 09	2.45e − 09	8.02e − 10	2.37e − 09	0.00e + 00	0.00e + 00	3.74e − 25	0.00e + 00	2.26e − 09	2.26e − 09	5.80e − 10	2.27e − 16	9.06e − 16
MA-BLX-T	Mean	3.05e − 09	6.19e − 09	3.30e − 09	4.79e − 09	0.00e + 00	0.00e + 00	5.19e − 02	4.76e + 02	3.25e − 09	5.10e − 09	3.76e − 09	1.40e − 04	1.36e − 01
	STD	2.67e − 09	2.85e − 09	3.03e − 09	3.17e − 09	0.00e + 00	0.00e + 00	4.17e − 02	1.51e + 02	3.20e − 09	2.77e − 09	3.43e − 09	1.14e − 04	4.90e − 02
MA-BLX-V	Mean	3.56e − 09	4.86e − 09	3.95e − 09	5.51e − 09	0.00e + 00	0.00e + 00	5.97e − 02	4.15e + 02	2.93e − 09	3.63e − 09	2.20e − 09	7.05e − 04	1.25e − 01
	STD	3.22e − 09	2.59e − 09	3.18e − 09	2.86e − 09	0.00e + 00	0.00e + 00	4.32e − 02	1.06e + 02	2.59e − 09	2.81e − 09	2.77e − 09	1.76e − 03	3.87e − 02
MA-BLX-Q	Mean	2.78e − 09	4.83e − 09	3.42e − 09	5.19e − 09	0.00e + 00	0.00e + 00	6.09e − 02	4.31e + 02	4.72e − 09	5.42e − 09	2.99e − 09	5.28e − 04	1.13e − 01
	STD	2.87e − 09	2.86e − 09	3.28e − 09	3.03e − 09	0.00e + 00	0.00e + 00	4.54e − 02	8.73e + 01	3.45e − 09	3.12e − 09	2.81e − 09	2.14e − 03	4.54e − 02
MA-SBX-T	Mean	3.35e − 09	4.38e − 09	2.16e − 09	3.56e − 09	0.00e + 00	0.00e + 00	7.20e − 02	−8.5e − 10	2.70e − 09	3.95e − 09	2.68e − 09	4.43e − 01	1.00e − 01
	STD	3.36e − 09	2.85e − 09	2.63e − 09	2.95e − 09	0.00e + 00	0.00e + 00	4.06e − 02	6.98e − 09	2.85e − 09	2.60e − 09	2.67e − 09	1.16e − 01	1.83e − 03
MA-SBX-V	Mean	1.98e − 09	3.83e − 09	2.27e − 09	4.54e − 09	0.00e + 00	0.00e + 00	4.22e − 02	−3.00e − 10	2.73e − 09	3.46e − 09	2.32e − 09	4.41e − 01	1.00e − 01
	STD	2.16e − 09	2.56e − 09	2.61e − 09	3.42e − 09	0.00e + 00	0.00e + 00	3.06e − 02	6.30e − 09	2.64e − 09	2.76e − 09	2.97e − 09	1.78e − 01	9.08e − 05
MA-SBX-Q	Mean	2.32e − 09	5.05e − 09	2.38e − 09	4.92e − 09	0.00e + 00	0.00e + 00	6.60e − 02	−4.90e − 10	2.86e − 09	3.59e − 09	2.37e − 09	3.96e − 01	1.01e − 01
	STD	3.45e − 09	2.84e − 09	2.77e − 09	3.18e − 09	0.00e + 00	0.00e + 00	6.33e − 02	6.97e − 09	3.29e − 09	2.58e − 09	2.47e − 09	1.73e − 01	4.20e − 03

Table 5: Comparison of the experimental results of algorithms used for dimension 50.

50D	F	Fsph	Fs2.22	Fsl.2	Fs2.21	Fros	Fste	Fqua	Fs2.26	Fras	Fack	Fgri	Fpenl	Fpen2
GA-BLX	Mean	2.39e + 00	1.66e + 01	2.90e + 03	9.58e + 00	1.15e + 02	3.64e + 00	2.67e − 01	6.84e + 02	1.02e + 02	8.45e + 00	6.68e − 01	1.59e − 03	3.66e − 02
	STD	1.31e + 00	2.14e + 00	7.82e + 02	1.28e + 00	3.82e + 01	2.27e + 00	1.18e − 01	1.16e + 03	1.57e + 01	5.44e − 01	1.92e − 01	9.60e − 04	1.79e − 02
GA-SBX	Mean	4.34e + 00	9.50e + 00	7.14e + 02	4.81e + 00	1.08e + 02	9.96e + 00	6.52e − 01	7.37e − 04	7.08e + 01	1.41e + 01	2.16e − 01	1.27e − 02	7.39e − 02
	STD	2.53e + 00	1.58e + 00	2.55e + 02	1.05e + 00	4.53e + 01	5.78e + 00	2.41e − 01	2.98e − 03	12.2e + 01	1.02e + 00	9.26e − 02	2.86e − 02	2.44e − 02
SIMBO-T	Mean	3.39e − 09	4.25e − 09	2.10e − 09	5.04e − 09	4.90e + 01	0.00e + 00	1.80e − 10	2.09e + 04	2.60e − 09	6.30e − 09	2.09e − 09	2.45e + 00	5.00e + 00
	STD	3.59e − 09	3.02e − 09	2.22e − 09	2.82e − 09	0.00e + 00	0.00e + 00	8.12e − 10	7.43e − 12	2.82e − 09	2.63e − 09	2.79e − 09	9.06e − 16	9.06e − 16
SIMBO-V	Mean	3.10e − 09	5.71e − 09	2.98e − 09	5.15e − 09	4.90e + 01	0.00e + 00	8.36e − 13	2.09e + 04	1.70e − 09	3.48e − 09	3.77e − 09	2.45e + 00	5.00e + 00
	STD	3.18e − 09	2.69e − 09	3.00e − 09	2.64e − 09	0.00e + 00	0.00e + 00	4.18e − 12	7.43e − 12	2.08e − 09	2.94e − 09	3.03e − 09	9.06e − 16	9.06e − 16
SIMBO-Q	Mean	1.54e − 09	2.76e − 09	8.84e − 10	3.00e − 09	4.90e + 01	0.00e + 00	2.31e − 20	2.09e + 04	1.31e − 09	2.78e − 09	1.49e − 09	2.45e + 00	5.00e + 00
	STD	2.79e − 09	2.48e − 09	1.70e − 09	2.48e − 09	0.00e + 00	0.00e + 00	9.81e − 20	7.43e − 12	2.11e − 09	2.67e − 09	2.23e − 09	9.06e − 16	9.06e − 16
mSIMBO-T	Mean	2.44e − 09	4.50e − 09	3.14e − 09	4.22e − 09	4.90e + 01	0.00e + 00	7.92e − 20	2.09e + 04	3.03e − 09	4.96e − 09	2.93e − 09	2.45e + 00	5.00e + 00
	STD	2.12e − 09	2.98e − 09	3.05e − 09	3.07e − 09	0.00e + 00	0.00e + 00	3.96e − 19	7.43e − 12	3.07e − 09	2.65e − 09	2.63e − 09	9.06e − 16	9.06e − 16
mSIMBO-V	Mean	2.55e − 09	3.67e − 09	2.64e − 09	5.86e − 09	4.90e + 01	0.00e + 00	1.18e − 10	2.09e + 04	1.56e − 09	4.93e − 09	3.14e − 09	2.45e + 00	5.00e + 00
	STD	3.09e − 09	3.00e − 09	2.89e − 09	3.44e − 09	0.00e + 00	0.00e + 00	5.92e − 10	7.43e − 12	2.18e − 09	3.02e − 09	3.29e − 09	9.06e − 16	9.06e − 16
mSIMBO-Q	Mean	9.86e − 10	2.07e − 09	1.13e − 09	2.03e − 09	4.90e + 01	0.00e + 00	7.02e − 18	2.09e + 04	1.07e − 09	3.27e − 09	8.69e − 10	2.45e + 00	5.00e + 00
	STD	1.88e − 09	2.37e − 09	2.03e − 09	2.57e − 09	0.00e + 00	0.00e + 00	3.51e − 17	7.43e − 12	2.26e − 09	3.41e − 09	1.31e − 09	9.06e − 16	9.06e − 16
MA-BLX-T	Mean	3.79e − 09	3.91e − 09	3.08e − 09	5.11e − 09	0.00e + 00	0.00e + 00	1.06e − 01	4.23e + 02	3.38e − 09	4.76e − 09	2.60e − 09	4.48e − 01	1.01e − 01
	STD	3.34e − 09	2.60e − 09	2.88e − 09	3.22e − 09	0.00e + 00	0.00e + 00	9.46e − 02	3.80e + 01	2.80e − 09	3.05e − 09	2.52e − 09	1.44e − 01	1.45e − 02
MA-BLX-V	Mean	2.95e − 09	4.63e − 09	3.02e − 09	4.79e − 09	0.00e + 00	0.00e + 00	1.44e − 01	4.14e + 02	2.67e − 09	4.75e − 09	3.32e − 09	4.53e − 01	1.04e − 01
	STD	3.08e − 09	2.58e − 09	2.28e − 09	2.82e − 09	0.00e + 00	0.00e + 00	1.10e − 01	6.49e + 01	2.19e − 09	2.80e − 09	3.13e − 09	1.36e − 01	2.22e − 02
MA-BLX-Q	Mean	2.87e − 09	5.61e − 09	2.39e − 09	5.06e − 09	0.00e + 00	0.00e + 00	1.26e − 01	4.47e + 02	3.31e − 09	5.37e − 09	2.56e − 09	4.82e − 01	9.84e − 02
	STD	3.38e − 09	3.48e − 09	2.63e − 09	3.20e − 09	0.00e + 00	0.00e + 00	9.24e − 02	9.76e + 01	3.19e − 09	2.78e − 09	2.92e − 09	9.19e − 02	1.62e − 02
MA-SBX-T	Mean	3.36e − 09	4.98e − 09	2.47e − 09	4.41e − 09	0.00e + 00	0.00e + 00	1.16e − 01	1.30e − 03	2.95e − 09	5.33e − 09	3.66e − 09	5.20e − 01	1.00e − 01
	STD	3.21e − 09	2.75e − 09	2.88e − 09	3.19e − 09	0.00e + 00	0.00e + 00	1.08e − 01	5.18e − 03	3.07e − 09	2.46e − 09	3.48e − 09	2.17e − 02	1.10e − 01
MA-SBX-V	Mean	2.14e − 09	4.41e − 09	3.36e − 09	4.23e − 09	0.00e + 00	0.00e + 00	1.09e − 01	5.52e − 04	2.55e − 09	5.53e − 09	3.91e − 09	4.99e − 01	1.00e − 01
	STD	2.78e − 09	2.56e − 09	2.80e − 09	3.05e − 09	0.00e + 00	0.00e + 00	1.09e − 01	5.25e − 03	2.74e − 09	3.26e − 09	3.24e − 09	9.90e − 02	2.77e − 04
MA-SBX-Q	Mean	2.32e − 09	4.15e − 09	2.30e − 09	5.27e − 09	0.00e + 00	0.00e + 00	1.27e − 01	9.30e − 04	2.86e − 09	5.43e − 09	2.49e − 09	5.11e − 01	1.00e − 01
	STD	2.36e − 09	2.88e − 09	2.34e − 09	3.10e − 09	0.00e + 00	0.00e + 00	1.04e − 01	4.38e − 03	3.18e − 09	2.72e − 09	2.29e − 09	6.54e − 02	9.95e − 07

Table 6: Comparison of the experimental results of algorithms using statistical test.

			Fsph	Fs2.22	Fs1.2	Fs2.21	Fros	Fste	Fqua	Fs2.26	Fras	Fack	Fgri	Fpen1	Fpen2
	10D	*P*-value	7.30*e* − 01	8.52*e* − 05	8.20*e* − 01	3.21*e* − 02	NA	NA	3.27*e* − 01	NA	1.39*e* − 03	1.90*e* − 05	1.23*e* − 05	NA	NA
	10D	SL = 0.05	0	1	0	1	—	—	0	—	1	1	1	—	—
	10D	SL = 0.01	0	1	0	0	—	—	0	—	1	1	1	—	—
	30D	*P*-value	5.40*e* − 05	1.60*e* − 01	1.69*e* − 03	4.77*e* − 08	NA	NA	3.27*e* − 01	NA	6.08*e* − 05	5.99*e* − 01	1.43*e* − 04	NA	NA
m-SIMBO-T versus SIMBO-T	30D	SL = 0.05	1	0	1	1	—	—	0	—	1	0	1	—	—
	30D	SL = 0.01	1	0	1	1	—	—	0	—	1	0	1	—	—
	50D	*P*-value	7.91*e* − 03	2.92*e* − 02	3.10*e* − 04	1.20*e* − 05	NA	NA	2.80*e* − 01	NA	1.87*e* − 03	4.45*e* − 08	8.51*e* − 05	NA	NA
	50D	SL = 0.05	1	1	1	1	—	—	0	—	1	1	1	—	—
	50D	SL = 0.01	1	0	1	1	—	—	0	—	1	1	1	—	—
	10D	*P*-value	2.51*e* − 04	7.16*e* − 06	4.91*e* − 02	2.38*e* − 07	NA	NA	1.27*e* − 01	NA	3.27*e* − 03	7.99*e* − 04	7.07*e* − 01	NA	NA
	10D	SL = 0.05	1	1	1	1	—	—	0	—	1	1	0	—	—
	10D	SL = 0.01	1	1	0	1	—	—	0	—	1	1	0	—	—
	30D	*P*-value	7.57*e* − 01	2.48*e* − 02	7.50*e* − 03	7.65*e* − 01	NA	NA	3.27*e* − 01	NA	9.09*e* − 01	7.57*e* − 01	2.03*e* − 02	NA	NA
m-SIMBO-V versus SIMBO-V	30D	SL = 0.05	0	1	1	0	—	—	0	—	0	0	1	—	—
	30D	SL = 0.01	0	0	1	0	—	—	0	—	0	0	0	—	—
	50D	*P*-value	1.15*e* − 02	1.21*e* − 05	2.00*e* − 02	3.01*e* − 05	NA	NA	3.28*e* − 01	NA	1.98*e* − 01	1.02*e* − 02	3.72*e* − 03	NA	NA
	50D	SL = 0.05	1	1	1	1	—	—	0	—	0	1	1	—	—
	50D	SL = 0.01	0	1	0	1	—	—	0	—	0	0	1	—	—
	10D	*P*-value	5.57*e* − 03	6.69*e* − 01	7.92*e* − 01	1.64*e* − 07	NA	NA	3.27*e* − 01	NA	8.28*e* − 01	1.66*e* − 04	1.96*e* − 02	NA	NA
	10D	SL = 0.05	1	0	0	1	—	—	0	—	0	1	1	—	—
	10D	SL = 0.01	1	0	0	1	—	—	0	—	0	1	0	—	—
	30D	*P*-value	8.20*e* − 01	5.17*e* − 02	2.21*e* − 03	1.70*e* − 10	NA	NA	1.89*e* − 01	NA	1.18*e* − 02	7.50*e* − 06	4.15*e* − 04	NA	NA
m-SIMBO-Q versus SIMBO-Q	30D	SL = 0.05	0	0	1	1	—	—	0	—	1	1	1	—	—
	30D	SL = 0.01	0	0	1	1	—	—	0	—	0	1	1	—	—
	50D	*P*-value	6.87*e* − 03	4.06*e* − 10	2.47*e* − 02	6.48*e* − 03	NA	NA	3.27*e* − 01	NA	2.08*e* − 01	2.00*e* − 10	4.01*e* − 03	NA	NA
	50D	SL = 0.05	1	1	1	1	—	—	0	—	0	1	1	—	—
	50D	SL = 0.01	1	1	0	1	—	—	0	—	0	1	1	—	—

TABLE 7: Comparison of the experimental results of algorithms used for dimension 02.

02D	F	F1	F2	F3	F4	F5	F6	F7	F8	F9	F10	F11	F12	F13
GA-BLX	Mean	3.47e − 03	3.78e − 01	1.78e + 03	1.94e + 00	1.61e + 01	3.81e + 01	5.98e − 01	1.05e + 01	1.22e − 03	2.72e − 01	1.39e − 02	2.45e + 00	2.14e − 02
	STD	1.32e − 02	8.35e − 01	3.38e + 03	5.03e + 00	2.10e + 01	4.55e + 01	3.80e − 01	6.98e + 00	2.54e − 03	5.17e − 01	2.88e − 02	5.45e + 00	3.66e − 02
GA-SBX	Mean	2.38e − 08	6.18e − 09	8.54e + 01	1.29e − 08	7.59e − 09	1.99e + 01	1.70e − 01	9.79e + 00	6.69e − 06	3.83e − 01	2.80e − 03	8.92e − 01	2.25e − 02
	STD	8.86e − 08	2.61e − 09	2.69e + 02	3.54e − 08	2.26e − 09	3.85e + 01	1.81e − 01	6.67e + 00	3.34e − 05	5.34e − 01	1.16e − 02	2.16e + 00	2.82e − 02
MA-BLX-T	Mean	7.10e − 04	5.89e − 02	4.30e + 03	8.97e − 01	1.33e + 01	3.15e + 01	8.23e − 01	8.99e + 00	1.53e − 03	2.66e − 01	2.62e − 02	7.27e + 00	1.82e − 02
	STD	2.83e − 03	9.01e − 02	5.28e + 03	1.70e + 00	2.30e + 01	4.08e + 01	4.56e − 01	6.55e + 00	3.39e − 03	4.38e − 01	4.83e − 02	1.43e + 01	3.28e − 02
MA-BLX-V	Mean	1.57e − 04	1.64e − 01	3.71e + 03	6.29e − 01	1.06e + 01	1.64e + 01	8.35e − 01	1.03e + 01	3.71e − 04	3.67e − 01	2.91e − 02	2.01e + 00	1.41e − 02
	STD	4.03e − 04	4.29e − 01	4.10e + 03	1.42e + 00	1.70e + 01	2.17e + 01	5.33e − 01	6.85e + 00	5.20e − 04	4.75e − 01	6.57e − 02	4.85e + 00	2.43e − 02
MA-BLX-Q	Mean	1.39e − 04	7.79e − 02	2.59e + 03	7.54e − 01	1.23e + 01	2.87e + 01	8.48e − 01	8.51e + 00	2.89e − 04	3.19e − 01	2.60e − 02	2.83e + 00	7.56e − 03
	STD	4.10e − 04	1.37e − 01	3.82e + 03	9.81e − 01	2.29e + 01	3.49e + 01	5.52e − 01	7.20e + 00	5.91e − 04	4.73e − 01	5.87e − 02	8.15e + 00	8.23e − 03
MA-SBX-T	Mean	5.61e − 09	1.65e − 06	2.01e + 01	4.88e − 05	6.99e − 09	1.69e + 01	1.18e − 01	7.60e + 00	3.76e − 08	4.96e − 01	7.89e − 04	6.60e − 01	1.16e − 02
	STD	2.83e − 09	5.59e − 06	1.00e + 02	2.40e − 04	1.96e − 09	2.31e + 01	1.29e − 01	6.58e + 00	1.11e − 07	4.97e − 01	2.29e − 03	2.25e + 00	2.41e − 02
MA-SBX-V	Mean	4.42e − 05	1.39e − 06	5.23e + 01	6.46e − 09	7.28e − 09	1.27e + 01	1.15e − 01	7.51e + 00	3.17e − 04	3.18e − 01	4.65e − 03	3.91e − 01	7.12e − 03
	STD	2.20e − 04	6.89e − 06	2.40e + 02	4.94e − 09	2.22e − 09	1.61e + 01	1.13e − 01	7.17e + 00	1.58e − 03	4.74e − 01	2.32e − 02	1.46e + 00	7.28e − 03
MA-SBX-Q	Mean	4.12e − 04	6.79e − 09	5.31e + 02	1.28e − 05	6.30e − 09	7.20e + 00	1.58e − 01	7.98e + 00	6.89e − 05	3.98e − 01	7.98e − 03	1.34e − 01	1.44e − 02
	STD	2.06e − 03	2.31e − 09	2.52e + 03	5.01e − 05	2.49e − 09	1.15e + 01	1.77e − 01	7.84e + 00	3.17e − 04	4.97e − 01	2.90e − 02	3.34e − 01	2.24e − 02

02D	F	F14	F15	F16	F17	F18	F19	F20	F21	F22	F23	F24	F25
GA-BLX	Mean	4.63e − 02	4.07e + 00	2.26e + 01	2.74e + 01	2.19e + 02	2.40e + 02	2.54e + 02	2.06e + 02	2.66e + 02	2.65e + 02	2.00e + 02	2.00e + 02
	STD	4.95e − 02	2.00e + 01	2.90e + 01	2.80e + 01	8.90e + 01	8.00e + 01	7.30e + 01	1.19e + 02	7.71e + 01	1.40e + 02	6.32e − 05	2.69e − 05
GA-SBX	Mean	2.53e − 02	8.00e + 00	3.50e + 01	3.36e + 01	1.78e + 02	2.56e + 02	2.37e + 02	2.12e + 02	2.58e + 02	2.83e + 02	1.98e + 02	1.98e + 02
	STD	1.89e − 02	2.77e + 01	3.84e + 01	3.55e + 01	1.47e + 02	8.70e + 01	1.24e + 02	7.55e + 01	8.62e + 01	8.20e + 01	7.61e + 00	9.85e + 00
MA-BLX-T	Mean	5.78e − 02	2.28e − 02	2.61e + 01	2.83e + 01	1.73e + 02	2.44e + 02	1.90e + 02	1.79e + 02	2.72e + 02	2.30e + 02	2.00e + 02	1.85e + 02
	STD	4.65e − 02	3.43e − 02	2.67e + 01	3.27e + 01	1.13e + 02	7.12e + 01	9.95e + 01	1.09e + 02	6.73e + 01	1.16e + 02	9.37e − 05	3.25e + 01
MA-BLX-V	Mean	5.19e − 02	4.05e + 00	3.71e + 01	2.62e + 01	1.85e + 02	2.60e + 02	2.16e + 02	2.11e + 02	2.37e + 02	2.48e + 02	1.96e + 02	1.94e + 02
	STD	5.46e − 02	2.00e + 01	3.41e + 01	3.41e + 01	1.30e + 02	6.45e + 01	8.89e + 01	1.15e + 02	1.00e + 02	1.56e + 02	1.40e + 01	2.11e + 01
MA-BLX-Q	Mean	5.95e − 02	4.16e + 00	3.55e + 01	3.09e + 01	1.77e + 02	2.20e + 02	2.50e + 02	2.09e + 02	2.33e + 02	2.16e + 02	1.98e + 02	1.88e + 02
	STD	6.52e − 02	2.00e + 01	3.66e + 01	2.91e + 01	1.19e + 02	8.92e + 01	7.11e + 01	1.28e + 02	6.10e + 01	1.27e + 02	1.13e + 01	2.94e + 01
MA-SBX-T	Mean	3.15e − 02	4.00e + 00	2.66e + 01	3.84e + 01	2.16e + 02	2.88e + 02	2.13e + 02	1.90e + 02	2.20e + 02	2.81e + 02	2.00e + 02	1.86e + 02
	STD	3.17e − 02	2.00e + 01	3.17e + 01	3.69e + 01	1.46e + 02	8.33e + 01	1.30e + 02	9.33e + 01	4.08e + 01	1.06e + 02	2.81e − 05	3.45e + 01
MA-SBX-V	Mean	4.47e − 02	2.34e − 02	1.73e + 01	2.47e + 01	1.50e + 02	2.51e + 02	2.09e + 02	1.75e + 02	2.35e + 02	2.44e + 02	2.00e + 02	2.00e + 02
	STD	4.67e − 02	1.11e − 01	2.65e + 01	3.22e + 01	1.19e + 02	8.93e + 01	9.97e + 01	1.10e + 02	4.78e + 01	1.18e + 02	3.04e − 05	1.65e − 04
MA-SBX-Q	Mean	3.81e − 02	3.95e − 03	3.18e + 01	2.31e + 01	1.67e + 02	2.59e + 02	1.99e + 02	1.92e + 02	1.92e + 02	2.01e + 02	2.00e + 02	1.92e + 02
	STD	3.43e − 02	1.60e − 02	3.16e + 01	2.58e + 01	1.31e + 02	6.80e + 01	1.24e + 02	1.29e + 02	8.12e + 01	1.22e + 02	3.86e − 05	2.36e + 01

Table 8: Comparison of the experimental results of algorithms used for dimension 10.

10D	F	F1	F2	F3	F4	F5	F6	F7	F8	F9	F10	F11	F12	F13
GA-BLX	Mean	7.00e − 03	1.37e + 02	3.51e + 05	5.88e + 02	4.28e + 02	2.48e + 02	3.95e + 00	2.04e + 01	2.31e − 03	1.72e + 01	4.36e + 00	2.60e + 03	3.63e − 01
	STD	6.02e − 03	5.37e + 01	1.71e + 05	2.50e + 02	1.84e + 02	6.58e + 02	2.49e + 00	4.82e − 02	1.90e − 03	6.74e + 00	1.91e + 00	1.61e + 03	1.08e − 01
GA-SBX	Mean	3.40e − 03	2.85e + 00	3.04e + 05	2.71e + 01	4.42e + 02	3.98e + 02	2.39e + 00	2.03e + 01	1.28e − 03	2.18e + 01	7.26e + 00	1.23e + 03	3.85e − 01
	STD	5.05e − 03	3.33e + 00	3.40e + 05	2.15e + 01	2.52e + 02	1.05e + 03	1.27e + 00	8.32e − 02	1.54e − 03	7.66e + 00	2.11e + 00	1.41e + 03	1.17e − 01
MA-BLX-T	Mean	9.53e − 03	1.34e + 02	4.51e + 05	6.82e + 02	3.35e + 02	3.69e + 02	3.45e + 00	2.04e + 01	3.46e − 03	2.15e + 01	5.09e + 00	2.29e + 03	3.66e − 01
	STD	9.93e − 03	5.36e + 01	3.14e + 05	2.59e + 02	1.45e + 02	1.05e + 03	2.56e + 00	7.76e − 02	2.64e − 03	7.31e + 00	1.90e + 00	1.21e + 03	1.03e − 01
MA-BLX-V	Mean	8.62e − 03	1.63e + 02	4.98e + 05	6.19e + 02	3.73e + 02	1.32e + 02	4.40e + 00	2.04e + 01	3.77e − 03	2.29e + 01	4.94e + 00	3.03e + 03	4.32e − 01
	STD	4.90e − 03	5.40e + 01	4.47e + 05	2.55e + 02	1.72e + 02	3.09e + 02	2.96e + 00	7.81e − 02	3.77e − 03	7.65e + 00	2.05e + 00	1.89e + 03	1.05e − 01
MA-BLX-Q	Mean	9.22e − 03	1.64e + 02	4.94e + 05	7.23e + 02	4.00e + 02	3.63e + 02	3.22e + 00	2.04e + 01	2.81e − 03	2.19e + 01	4.70e + 00	2.26e + 03	3.49e − 01
	STD	4.85e − 03	6.72e + 01	3.47e + 05	2.51e + 02	1.91e + 02	1.29e + 03	2.08e + 00	6.68e − 02	2.38e − 03	6.32e + 00	1.82e + 00	1.74e + 03	1.24e − 01
MA-SBX-T	Mean	5.59e − 03	2.10e + 00	5.58e + 05	2.60e + 01	5.17e + 02	9.17e + 01	2.78e + 00	2.04e + 01	2.87e − 03	2.26e + 01	6.97e + 00	9.92e + 02	4.04e − 01
	STD	4.53e − 03	1.80e + 00	6.25e + 05	2.32e + 01	4.44e + 02	2.26e + 02	1.41e + 00	6.86e − 02	2.79e − 03	9.73e + 00	2.06e + 00	9.02e + 02	1.34e − 01
MA-SBX-V	Mean	5.07e − 03	2.09e + 00	4.75e + 05	2.33e + 01	4.56e + 02	9.06e + 01	2.11e + 00	2.04e + 01	3.39e − 03	2.42e + 01	6.49e + 00	1.29e + 03	3.85e − 01
	STD	2.83e − 03	2.25e + 00	3.95e + 05	1.48e + 01	3.68e + 02	1.36e + 02	7.10e − 01	6.26e − 02	2.16e − 03	9.16e + 00	2.45e + 00	1.28e + 03	1.02e − 01
MA-SBX-Q	Mean	4.26e − 03	2.81e + 00	4.08e + 05	2.90e + 01	4.41e + 02	6.98e + 01	1.88e + 00	2.04e + 01	3.18e − 03	2.59e + 01	7.49e + 00	1.18e + 03	3.73e − 01
	STD	3.11e − 03	2.29e + 00	4.53e + 05	1.80e + 01	3.39e + 02	6.86e + 01	7.89e − 01	7.73e − 02	3.30e − 03	1.00e + 01	1.75e + 00	1.18e + 03	1.23e − 01

10D	F	F14	F15	F16	F17	F18	F19	F20	F21	F22	F23	F24	F25
GA-BLX	Mean	3.66e + 00	4.42e + 01	1.25e + 02	1.88e + 02	8.10e + 02	8.10e + 02	8.10e + 02	1.15e + 03	4.25e + 02	1.17e + 03	3.59e + 02	9.60e + 02
	STD	2.08e − 01	1.45e + 02	1.50e + 01	5.92e + 01	5.27e + 00	5.22e + 00	4.81e + 00	2.01e + 02	2.26e + 02	1.91e + 02	3.17e + 02	4.81e + 02
GA-SBX	Mean	4.10e + 00	2.07e + 01	1.31e + 02	1.33e + 02	7.82e + 02	7.52e + 02	8.04e + 02	6.75e + 02	7.21e + 02	7.97e + 02	4.24e + 02	4.19e + 02
	STD	1.92e − 01	8.47e + 01	1.71e + 01	2.69e + 01	2.28e + 02	2.48e + 02	1.96e + 02	3.31e + 02	1.91e + 02	2.39e + 02	1.88e + 01	1.24e + 01
MA-BLX-T	Mean	3.59e + 00	8.24e + 01	1.26e + 02	1.59e + 02	8.10e + 02	7.95e + 02	8.30e + 02	1.06e + 03	4.78e + 02	1.19e + 03	4.31e + 02	9.84e + 02
	STD	1.81e − 01	1.93e + 02	2.40e + 01	4.44e + 01	3.13e + 01	9.60e + 01	6.95e + 01	2.72e + 02	2.47e + 02	1.39e + 02	3.66e + 02	4.49e + 02
MA-BLX-V	Mean	3.60e + 00	1.87e + 02	1.41e + 02	1.49e + 02	8.04e + 02	7.84e + 02	7.96e + 02	1.10e + 03	6.09e + 02	1.17e + 03	4.12e + 02	1.03e + 03
	STD	1.60e − 01	2.47e + 02	2.47e + 01	5.71e + 01	7.04e + 00	1.01e + 02	1.12e + 02	2.63e + 02	2.57e + 02	1.41e + 02	3.16e + 02	4.45e + 02
MA-BLX-Q	Mean	3.53e + 00	1.55e + 02	1.23e + 02	1.69e + 02	7.81e + 02	8.27e + 02	8.20e + 02	1.14e + 03	4.75e + 02	1.20e + 03	3.86e + 02	9.92e + 02
	STD	1.93e − 01	2.49e + 02	1.76e + 01	5.68e + 01	1.27e + 02	4.18e + 01	7.82e + 01	1.58e + 02	2.41e + 02	1.13e + 02	3.45e + 02	4.80e + 02
MA-SBX-T	Mean	4.00e + 00	4.02e + 01	1.31e + 02	1.37e + 02	7.89e + 02	7.61e + 02	7.31e + 02	7.47e + 02	7.21e + 02	8.46e + 02	4.24e + 02	4.45e + 02
	STD	2.21e − 01	1.25e + 02	1.53e + 01	2.65e + 01	2.21e + 02	2.56e + 02	2.42e + 02	3.38e + 02	1.89e + 02	2.77e + 02	2.44e + 01	7.77e + 01
MA-SBX-V	Mean	4.03e + 00	2.30e + 01	1.36e + 02	1.42e + 02	8.27e + 02	8.08e + 02	7.86e + 02	7.27e + 02	6.58e + 02	7.26e + 02	4.18e + 02	4.35e + 02
	STD	1.63e − 01	9.61e + 01	2.10e + 01	2.79e + 01	1.62e + 02	1.74e + 02	2.14e + 02	3.41e + 02	2.29e + 02	2.03e + 02	1.97e + 01	7.67e + 01
MA-SBX-Q	Mean	3.97e + 00	6.22e + 01	1.34e + 02	1.36e + 02	7.96e + 02	7.81e + 02	8.07e + 02	7.30e + 02	7.16e + 02	8.42e + 02	4.16e + 02	4.25e + 02
	STD	2.18e − 01	1.51e + 02	2.23e + 01	1.82e + 01	2.20e + 02	2.15e + 02	1.69e + 02	3.21e + 02	1.86e + 02	2.60e + 02	1.01e + 01	9.00e + 01

Table 9: Comparison of the experimental results of algorithms using statistical test.

F	MABLX_T versus GABLX						MABLX_V versus GABLX						MABLX_Q versus GABLX					
	02D P value	02D SL = 0.05	02D SL = 0.01	10D P value	10D SL = 0.05	10D SL = 0.01	02D P value	02D SL = 0.05	02D SL = 0.01	10D P value	10D SL = 0.05	10D SL = 0.01	02D P value	02D SL = 0.05	02D SL = 0.01	10D P value	10D SL = 0.05	10D SL = 0.01
F1	1.97e − 01	0	0	9.07e − 03	1	1	2.11e − 01	0	0	2.46e − 04	1	1	2.05e − 01	0	0	1.95e − 05	1	1
F2	4.58e − 02	1	0	3.73e − 01	0	0	1.78e − 02	1	0	3.30e − 07	1	1	4.55e − 02	1	0	1.14e − 07	1	1
F3	2.29e − 03	1	1	2.84e − 03	1	1	1.29e − 03	1	1	1.80e − 02	1	0	4.23e − 02	1	0	5.48e − 04	1	1
F4	1.40e − 01	0	0	1.08e − 14	1	1	9.11e − 02	0	0	2.94e − 02	1	0	1.79e − 01	0	0	2.87e − 16	1	1
F5	6.42e − 02	0	0	1.26e − 07	1	1	2.64e − 04	1	1	3.28e − 07	1	1	2.31e − 02	1	0	3.54e − 03	1	1
F6	9.39e − 03	1	1	1.90e − 01	0	0	1.77e − 04	1	1	1.22e − 01	0	0	2.68e − 03	1	1	4.06e − 01	0	0
F7	1.97e − 09	1	1	1.02e − 01	0	0	5.57e − 06	1	1	2.72e − 03	1	1	3.07e − 05	1	1	2.71e − 03	1	1
F8	9.97e − 06	1	1	8.16e − 03	1	1	4.50e − 01	0	0	4.06e − 01	0	0	1.25e − 07	1	1	2.03e − 01	0	0
F9	3.57e − 01	0	0	1.95e − 07	1	1	5.37e − 02	0	0	3.13e − 03	1	1	3.39e − 02	1	0	9.51e − 03	1	1
F10	8.80e − 01	0	0	4.14e − 13	1	1	1.79e − 01	0	0	2.93e − 14	1	1	4.84e − 01	0	0	2.59e − 13	1	1
F11	1.32e − 02	1	0	1.18e − 08	1	1	5.17e − 02	0	0	2.03e − 08	1	1	5.92e − 02	0	0	2.23e − 04	1	1
F12	2.88e − 02	1	0	8.65e − 03	1	1	3.75e − 03	1	1	1.18e − 05	1	1	6.47e − 01	0	0	5.14e − 03	1	1
F13	5.19e − 01	0	0	6.72e − 01	0	0	1.35e − 01	0	0	2.29e − 07	1	1	3.26e − 02	1	0	1.16e − 01	0	0
F14	5.49e − 02	0	0	4.16e − 06	1	1	4.18e − 01	0	0	3.85e − 03	1	1	2.73e − 02	1	0	5.32e − 07	1	1
F15	3.21e − 01	0	0	1.64e − 01	0	0	3.64e − 02	1	0	4.17e − 03	1	1	2.04e − 01	0	0	9.74e − 03	1	1
F16	1.55e − 01	0	0	5.47e − 01	0	0	1.04e − 03	1	1	1.41e − 06	1	1	3.15e − 03	1	1	4.57e − 01	0	0
F17	7.21e − 01	0	0	8.05e − 07	1	1	7.37e − 01	0	0	1.70e − 11	1	1	2.20e − 01	0	0	7.67e − 06	1	1
F18	1.26e − 03	1	1	9.29e − 01	0	0	3.51e − 02	1	0	8.32e − 08	1	1	2.49e − 03	1	1	2.55e − 01	0	0
F19	5.13e − 01	0	0	4.26e − 01	0	0	1.33e − 02	1	0	2.00e − 01	0	0	3.94e − 02	1	0	3.09e − 02	1	0
F20	1.22e − 05	1	1	1.38e − 01	0	0	6.04e − 04	1	1	5.44e − 01	0	0	5.62e − 01	0	0	4.89e − 01	0	0
F21	2.05e − 03	1	1	1.08e − 02	1	0	4.50e − 01	0	0	2.98e − 02	1	0	7.40e − 01	0	0	5.29e − 01	0	0
F22	3.54e − 01	0	0	7.92e − 02	0	0	1.22e − 02	1	0	1.04e − 03	1	1	5.35e − 03	1	1	9.59e − 02	0	0
F23	2.88e − 02	1	0	4.20e − 01	0	0	3.34e − 01	0	0	9.56e − 01	0	0	2.34e − 02	1	0	1.80e − 01	0	0
F24	1.34e − 03	1	1	6.78e − 03	1	1	1.73e − 01	0	0	3.36e − 02	1	0	3.27e − 01	0	0	3.26e − 01	0	0
F25	3.51e − 02	1	0	3.99e − 01	0	0	1.69e − 01	0	0	7.23e − 03	1	1	5.40e − 02	0	0	1.48e − 03	1	1

Table 10: Comparison of the experimental results of algorithms using statistical test.

F	MASBX_T versus GASBX						MASBX_V versus GASBX						MASBX_Q versus GASBX					
	02D P value	02D SL = 0.05	02D SL = 0.01	10D P value	10D SL = 0.05	10D SL = 0.01	02D P value	02D SL = 0.05	02D SL = 0.01	10D P value	10D SL = 0.05	10D SL = 0.01	02D P value	02D SL = 0.05	02D SL = 0.01	10D P value	10D SL = 0.05	10D SL = 0.01
F1	3.09e − 01	0	0	1.73e − 04	1	1	3.27e − 01	0	0	2.64e − 02	1	0	3.27e − 01	0	0	1.14e − 01	0	0
F2	1.53e − 01	0	0	4.23e − 02	1	0	3.27e − 01	0	0	4.72e − 03	1	1	8.06e − 08	1	1	8.85e − 01	0	0
F3	1.04e − 01	0	0	2.55e − 04	1	1	2.32e − 01	0	0	1.61e − 09	1	1	3.43e − 01	0	0	5.91e − 03	1	1
F4	3.19e − 01	0	0	5.92e − 01	0	0	3.18e − 01	0	0	8.56e − 02	0	0	2.14e − 01	0	0	6.91e − 02	0	0
F5	4.26e − 03	1	1	9.73e − 02	0	0	4.54e − 02	1	0	6.92e − 01	0	0	9.03e − 08	1	1	9.76e − 01	0	0
F6	4.33e − 01	0	0	9.45e − 02	0	0	1.49e − 01	0	0	1.14e − 01	0	0	3.20e − 02	1	0	1.11e − 01	0	0
F7	1.76e − 04	1	1	2.26e − 05	1	1	7.61e − 04	1	1	3.37e − 02	1	0	2.47e − 01	0	0	1.51e − 04	1	1
F8	2.29e − 05	1	1	2.44e − 02	1	0	1.06e − 04	1	1	3.03e − 08	1	1	5.82e − 03	1	1	3.24e − 04	1	1
F9	3.28e − 01	0	0	1.44e − 05	1	1	3.26e − 01	0	0	1.99e − 08	1	1	2.84e − 01	0	0	3.17e − 05	1	1
F10	1.36e − 01	0	0	2.55e − 01	0	0	1.74e − 01	0	0	2.80e − 05	1	1	7.50e − 01	0	0	6.51e − 06	1	1
F11	3.21e − 01	0	0	1.34e − 01	0	0	4.41e − 01	0	0	1.08e − 03	1	1	1.76e − 01	0	0	1.69e − 01	0	0
F12	4.99e − 02	1	0	8.57e − 02	0	0	9.26e − 03	1	1	4.50e − 01	0	0	4.98e − 02	1	0	5.51e − 01	0	0
F13	7.54e − 04	1	1	2.84e − 03	1	1	2.42e − 03	1	1	9.34e − 01	0	0	1.30e − 04	1	1	7.76e − 03	1	1
F14	1.22e − 01	0	0	9.69e − 06	1	1	4.64e − 03	1	1	2.01e − 06	1	1	1.70e − 02	1	0	1.55e − 07	1	1
F15	3.27e − 01	0	0	1.88e − 01	0	0	1.61e − 01	0	0	3.21e − 01	0	0	1.61e − 01	0	0	8.09e − 02	0	0
F16	1.98e − 02	1	0	7.62e − 01	0	0	4.71e − 04	1	1	7.11e − 04	1	1	3.03e − 01	0	0	1.29e − 02	1	0
F17	3.78e − 02	1	0	1.30e − 02	1	0	6.25e − 03	1	1	8.87e − 03	1	1	5.54e − 03	1	1	4.27e − 01	0	0
F18	1.70e − 03	1	1	5.11e − 01	0	0	1.21e − 02	1	0	6.98e − 02	0	0	2.14e − 01	0	0	1.87e − 01	0	0
F19	2.60e − 03	1	1	7.27e − 01	0	0	6.02e − 01	0	0	7.54e − 02	0	0	7.36e − 01	0	0	1.61e − 01	0	0
F20	9.68e − 03	1	1	4.39e − 03	1	1	1.37e − 02	1	0	1.18e − 01	0	0	5.85e − 04	1	1	8.69e − 01	0	0
F21	1.15e − 01	0	0	1.13e − 02	1	0	1.35e − 02	1	0	8.12e − 02	0	0	2.24e − 01	0	0	4.44e − 02	1	0
F22	5.88e − 03	1	1	9.39e − 01	0	0	5.40e − 02	0	0	5.46e − 02	0	0	3.59e − 04	1	1	9.71e − 02	0	0
F23	8.80e − 01	0	0	1.24e − 03	1	1	4.38e − 02	1	0	5.66e − 04	1	1	1.57e − 03	1	1	6.18e − 03	1	1
F24	1.64e − 01	0	0	7.67e − 01	0	0	1.64e − 01	0	0	2.07e − 10	1	1	1.64e − 01	0	0	2.39e − 03	1	1
F25	5.25e − 02	0	0	8.21e − 02	0	0	3.27e − 01	0	0	2.85e − 01	0	0	9.08e − 02	0	0	7.23e − 01	0	0

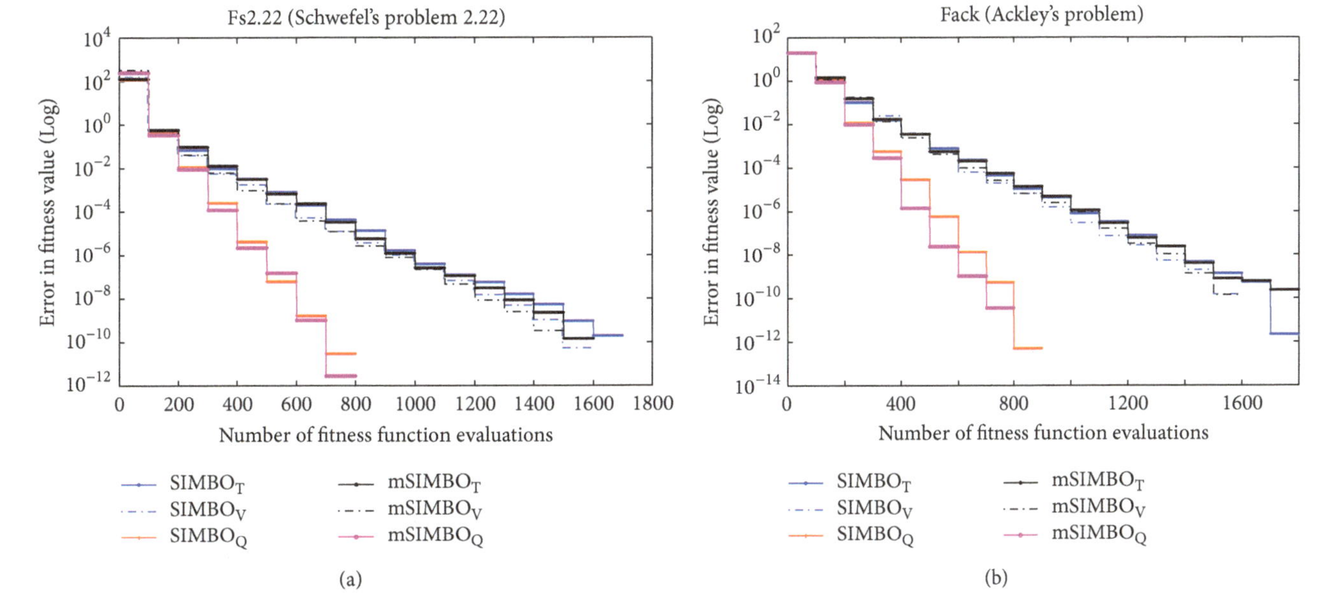

FIGURE 1: Fitness error versus number of function evaluation calls of test functions averaged over 25 runs in cases of SIMBOs and mSIMBOs for functions (a) Fs2.22, (b) Fack.

Fpen2 (generalized penalized function no. 2) are multimodal functions in which the number of local minima increases exponentially with the problem dimension [7]. They are the most difficult class of problems for many optimization algorithms [7]. For unimodal functions, the convergence rates are more interesting than the final results of optimization. For unimodal functions, there are other methods like gradient based methods which are specifically designed to optimize these. In case of multimodal functions, the final results are important since they reflect an algorithm's ability to escape from poor local optima and locating a good near-global optimum.

From Tables 3, 4 and 5, it is clearly seen that the proposed algorithms all modified SIMBOs and all MAs with the modified SIMBOs as local searches generate better results compared to all the competitor algorithms used for comparison. The modified SIMBOs as well as MAs based on modified SIMBOs generally demonstrate the superiority than all GAs and basic SIMBOs in solving the standard benchmark test problems. In case of all unimodal functions (Fsph, Fs2.22, Fs1.2, Fs2.21, and Fros), all the MAs converge successfully for all the dimensions of 10, 30, and 50, whereas basic SIMBO family and modified SIMBO family converge successfully in all unimodal functions for all dimensions except Rosenbrock's function, that is, Fros. The basic as well as modified SIMBO families not at all successfully converge for any of the dimensions in case of Rosenbrock's function, that is, Fros. The GAs do not reach the required accuracy in stipulated NFFEs for any of the run and for any of the dimensions in cases of unimodal functions (Fsph, Fs2.22, Fs1.2, Fs2.21, and Fros). All algorithms with all dimensions under test in this experiment successfully converge for the discontinuous step function (Fste). In case of noisy quartic function (Fqua), all basic as well as modified SIMBOs successfully converge in all runs for all dimensions, whereas all GAs do not converge successfully and all MAs converge successfully in very few runs. For generalized Schwefel's problem 2.26 (Fs2.26) test all GAs as well as all MAs show successful convergence in many runs, whereas the basic as well as modified SIMBOs do not at all converge. In cases of multimodal functions, namely, generalized Rastrigin problem, Ackley's function, generalized Griewank function, that is, Fras, Fack, and Fgri, all the algorithms except the GAs successfully converge to set accuracy. For the generalized penalized functions nos. 1 and 2, that is, Fpen1 and Fpen2, the GAs as well as MAs are more promising than the basic and modified SIMBOs for convergence characteristics.

As seen from Figures 1(a) and 1(b), the SIMBO has the better characteristics of convergence and mSIMBO shows still better converging characteristic in general in almost all the types of numerical standard benchmark functions for all 10, 30, and 50 dimensions. The mSIMBOs have quite faster speed of convergence but when the mSIMBOs are embedded in the evolutionary algorithm GA, the developed MAs show slightly lesser speed of convergence than the mSIMBOs as seen from Figures 2(a), 2(b), 2(c), and 2(d). The very small tradeoffs are repaid by solving the larger number of problem functions of a variety of domains for all MAs as compared to all SIMBOs. Along with SIMBOs, the MAs also successfully converge in cases of unimodal (Fsph, Fs2.22, Fs1.2, and Fs2.21) and multimodal (Fras, Fack, and Fgri). Additionally, the success in convergence is attained by MAs in cases of Fs2.26, Fpen1, and Fpen2 along with promising characteristics of GAs in the same cases. The SIMBOs have quite large drag in origin centric solutions and fail in other cases which are overcome by MAs with heavy explorations along with the exploitations by mSIMBOs. Among the SIMBOs and mSIMBOs, treatment based are the slowest and

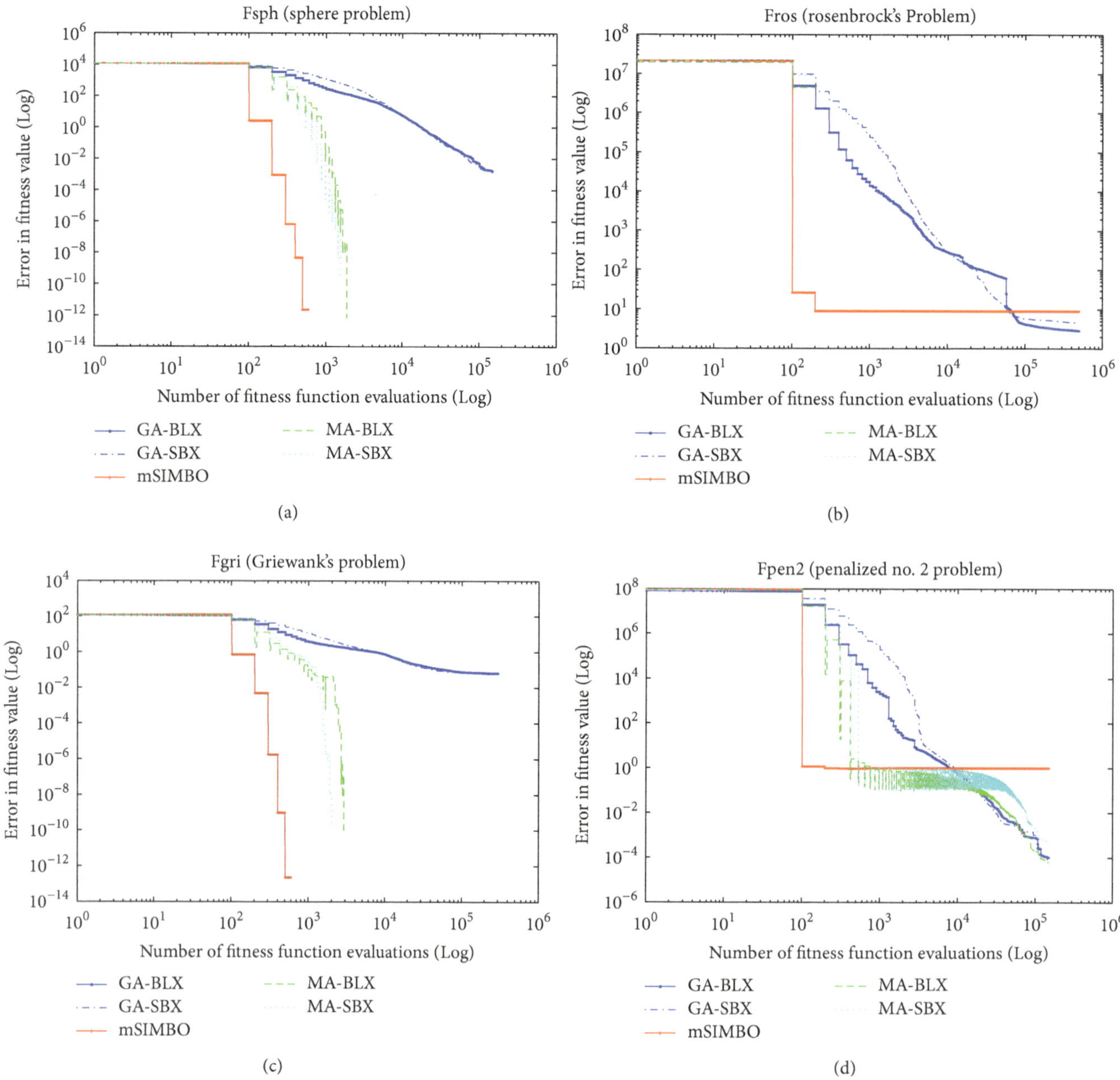

FIGURE 2: Fitness error versus number of function evaluation calls of test functions averaged over 25 runs in cases of GAs, mSIMBOs, and MAs for functions (a) Fsph, (b) Fros, (c) Fgri, and (d) Fpen2.

quarantine based are the fastest; similarly, the MAs based on mSIMBOs have the same characteristics, that is, treatment based are the slowest and quarantine based,the are fastest.

A statistical test, Student's t-test, is performed for statistical validity of the obtained results [33]. Here, N.A. denotes "not applicable." From Table 6, it is seen that the proposed algorithm modified SIMBOs (mSIMBO_T, mSIMBO_V, and mSIMBO_Q) show better performance than basic SIMBOs (SIMBO_T, SIMBO_V, and SIMBO_Q) in their respective comparison. Both basic as well as proposed algorithms show better performances as per statistical comparison for many cases of basic standard benchmark functions of 10 dimensions and considering a significance level of 0.05. As the significance level becomes 0.01, the superiority of the modified SIMBOs becomes clearly distinct. The modified SIMBOs also show the consistent performance for higher dimensions of 30 and 50, and therefore, the proposed algorithm wins in the statistical tests in these cases. Overall, it is seen that the proposed SIMBOs are better algorithms than basic SIMBOs.

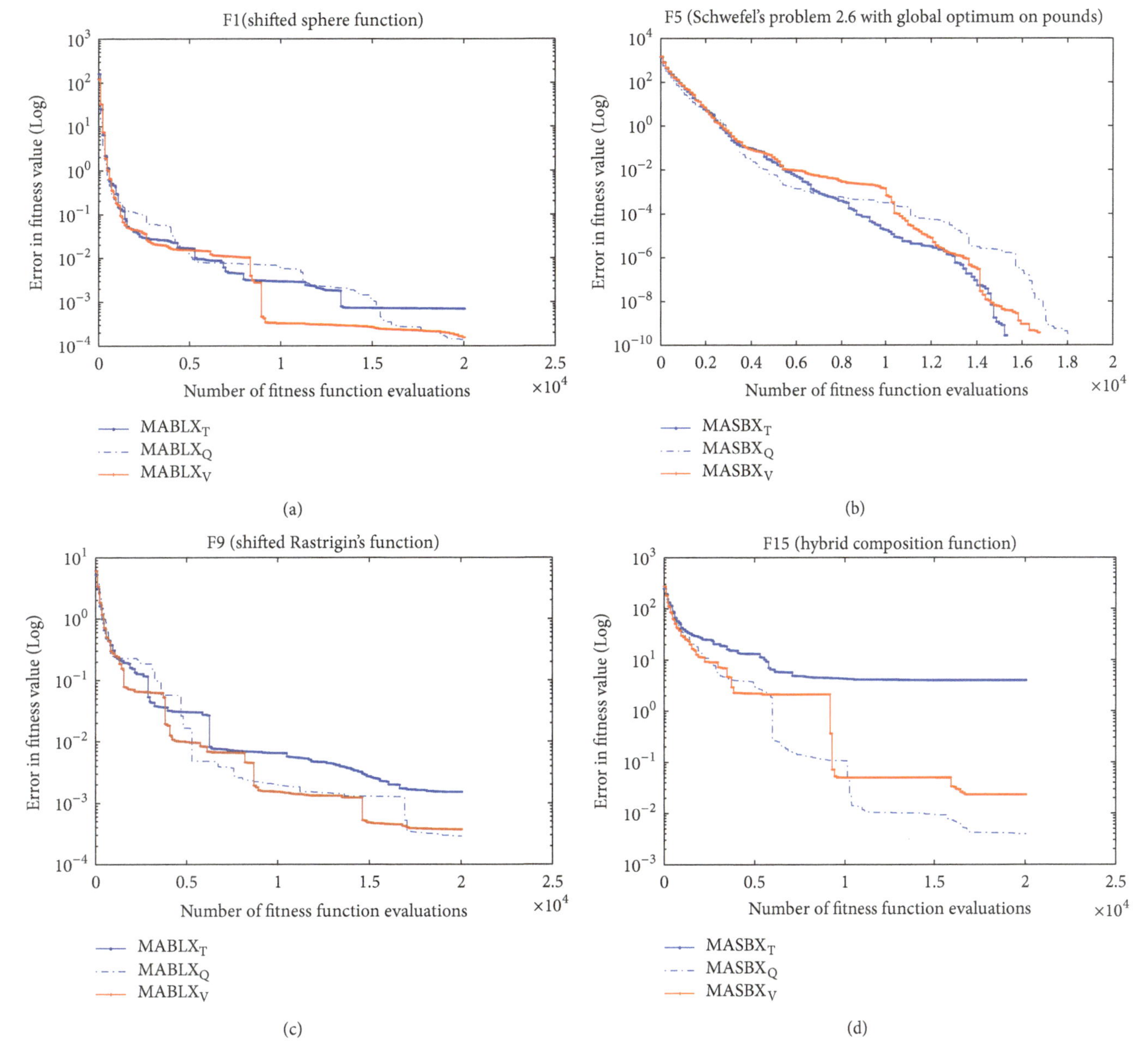

FIGURE 3: Fitness error versus number of function evaluation calls of test functions averaged over 25 runs in cases of MAs for functions (a) F1, (b) F5, (c) F9, and (d) F15.

5.2.2. CEC 2005 Standard Benchmark Problems [29]. There are various types of optimization algorithms available and used to solve real-parameter function optimization problems. There is a variety of mechanisms having variations in their operators and working principles. Catering to the need of evaluating the algorithms in a more systematic manner by specifying a common termination criterion, size of problems, initialization scheme, linkages/rotation, and so forth, the CEC 2005 having 25 test problems is designed [29].

All functions are nonseparable and scalable except F1 and F9, which are separable, whereas F15 is separable near the global optimum (Rastrigin). The functions in the test suit are shifted and rotated. Functions F1–F5 are unimodal and functions F6–F25 are multimodal in which F6–F12 are single functions, F13-F14, are expanded functions and F15–F25 are hybrid composition functions. Functions F9, F10, and F15–F25 have a huge number of local optima in their fitness landscape. In functions F15–F25, different function's properties are mixed together to make hybrid composition functions. Functions F4 and F17 have Gaussian noise in fitness and they are similar to functions F2 and F16, respectively. Functions F15–F20 have effect due to sphere functions giving two flat areas for the function and functions F24-F25 have effect due to unimodal functions giving flat areas for the function. Functions F18–F20 have a local optimum and are set on the origin. Functions F8, F20, F22-F23, and F25 have Global optimum on the bound. For the performance evaluation, the function error, NFFEs, and convergence graphs for the

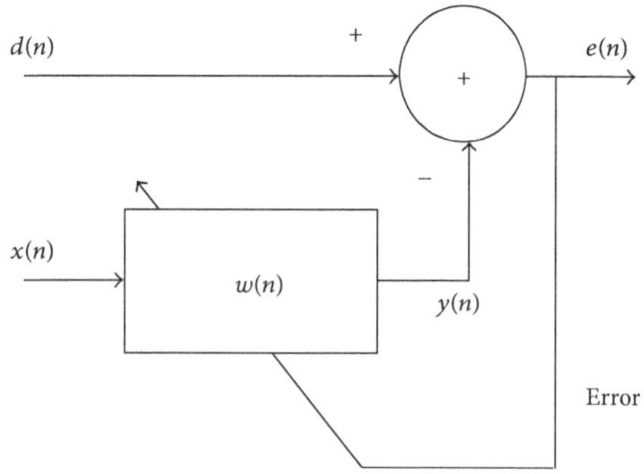

FIGURE 4: Adaptive filter structure.

problems are used [29]. Function error value is $(f(x)-f(x^*))$, where x^* = global optimum. Minimum error is recorded when the maximum number of fitness function evaluation is reached or NFFEs is recorded when error of 10^{-8} is attained, each in all 25 runs [29]. Mean and standard deviation of the error are also calculated. Convergence graphs (or run-length distribution graphs) are the semilog graphs that should show $\log_{10}(f(x) - f(x^*))$ versus NFFEs for each problem [29].

From Tables 7 and 8, the convergence properties GABLX, GASBX, and MAs with various modified SIMBOs as local searches are tested for convergence on CEC 2005 functions. All the SBX crossover algorithms, that is, GASBX and MASBX, with all types of mSIMBOs as local search and with 2 dimensions are 100% successfully converging in all of the runs for function F5. All the algorithms (GABLX, GASBX, MABLX, and MASBX with all mSIMBOs as local searches) are successfully attaining optimum with error less than 10^{-8} in all of the runs or some of the runs with 2 dimensions for functions F1, F2, F9–F11, F13, F15, and F16. All the algorithms (GABLX, GASBX, MABLX, and MASBX with all mSIMBOs as local searches) show promising characteristics in convergence for function F14 with 2 dimensions and for functions F1, F9, and F13 with 10 dimensions. All the SBX crossover algorithms that is, GASBX and MASBX, with all types of mSIMBOs as local search and with 2 dimensions show achievement of success with error less than 10^{-8} with more than 92% successful runs for functions F1, F2, F4, and F15; 52% to 79% successful runs for function F3; 20% to 24% successful runs for functions F6, and F8; 32% to 48% successful runs for functions F12-F13, F16–F18; 4% to 12% successful runs for functions F20, F21, and F23. Also, all SBX algorithms promisingly converge in case of the function F7.

As per Figures 3(a), 3(b), 3(c), and 3(d), MA with mSIMBO_Q as local search shows more promising average convergence property, and after that, MAs with mSIMBO_V and mSIMBO_T as local searches follow in the convergence performance. For some functions, MAs with mSIMBO_V and mSIMBO_T also show better average performance characteristics than MA with mSIMBO_Q. In general, among the MAs with treatment based mSIMBOs as local searches are the slowest and those of quarantine based are the fastest converging characteristics for CEC 2005 functions. The averaged error and convergence characteristics show the success in a variety of functions with unimodal, multimodal, shifted, rotated, and composite properties.

When the statistical test, Student's t-test, is performed in comparison of the results of MAs and GAs for CEC 2005 problems, it is seen that MAs behave better in generality for convergences with 2 dimensions and also show promising performance in many cases with dimensions 10. As per Tables 9 and 10 for 2 dimensions, both GA and MAs promisingly perform and also perform similar in some cases, but for 10 dimensions MAs show consistency in performance. Therefore, with higher dimensions, MAs win in the statistical tests in generality.

6. The LMS Filter Weight Optimization for PLI Removal from ECG

The LMS-based adaptive filter having a length L is depicted in Figure 4. It generally uses the gradient-based technique to update the weights with, $w(n+1) = w(n) + \mu x(n)e(n)$, where $w(n) = [w_0(n)\ w_1(n)\ \cdots\ w_{L-1}(n)]^t$ is the weight vector. For nth index, $x(n) = [x(n)\ x(n-1)\ \cdots\ x(n-L+1)]^t$, error $e(n) = d(n) - w^t(n)x(n)$, having $d(n)$ as the desired response available in starting training period and μ indicating step-size parameter, and $x(n)$ is reference or inference signal necessary for the adaptive filter [34].

The stochastic optimization with proposed MAs is used to search for the optimum filter weight vector giving error. In this case to remove the noise from the ECG signal, the ECG signal $s_1(n)$ contaminated by the PLI signal $p_1(n)$ is used in case of the desired response $d(n)$ in the adaptive filter shown in Figure 4. The required inference signal is obtained with the help of selective usage of the intrinsic mode functions (IMFs) of empirical mode decomposition (EMD) in generating the signal and is used in the adaptive filter [34]. Where the inference signal obtained from the EMD process is named as $p_2(n)$ that is in the correlation with $p_1(n)$ and is applied to the filter, that is, $x(n) = p_2(n)$. The filter error becomes $e(n) = [s_1(n) + p_1(n)] - y(n)$, where, $y(n)$ is the filter output given by $y(n) = w^t(n)x(n)$.

As the signal and noise are uncorrelated, then mean squared error (MSE) becomes:

$$E\left[e^2(n)\right] = E\left\{\left[s_1(n) - y(n)\right]^2\right\} + E\left[p_1^2(n)\right]. \quad (11)$$

In this method of MA-based optimization, the filter weights are generated with minimization of the MSE results in a filter output which is the best least-squares estimate of the signal $s_1(n)$.

The ECG contaminated by the artificial PLI with 50 Hz frequency is put forward for the PLI removal, to the LMS adaptive filter for its weights optimization by the MA using fitness function articulated in (11). The MSE is used to measure the difference between the original "clean" ECG and the reconstructed ECG. The MSE between original and recovered ECG indicates that MA optimizes the filter weights effectively with better mean square error (MSE) as in Table 11. Frequency spectrum of ECG having PLI and after filtering

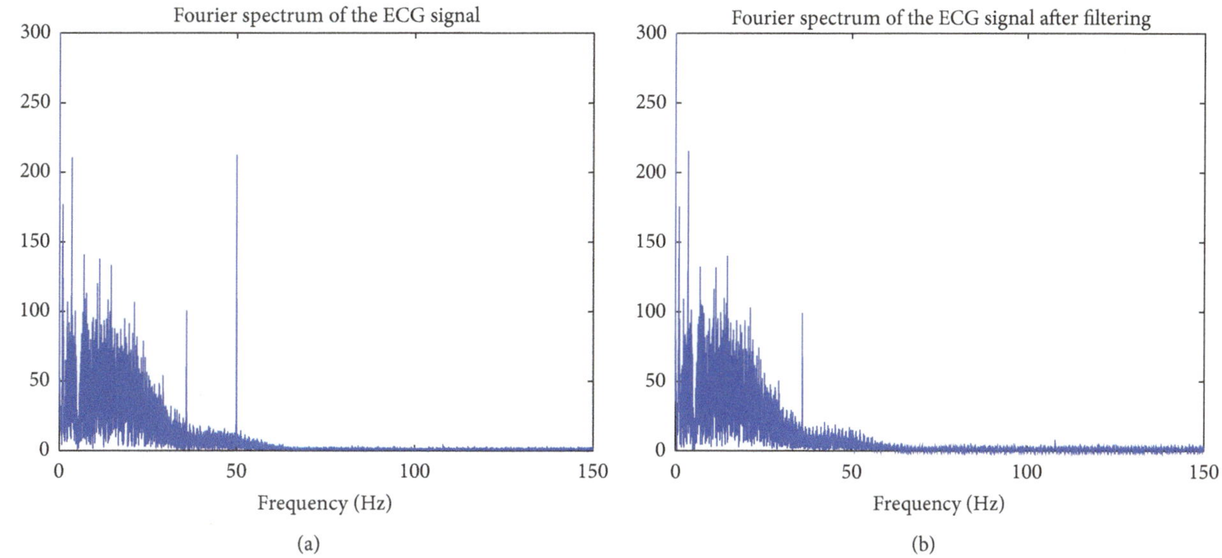

FIGURE 5: Frequency spectrum of the ECG signal (a) having power line interference (PLI) at 50 Hz and (b) after filtering with MA-based LMS adaptive filter.

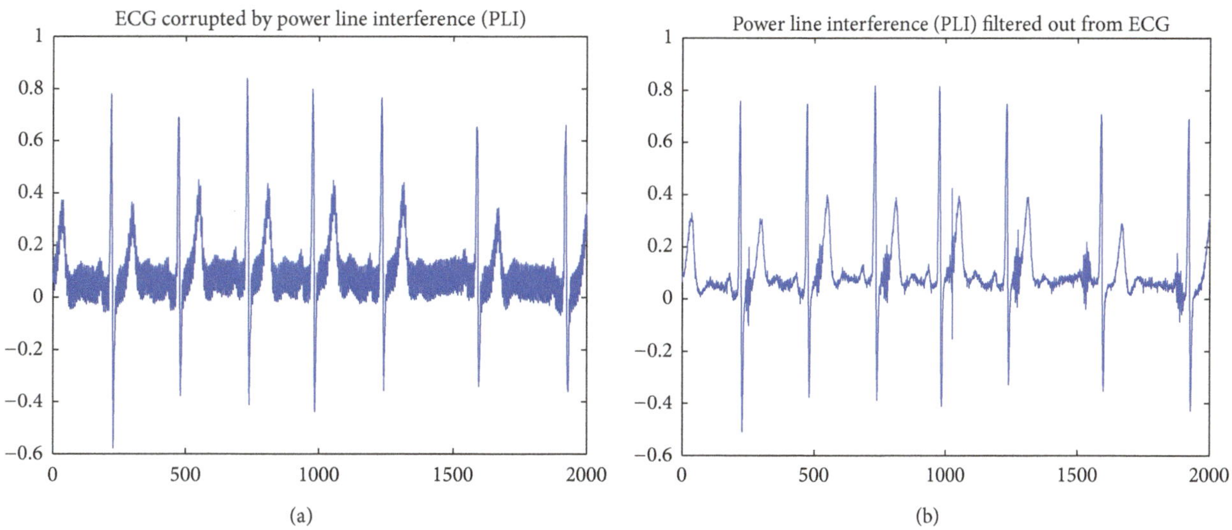

FIGURE 6: ECG—(a) corrupted by power line interference (PLI) of 50 Hz and (b) recovered.

with LMS adaptive filter are shown in Figures 5(a) and 5(b), respectively, which indicates the absence of only power line inference (PLI) with other frequency spectrum remaining unaffected after filtering. In Figure 6(a), the corrupted ECG and in Figure 6(b) the recovered ECG are shown. The ECG is used from the benchmark MIT-BIH database [35, 36].

7. Conclusion

The Swine Influenza Model-Based Optimization (SIMBO) is a faster optimization technique having the adaptive nature in its process. With the help of the three-fold modifications, the algorithm is made still faster in this work. The SIMBO family is very good in searching the basin under consideration, so it is utilized in local searches to develop the memetic algorithms (MAs). The SIMBO family has high pressure in the direction of origin and hence demonstrates better results in noisy function also. But this makes it weak in case of an other type of the solution that requires more explorations. This is achieved with help of MA having balance in exploration and exploitation. So, MAs have little tradeoffs in convergence speed but are more comprehensive in solving the numerical standard benchmark test bed of functions having the different properties. The modifications in SIMBO and application of it as the local searchers in making MAs as a process exploit the

Table 11: Mean square error - MSEo-r is between original and recovered ECG signal; MSEo-c is between original and corrupted ECG signal by power line interference (PLI).

Record No.	MSEo-c	MSEo-r
sel123	$4.41e-3$	$2.23e-3$
sel123	$6.89e-4$	$3.73e-4$
100m	$6.18e-3$	$4.15e-3$
100m	$9.85e-4$	$6.64e-4$
118e00m	$5.02e-3$	$3.31e-3$
118e00m	$9.14e-4$	$6.78e-4$

properties of both the algorithms to enhance the efficiency. The ensemble strategy to tune parameters and operators can be useful to develop further the more adaptive techniques.

The developed MA is also applied to remove the power line interference (PLI) from the biomedical signal ECG with the help of adaptive filter whose weights are optimized by the MA. In this, the adaptive filter inference signal is generated with the selective reconstruction of ECG from the intrinsic mode functions (IMFs) of empirical mode decomposition (EMD). The adaptive filter with the help of the optimized weights filters the PLI from ECG effectively with reduced error (mean square error (MSE)).

Conflict of Interests

The authors declare that there is no conflict of interests regarding the publication of this paper.

References

[1] Q. H. Nguyen, Y. S. Ong, and N. Krasnogor, "A study on the design issues of memetic algorithm," in *Proceedings of IEEE Congress on Evolutionary Computation (CEC '07)*, pp. 2390–2397, September 2007.

[2] Z. Michalewicz, *Genetic Algorithms + Data Structures = Evolution Programs*, Springer-, 1996.

[3] S. Das and P. N. Suganthan, "Differential evolution: a survey of the state-of-the-art," *IEEE Transactions on Evolutionary Computation*, vol. 15, no. 1, pp. 4–31, 2011.

[4] J. J. Liang, A. K. Qin, P. N. Suganthan, and S. Baskar, "Comprehensive learning particle swarm optimizer for global optimization of multimodal functions," *IEEE Transactions on Evolutionary Computation*, vol. 10, no. 3, pp. 281–295, 2006.

[5] N. V. George and G. Panda, "A particle-swarm-optimization-based decentralized nonlinear active noise control system," *IEEE Transactions on Instrumentation and Measurement*, vol. 61, no. 12, pp. 3378–3386, 2012.

[6] K. M. Passino, "Biomimicry of bacterial foraging for distributed optimization and control," *IEEE Control Systems Magazine*, vol. 22, no. 3, pp. 52–67, 2002.

[7] X. Yao, Y. Liu, and G. Lin, "Evolutionary programming made faster," *IEEE Transactions on Evolutionary Computation*, vol. 3, no. 2, pp. 82–102, 1999.

[8] S. Mishra, "A hybrid least square-fuzzy bacterial foraging strategy for harmonic estimation," *IEEE Transactions on Evolutionary Computation*, vol. 9, no. 1, pp. 61–73, 2005.

[9] D. H. Kim, A. Abraham, and J. H. Cho, "A hybrid genetic algorithm and bacterial foraging approach for global optimization," *Information Sciences*, vol. 177, no. 18, pp. 3918–3937, 2007.

[10] B. K. Panigrahi and V. R. Pandi, "Bacterial foraging optimisation: nelder-mead hybrid algorithm for economic load dispatch," *IET Generation, Transmission & Distribution*, vol. 2, no. 4, pp. 556–565, 2008.

[11] D. P. Acharya, G. Panda, S. Mishra, and Y. V. S. Lakshmi, "Bacteria foraging based independent component analysis," in *Proceedings of the International Conference on Computational Intelligence and Multimedia Applications (ICCIMA '07)*, pp. 527–531, December 2007.

[12] B. Li, Y. S. Ong, M. N. Le, and C. K. Goh, "Memetic gradient search," in *Proceedings of IEEE Congress on Evolutionary Computation (CEC '08)*, pp. 2894–2901, June 2008.

[13] R. Akbari and K. Ziarati, "Combination of particle swarm optimization and stochastic local search for multimodal function optimization," in *Proceedings of the Pacific-Asia Workshop on Computational Intelligence and Industrial Application (PACIIA '08)*, pp. 388–392, December 2008.

[14] D. G. Jadhav, S. S. Pattnaik, S. Devi, M. R. Lohokare, and K. M. Bakwad, "Approximate memetic algorithm for consistent convergence," in *Proceedings of the National Conference on Computational Instrumentation CSIO Chandigarh (NCCI '10)*, pp. 118–122, India, March 2010.

[15] P. A. Moscato, "On evolution, search, optimization, genetic algorithms and martial arts: towards memetic algorithms," Caltech Concurrent Computation Program, Report 826, California Institute of Technology, Pasadena, Calif, USA, 1989.

[16] S. S. Pattnaik, K. M. Bakwad, B. S. Sohi, R. K. Ratho, and S. Devi, "Swine influenza models based optimization (SIMBO)," *Applied Soft Computing*, vol. 13, no. 1, pp. 628–653, 2013.

[17] B. J. Coburn, B. G. Wagner, and S. Blower, "Modeling influenza epidemics and pandemics: insights into the future of swine flu (H1N1)," *BMC Medicine*, vol. 7, article 30, 2009.

[18] A. Biswas, K. K. Mishra, S. Tiwari, and A. K. Misra, "Physics-inspired optimization algorithms: a survey," *Journal of Optimization*, vol. 2013, Article ID 438152, 16 pages, 2013.

[19] D. Dasgupta, "Advances in artificial immune systems," *IEEE Computational Intelligence Magazine*, vol. 1, no. 4, pp. 40–49, 2006.

[20] D. Dasgupta, S. Yu, and F. Nino, "Recent advances in artificial immune systems: models and applications," *Applied Soft Computing*, vol. 11, no. 2, pp. 1574–1587, 2011.

[21] S. Knobler, A. Mack, A. Mahmoud, and S. Lemon, *The Story of Influenza: The Threat of Pandemic Influenza: Are We Ready? Workshop Summary*, The National Academies Press, Washington, DC, USA, 2005, http://books.nap.edu/openbook.php?isbn=0309095042&page=75.

[22] Q. H. Nguyen, Y.-S. Ong, and M. H. Lim, "A probabilistic memetic framework," *IEEE Transactions on Evolutionary Computation*, vol. 13, no. 3, pp. 604–623, 2009.

[23] M. Lozano, F. Herrera, N. Krasnogor, and D. Molina, "Real-coded memetic algorithms with crossover hill-climbing," *Evolutionary Computation*, vol. 12, no. 3, pp. 273–302, 2004.

[24] D. E. Goldberg, *Genetic Algorithms in Search, Optimization and Machine Learning*, Addison-Wesley, Reading, Mass, USA, 1989.

[25] D. E. Goldberg and K. Deb, "A comparative analysis of selection schemes used in genetic algorithms," in *Proceedings of the Foundations of Genetic Algorithms*, pp. 69–93, 1991.

[26] J. A. Vasconcelos, J. A. Ramírez, R. H. C. Takahashi, and R. R. Saldanha, "Improvements in genetic algorithms," *IEEE Transactions on Magnetics*, vol. 37, no. 5, pp. 3414–3417, 2001.

[27] P. Moscato and C. Cotta, "A gentle introduction to memetic algorithms," in *Handbook of Metaheuristics*, chapter 5, pp. 105–144, Springer, New York, NY, USA, 2003.

[28] D. Garrett and D. Dasgupta, "An empirical comparison of memetic algorithm strategies on the multiobjective quadratic assignment problem," in *Proceedings of IEEE Symposium on Computational Intelligence in Multi-Criteria Decision-Making (MCDM '09)*, pp. 80–87, March-April 2009.

[29] P. N. Suganthan, N. Hansen, J. J. Liang et al., "Problem definition and evaluation criteria for the CEC, 2005 special session on real parameter optimization," Nanyang Technological University and KanGAL Report 2005005, IIT Kanpur, 2005, http://www.ntu.edu.sg/home/epnsugan.

[30] L. J. Eshelman and J. D. Schaffer, "Real-coded genetic algorithms and interval-shemata," in *Foundation of Genetic Algorithms 2*, L. Darrell Whitley, Ed., pp. 187–202, Morgan Kaufmann, San Mateo, Calif, USA, 1993.

[31] K. Deb and R. B. Agrawal, "Simulated binary crossover for continuous search space," *Complex System*, vol. 9, no. 2, pp. 115–148, 1995.

[32] R. Mallipeddi, P. N. Suganthan, Q. K. Pan, and M. F. Tasgetiren, "Differential evolution algorithm with ensemble of parameters and mutation strategies," *Applied Soft Computing*, vol. 11, no. 2, pp. 1679–1696, 2011.

[33] J. Derrac, S. García, D. Molina, and F. Herrera, "A practical tutorial on the use of nonparametric statistical tests as a methodology for comparing evolutionary and swarm intelligence algorithms," *Swarm and Evolutionary Computation*, vol. 1, no. 1, pp. 3–18, 2011.

[34] Z. Zhidong and M. Chan, "A novel cancellation method of power line interference in ECG signal based on EMD and adaptive filter," in *Proceedings of the 11th IEEE International Conference on Communication Technology (ICCT '08)*, pp. 517–520, Hangzhou, China, November 2008.

[35] G. B. Moody and R. G. Mark, "The impact of the MIT-BIH arrhythmia database," *IEEE Engineering in Medicine and Biology Magazine*, vol. 20, no. 3, pp. 45–50, 2001.

[36] A. L. Goldberger, L. A. N. Amaral, L. Glass et al., "PhysioBank, PhysioToolkit, and PhysioNet: components of a new researchresource for complex physiologic signals," *Circulation*, vol. 101, no. 23, pp. e215–e220, 2000.

Permissions

The contributors of this book come from diverse backgrounds, making this book a truly international effort. This book will bring forth new frontiers with its revolutionizing research information and detailed analysis of the nascent developments around the world.

We would like to thank all the contributing authors for lending their expertise to make the book truly unique. They have played a crucial role in the development of this book. Without their invaluable contributions this book wouldn't have been possible. They have made vital efforts to compile up to date information on the varied aspects of this subject to make this book a valuable addition to the collection of many professionals and students.

This book was conceptualized with the vision of imparting up-to-date information and advanced data in this field. To ensure the same, a matchless editorial board was set up. Every individual on the board went through rigorous rounds of assessment to prove their worth. After which they invested a large part of their time researching and compiling the most relevant data for our readers. Conferences and sessions were held from time to time between the editorial board and the contributing authors to present the data in the most comprehensible form. The editorial team has worked tirelessly to provide valuable and valid information to help people across the globe.

Every chapter published in this book has been scrutinized by our experts. Their significance has been extensively debated. The topics covered herein carry significant findings which will fuel the growth of the discipline. They may even be implemented as practical applications or may be referred to as a beginning point for another development. Chapters in this book were first published by Hindawi Publishing Corporation; hereby published with permission under the Creative Commons Attribution License or equivalent.

The editorial board has been involved in producing this book since its inception. They have spent rigorous hours researching and exploring the diverse topics which have resulted in the successful publishing of this book. They have passed on their knowledge of decades through this book. To expedite this challenging task, the publisher supported the team at every step. A small team of assistant editors was also appointed to further simplify the editing procedure and attain best results for the readers.

Our editorial team has been hand-picked from every corner of the world. Their multi-ethnicity adds dynamic inputs to the discussions which result in innovative outcomes. These outcomes are then further discussed with the researchers and contributors who give their valuable feedback and opinion regarding the same. The feedback is then collaborated with the researches and they are edited in a comprehensive manner to aid the understanding of the subject.

Apart from the editorial board, the designing team has also invested a significant amount of their time in understanding the subject and creating the most relevant covers. They scrutinized every image to scout for the most suitable representation of the subject and create an appropriate cover for the book.

The publishing team has been involved in this book since its early stages. They were actively engaged in every process, be it collecting the data, connecting with the contributors or procuring relevant information. The team has been an ardent support to the editorial, designing and production team. Their endless efforts to recruit the best for this project, has resulted in the accomplishment of this book. They are a veteran in the field of academics and their pool of knowledge is as vast as their experience in printing. Their expertise and guidance has proved useful at every step. Their uncompromising quality standards have made this book an exceptional effort. Their encouragement from time to time has been an inspiration for everyone.

The publisher and the editorial board hope that this book will prove to be a valuable piece of knowledge for researchers, students, practitioners and scholars across the globe.

List of Contributors

Maneesha Gupta and Richa Yadav
Advanced Electronics Lab, Department of ECE, Netaji Subhas Institute of Technology, Dwarka, New Delhi 110078, India

Anupam Biswas, K. K. Mishra, Shailesh Tiwari and A. K. Misra
Department of Computer Science & Engineering, Motilal Nehru National Institute of Technology Allahabad, Allahabad 211004, India

Mohamed E. Seliaman
Department of Information Systems, King Faisal University, P.O. Box 78, Al-Hufof 31982, Saudi Arabia

Morteza Abbasi, Reza Hosnavi and Mehrdad Mohammadi
Department of Management and Soft Technologies, Malek Ashtar University of Technology, P.O. Box 1774/15875, Tehran, Iran

P. Vijaya Laxmi and K. Jyothsna
Department of Applied Mathematics, Andhra University, Visakhapatnam 530 003, India

V. Goswami
School of Computer Application, KIIT University, Bhubaneswar 751 024, India

Abd-Elmoneim Anwar Mohamed and Mohamed Abd Allah El-Hadidy
Department of Mathematics, Faculty of Science, Tanta University, Tanta 31527, Egypt

Khaled Loukhaoukha
Department of Electrical and Computer Engineering, Laval University, Quebec, QC, Canada 1K 7P4

Abbas Mahmoudabadi
Department of Industrial Engineering, Payame Noor University (PNU), Shahnaz Alley, Nourian Street, North Dibagi Avenue, Tehran, Iran

Ali Ghazizadeh
Technical and Engineering Department, Payame Noor University (PNU), Firouzbakhsh Street, Movahed Danesh Avenue, Aqdasieh, Tehran, Iran

Balamati Choudhury, Sangeetha Manickam and R. M. Jha
Centre for Electromagnetics, CSIR-National Aerospace Laboratories, Bangalore 560017, India

Xinbo Zhang, Feng Zhang and Zhong Wan
School of Mathematics and Statistics, Central South University, Changsha, Hunan 410083, China

Xiaohong Chen
School of Business, Central South University, Changsha, Hunan 410083, China

Ney José Luiggi and María Valera
Grupo Fısica de Metales, Departamento de Fısica, Nucleo de Sucre, Universidad de Oriente, Cumana 6101, Venezuela

Jesús P. Rodriguez
IDECYT, Universidad Simon Rodrıguez, Caracas 1060, Venezuela

José Prin
Instituto de Investigaciones de Biomedicina y Ciencias Aplicadas (IIBCA), Universidad de Oriente, Cumana 6101, Venezuela

Zhaocheng Cui
Department of Mathematics and Physics, Shandong Jiaotong University, Jinan, Shandong Province 250023, China

Mehdi Ghamari Adian
Electrical Engineering Department, University of Zanjan, Zanjan 45371-38791, Iran

Neal M. Bengtson
School of Business, Barton College, Wilson, NC 27893, USA

Devidas G. Jadhav and Shyam S. Pattnaik
National Institute of Technical Teachers Training & Research (NITTTR), Chandigarh 160019, India

Sanjoy Das
Kansas State University, Manhattan, KS 66506, USA